Lecture Notes in Mathematics

Edited by A. Dold, F. Takens and B. Teissier

Editorial Policy
for the publication of monographs

1. Lecture Notes aim to report new developments in all areas of mathematics – quickly, informally and at a high level. Monograph manuscripts should be reasonably self-contained and rounded off. Thus they may, and often will, present not only results of the author but also related work by other people. They may be based on specialized lecture courses. Furthermore, the manuscripts should provide sufficient motivation, examples and applications. This clearly distinguishes Lecture Notes from journal articles or technical reports which normally are very concise. Articles intended for a journal but too long to be accepted by most journals, usually do not have this "lecture notes" character. For similar reasons it is unusual for doctoral theses to be accepted for the Lecture Notes series.

2. Manuscripts should be submitted (preferably in duplicate) either to one of the series editors or to Springer-Verlag, Heidelberg. In general, manuscripts will be sent out to 2 external referees for evaluation. If a decision cannot yet be reached on the basis of the first 2 reports, further referees may be contacted: the author will be informed of this. A final decision to publish can be made only on the basis of the complete manuscript, however a refereeing process leading to a preliminary decision can be based on a pre-final or incomplete manuscript. The strict minimum amount of material that will be considered should include a detailed outline describing the planned contents of each chapter, a bibliography and several sample chapters.
Authors should be aware that incomplete or insufficiently close to final manuscripts almost always result in longer refereeing times and nevertheless unclear referees' recommendations, making further refereeing of a final draft necessary.
Authors should also be aware that parallel submission of their manuscript to another publisher while under consideration for LNM will in general lead to immediate rejection.

3. Manuscripts should in general be submitted in English.
Final manuscripts should contain at least 100 pages of mathematical text and should include
– a table of contents;
– an informative introduction, with adequate motivation and perhaps some
  historical remarks: it should be accessible to a reader not intimately familiar
  with the topic treated;
– a subject index: as a rule this is genuinely helpful for the reader.

Continued on back inside cover

Lecture Notes in Mathematics 1734

Springer
Berlin
Heidelberg
New York
Barcelona
Hong Kong
London
Milan
Paris
Singapore
Tokyo

M. S. Espedal   A. Fasano   A. Mikelić

Filtration in Porous Media and Industrial Application

Lectures given at the 4th Session of the
Centro Internazionale Matematico Estivo
(C.I.M.E.) held in Cetraro, Italy
August 24–29, 1998

Editor: A. Fasano

Fondazione
C.I.M.E.

Springer

Authors and Editor

Magne Espedal
Department of Mathematics
University of Bergen
Johs. Brunsgt. 12
5008 Bergen, Norway

magne.espedal@mi.uib.no

Antonio Fasano
Department of Mathematics "U. Dini"
University of Florence
Viale Morgagni 67a
50134 Florence, Italy

fasano@math.unifi.it

Andro Mikelić
Laboratoire d' Analyse Numérique, Bât.
101
Université Lyon I
43 Bd. du 11 novembre 1918
69622 Villeurbanne Cedex, Frace

andro@lan.univ-lyon1.fr

Cataloging-in-Publication Data applied for

Die Deutsche Bibliothek - CIP-Einheitsaufnahme

Filtration in porous media and industrial application : held in
Cetraro, Italy, August 24 - 29, 1998 / Fondazione CIME. M. S. Espedal.
... Ed.: A. Fasano. - Berlin ; Heidelberg ; New York ; Barcelona ;
Hong Kong ; London ; Milan ; Paris ; Singapore ; Tokyo : Springer,
2000 (Lectures given at the ... session of the Centro Internazionale
Matematico Estivo (CIME) ... ; 1998,4)
(Lecture notes in mathematics ; Vol. 1734 : Subseries: Fondazione
CIME) ISBN 3-540-67868-9

Mathematics Subject Classification (2000): 76S05, 76M50, 74A, 35R35, 76M10

ISSN 0075-8434
ISBN 3-540-67868-9 Springer-Verlag Berlin Heidelberg New York

Springer-Verlag Berlin Heidelberg New York
a member of BertelsmannSpringer Science+Business Media GmbH

© Springer-Verlag Berlin Heidelberg 2000
Printed in Germany

Typesetting: Camera-ready T_EX output by the author
SPIN: 10725018   41/3142-543210 - Printed on acid-free paper

Preface

This volume contains the notes of some of the lectures given at the CIME Course held in Cetraro during the week 24-29 August, 1998.

The course was organized by myself and by H. Van Duijn (CWI, Amsterdam) and was addressed to an audience of 33 people from various countries. The subjects covered a large spectrum in the area of filtration theory, emphasizing its importance for applications and its specific interest for mathematicians.

The course included the following series of Lectures:

1. Magne Espedal (University of Bergen, Norwey): Reservoirs flow models (3 lectures)

2. Antonio Fasano (University of Florence, Italy): Filtration problems in various industrial processes (6 lectures)

3. Peter Knabner (University of Erlangen, Germany): Reactive transport in porous media (4 lectures)

4. Andro Mikelić (University of Lyon, France): Homogenization theory in porous media (6 lectures)

5. Hans Van Duijn (CWI, Amsterdam, The Netherlands): Nonlinear model in subsurface transport (6 lectures).

The idea was to present a substantial overview of this very diversified research subject, giving at the same time an idea of its impact on industry and of its rich mathematical content. Therefore the stress was not only on fine mathematical problems, but also on modeling and on numerical analysis.

Here we give a short account of each series of lectures.

1) M. Espedal (lecture notes included). This is a remarkable overview of numerical methods widely applied in the engineering of oil recovery from reservoirs, with particular reference to the extraction technique consisting in the displacement of oil by another liquid. A self-contained exposition having a great value as a reference point for the study of this class of problems, including the mathematical background.

2) A. Fasano (lecture notes included). Three main themes were addressed: the so-called espresso-coffee problem (a familiar experience but a formidably complicated problem), various questions in the manufacturing of composite materials, and a very recent subject concerning injections of liquid in diapers.

3) P. Knabner. A series of lectures devoted to groundwater accompanied by chemical reactions with two main (opposite) applications: soil contamination (organic, inorganic, radioactive), soil remediation (flushing, venting, microbial degradation). Typical situations have been sketched and the main theoretical results summarized, together with the illustration of numerical schemes.

4) A. Mikelić (lecture notes included). A fundamental question in filtration theory is the following: how to justify the empirical laws (e.g. Darcy's law relating the fluid velocity to the pressure gradient) on the basis of the general laws of fluid dynamics. The answer to this question lies in homogenenization theory: solve the fluid dynamical problem in a cell which, repeated with periodicity provides an idealized model of the porous medium and then pass to the limit when the size of cell tends to zero. Here we have an extended explanation of this theory of remarkable clarity.

5) H. Van Duijn. These lectures were also dealing with underground transport of chemical substance, but with special emphasis on solutions. The ultimate aim is to study coastal aquifers and interfaces between fresh water and salt water. This problem has become classical, but there are new developments too (e.g. brine transport) that have been illustrated. Of particular relevance is the problem of cusps formation at fresh/salt water interface which may cause the breakthrough of salt water with obvious negative consequences when pumping drinkable water from the aquifer.

The lecture notes here collected are preceded by a short introduction to some basic facts concerning the flow of fluids through porous media, providing a general mathematical framework to this class of phenomena.

In conclusion I can say that this CIME Course was a unique opportunity of presenting classical and modern problems in a field that is rich of real applications and is also a great source of nice and profound mathematical problems.

I wish to thank the Scientific Council of CIME for inviting me to organize this Course and to express my gratitude to the other lecturers for their enthusiastic response and their extremely valuable contribution.

A. Fasano (Director of the Course)

Table of Contents

Antonio Fasano

Some general facts about filtration through porous media

Filtration of fluids through porous media is a phenomenon occurring in several natural processes, as well as in an incredible variety of technological applications. A porous medium is typically composed by solid particles (*grains*) and interstitial spaces (*pores*) that are connected, permitting a fluid to flow. Such a definition is quite generic, since it includes coarse materials, like granules, and soils with extremely fine texture, like clays, with a great range of intermediate possibilities.

However, the solid component is not necessarily a loose pack of grains, but it can also be connected, forming a skeleton. Examples can be found among artificial media (e.g. ceramics) and material products. Moreover the solid component may be rigid or it can be deformed to some extent (like in the extreme case of sponges).

In normal cases a porous medium can be modelled as a continuum by considering a representative volume element on which we define averaged quantities, next extended to the whole domain occupied by the system as regular functions of the space coordinates. The volume element must be large enough to contain a sufficiently high number of grains and pores, but much smaller than the typical size of the system. In this way we can define a basic geometric property, the *porosity* ϕ, i.e. the volume fraction occupied by the pores. When the flowing fluid is incompressible we may likewise introduce the *saturation* S, i.e. the fraction of the volume available to the flow (i.e. the pore volume) which is occupied by the fluid.

These are both numbers between 0 and 1. We say that the medium is saturated when $S = 1$, and unsaturated if $0 < S < 1$. When $S = 0$ the medium is dry.

We may have multicomponent flows (with a partial saturation defined for each component) and also quite complex porous structures (highly heterogeneous, with fissures, etc.). Even the case of grains which are in turn porous have been considered [19] [18].

Making an exhaustive list of filtration problems is a hopeless task, but we may roughly distinguish two large classes: filtration of gases and filtration of liquids, although this distinction is rather artificial, because flowing liquids quite often displace air and, when the media is not saturated, the presence of vapor may be important (like e.g. in drying processes). The most normal situation in the former class is the flow of a gas in a porous medium previously occupied by another gas (typically air) under small concentration gradients. This process is basically diffusive and we are not going to deal with it. Completely different is the problem of the expansion of a gas through a porous medium which is void. This is a celebrated problem that leads to the so-called porous media equation, a parabolic equation which degenerates at the propagation front. The degeneracy gives rise to typically non-parabolic features like waiting times and finite speed of the invasion front. Basic contributions in this field are the papers [2, 4], that have been the seed of a massive research work.

Filtration of liquids is the real subject of this CIME Course. By *liquids* we mean *Newtonian* incompressible fluids although theories for non-Newtonian flows through porous media have been developed [8].

For the basic theory we refer to classical books like [6] [7] [5], [21], [26] and to the survey papers [24] [25] [23] [16] [17].

It may be useful to expose some fundamental fact, about incompressible flows through porous media.

The fundamental experimental law, very well known and most frequently used to describe the flow of liquids through porous media is *Darcy's law*, which dates back to 1856 [15]. The experimental basis of Darcy's law is parallel to Fourier's law for heat conduction and Fick's law for diffusion: consider the flow through a homogeneous layer under a prescribed pressure gradient. The quantity to be measured is the volume of liquid passing through the unit section per unit time (*volumetric velocity*). For a vertical saturated flow in a horizontal layer, if $p(0) - p(b)$ is the driving pressure difference between the top surface ($z = 0$) and the bottom surface ($z = L$), L being the layer thickness, and ρ the liquid density, Darcy's law states that the volumetric velocity q is

$$q = k \left(\frac{p(0) - p(L)}{L} + \rho g \right), \tag{1}$$

where g is the gravity acceleration. We may have an upward flow ($q < 0$) if $\frac{\Delta p}{L} < -\rho g$.

The positive constant k is the *hydraulic conductivity* of the system. Experimenting with different liquids one realizes that k can be expressed as

$$k = \frac{K}{\mu}, \tag{2}$$

when K is a typical constant of the medium (*permeability*) and μ is the liquid *viscosity*.

Extending (1) to a general porous media, the volumetric velocity becomes a vector and (1) is replaced by

$$\mathbf{q} = -\frac{1}{\mu} \mathbf{K} \nabla (p - \rho g z), \tag{3}$$

where $\mathbf{K}$ is the *permeability tensor* and z the downward directed vertical coordinate. Of course we have $\mathbf{K} = K\mathbf{I}$ if and only if the medium is isotropic. The quantity $p - \rho g z$ plays the role of a potential and is called *piezometric head*. For an incompressible saturated flow on an indeformable porous medium, the conservation of mass is expressed by

$$\operatorname{div} \mathbf{q} = 0. \tag{4}$$

Typically, for an isotropic medium (3) and (4) take the respective forms

$$\mathbf{q} = -k\nabla(p - \rho g z), \tag{5}$$
$$\Delta p = 0, \tag{6}$$

Δ denoting the Laplacian operator.

When the flow is not saturated, then k is a decreasing function of saturation S (vanishing for $S = 0$) and mass conservation is no longer expressed by (4), but it becomes

$$\frac{\partial}{\partial t}(\phi S) + \operatorname{div} \mathbf{q} = 0 \tag{7}$$

(in the absence of sources or sinks).

In this case (5) and (7) are no longer sufficient to describe the flow, but we need a relationship between S and p, which is physically provided by *capillarity*. The function $S = S(p)$ is equal to 1 for $p \geq p_S$ (*saturation pressure*) and is monotonically decreasing for $p < p_S$. Once it has been wet a porous medium will contain at least a thin film of liquid surrounding the grains, no matter how far p is from p_S. For unsaturated flows, with $\phi =$constant, (5) and (7) lead to a parabolic equation for pressure, possibly degenerating at the interface with the saturated region.

One can also consider the limit case in which the effect of capillarity is totally disregarded (such an approximation makes sense in particular in the presence of a sufficiently fast *wetting front*). In this extreme situation the medium is either saturated or dry and in invasion problems the wetting front coincides with the saturation front.

The prototype of a penetration problem with no capillarity is the Green-Ampt model, developed in 1911 for the injection of water into a dry soil [20]. This is a one-dimensional model of a vertical flow in which the pressure penetration front $z = s(t)$ is the atmospheric pressure ($p = 0$), while at the inflow surface ($z = 0$) the injection pressure is specified as $p = p_0 > 0$. Since the advancing front is a material surface, its velocity is given by Darcy's law, hence

$$\phi \dot{s}(t) = -k\frac{\partial p}{\partial z} + k\rho g. \tag{8}$$

Because of incompressibility equation (6) applies, i.e. $\dfrac{\partial^2 p}{\partial z^2} = 0$, and therefore we find $p(z,t) = p_0\left(1 - \dfrac{z}{s(t)}\right)$, and (8) reduces to a simple first order o.d.e.

$$\phi \dot{s} = k\left(\frac{p_0}{s(t)} + \rho g\right), \qquad s(0) = 0. \tag{9}$$

Setting $\sigma = \dfrac{\rho g s}{p_0}$, we find

$$\sigma - \log(1 + \sigma) = \frac{k(\rho g)^2}{p_0 \phi}t. \tag{10}$$

For short times (r.h.s. $\ll 1$) we get $\sigma \sim \rho g \left(\dfrac{2kt}{p_0 \phi}\right)^{1/2}$, i.e. $s \sim \left(\dfrac{2kp_0 t}{\phi}\right)^{1/2}$ and the motion is dominated by pressure, while for large times $s \sim \dfrac{k\rho g}{\phi} t$ and the flow is driven by gravity.

A great deal of filtration processes taking place in many different technological applications are nontrivial variants of the Green-Ampt model. Not only just because of geometrical questions (anisotropy and in general multidimensional flows), but particularly for the simultaneous occurrence of other processes interacting with the flow, of mechanical, thermal and chemical nature.

Some of these cases will be considered in this course, but many more will not, like for instance most of the flows with deformations of the porous medium (see the review paper [23] and the recent studies on collapsing media [13, 14]), and flows accompanied by phase change (see the review paper [29]).

A naive aspect of Green-Ampt model is that it assumes that the simple 1-D geometry is preserved. However flows with a penetration front may exhibit instabilities leading to steadily amplifying deformations of the front (the well known *fingering* phenomenon). Such a situation is well known to occur also during the displacement of a liquid by another less viscous liquid. The problem of the motion of two immiscible incompressible fluids separated by an interface is indeed very important, in view of it applications to oil reservoir technology, and was first studied by Muskat [22]. This is clearly a free boundary problem in which the Laplace equation for pressure has to be solved in both phases and the interface conditions are continuity of pressure (neglecting capillarity), and of volumetric velocity, and Darcy's law expressing the interface velocity.

A variant of Muskat problem, considering a slight compressibility of the fluid, thus leading a parabolic equation for pressure, was proposed by Verigin [30].

There are formidable computational difficulties connected with this problem, mainly due to the natural instability of the moving interface, that have been discussed in this course.

The motion of two immiscible fluids in the ground under the effect of gravity has been extensively studied in connection with soil hydrology and oil reservoir engineering.

An extensively used procedure for recovering oil from reservoir not having a sufficient natural pressure is to inject water through a wall, thus pushing oil towards another wall. Much work has been done with different numerical methods in order to simulate this process, which is still studied very actively because of its economical relevance.

A typical problem in soil hydrology is the study of how fresh water is displaced underground by a denser liquid like salt water. On the large scale the two fluids can be treated as immiscible. It is clearly important to know where the interface is for fresh water recovery drilling.

A very classical problem belonging to the category of gravity driven flows is the extremely well known dam problem, about which a flourishing literature has been produced during the last three decades. In its basic version the problem consists in studying the stationary water flow between two basins at different levels, separated by a homogeneous porous wall with an inferior basis.

A fundamental contribution to the solution of this problem was given by C. Baiocchi [3] who, with his celebrated transformation (subsequently used in other contexts), reformulated it in the form of a variational inequality. A great number of extensions have been studied (time dependent problems, inhomogeneous and/or leaky dams, complicated geometries, etc.). The incredible viability of this problem is testified by recent surveys [11] (see also [1] [9]).

When fluids are not separated by a sharp interface, then quite different models are used in which partial saturations must be determined. A very useful accounts of this theory can be found in [28] (Chapt. 8). For the case of two liquids one introduces the two complementary saturations $\sigma, 1 - \sigma$ and writes the flow equations

$$\mathbf{v}_i = -\frac{k}{\mu_i} f_i(\sigma) \nabla p_i, \qquad i = 1, 2 \tag{11}$$

where $f_i(\sigma)$ are the phase permeabilities and p_1, p_2 are the partial pressures, linked by the capillarity relation

$$p_1 - p_2 = p_c(\sigma), \tag{12}$$

$p_c(\sigma)$ being the experimentally known capillarity pressure. The compound velocity $\mathbf{v} = \mathbf{v}_1 + \mathbf{v}_2$ is divergence free. In one space dimension, if one assumes that the volumetric velocity $v_1 + v_2 = q(t)$ is given, the overall mass balance equation takes the form

$$\phi \frac{\partial \sigma}{\partial t} - q(t) \frac{\partial}{\partial x} F(\sigma) = \frac{\partial}{\partial x} \left(\lambda(\sigma) \frac{\partial \sigma}{\partial x} \right), \tag{13}$$

where

$$F(\sigma) = \frac{f_1(\sigma)}{f_1(\sigma) + f_2(\sigma) \mu_2/\mu_1} \tag{14}$$

is known as the Buckley-Leverett function and

$$\lambda(\sigma) = k \frac{f_2}{\mu_2 f_1 + \mu_1 f_2} p_c'(\sigma) \tag{15}$$

should be called the Rapoport-Leas function (Rapoport and Leas [27] derived equation (13) adding the capillarity effect to the simpler model proposed much earlier by Buckley-Leverett [10]). Equation (13) has been used for long time in oil reservoirs engineering.

It must be stressed that the mobility of each species is considered zero when its concentration reaches some critical value. This is reflected in the model taking $f_1(\sigma) = 0$ for $\sigma \in (0, \sigma_1)$ and $f_2(\sigma) = 0$ for $\sigma \in (\sigma_2, 1)$. Moreover the capillary pressure becomes singular at σ_1 and tends rapidly to very small values as σ increases. Therefore from the mathematical point of view, even if we restrict our attention in the interval (σ_1, σ_2) we deal with the difficulty of the nonuniform parabolicity of equation (13).

Multiphase flow problems are always extremely delicate also from the point of view of modeling. We cannot develop further this question here, but we quote the paper [12], dealing with the process of filling a fibermat with resin in order to produce a composite material. The interesting aspect of the paper is that it does not neglect the flow of air, as it is frequently done (including the lectures in this course devoted to such a topic), which for this specific problem is critical, because the formation of air bubbles must be absolutely avoided. The basic point of such two-phase flow model is the correct definition of relative permeabilities of the medium for each species, which are related to respective partial saturations.

This short sketch has the scope of introducing a few basic concepts, but it certainly fails to show the impressive variety of problems in this vast research area. We did not mention for instance biological applications (the flow of fluids through body tissues, the remediation of soils by mean of bacteria colonies, etc.). However we hope these few pages can at least arise the curiosity of non-experts and stimulate further reading.

References

[1] H.W. Alt. A free boundary problem connected with the flow of ground water. *Arch. Rat. Mech. Anal.* **64** (1977), 111–126.

[2] D.G. Aronson. The Porous Media Equation. In "Nonlinear Diffusion Problems", Springer Lecture Notes in Mathematics 1224, A. Fasano, M. Primicerio eds. (1986), 1–46

[3] C. Baiocchi. Su un problema di frontiera libera connesso a questioni di idraulica. *Ann. Math. Pura Appl.* **92** (1972), 107–127.

[4] G.I. Barenblatt. On some unsteady motions of a liquid or a gas in a porous medium. *Prikl. Mat. Meh.* **16** (1952), 67–78.

[5] I. Barenblatt, V.M. Entov, and V.M. Ryzhik. *Theory of fluid flows through natural rocks.* Kluwer, 1990.

[6] J. Bear. *Dynamics of Fluids in Porous Media.* America Elsevier, New York, 1972.

[7] J. Bear and A. Verruijt. *Modelling Ground water Flow and Pollution.* Reidel, Dordrecht, N. Y., 1987.

[8] A. Bourgeat and A. Mikelić. Note on the homogenization of Bingham flows through porous medium. *Journ. Math. Pures et Appliquées,* **72** (1993), 405–414.

[9] H. Brezis. The dam problem revisited. In *Free Boundary Problems: Theory and Applications,* A. Fasano, M. Primicerio eds., Pitman Res. Notes in Math. **78** (1983), 77–87.

[10] S.F. Bukley and M.C. Leverett. Mechanism of fluid displacement in sands. *Trans. AIME* **146** (1942), 107.

[11] M. Chipot. Variational inequalities and flow in porous media. *Appl. Math. Sci.* **52**, Springer-Verlag (1984).

[12] W.K. Chui, J. Glimm, F.M. Tangerman, A.P. Jardine, J.S. Madsen, T.M. Donnellan and R. Leek. Process modeling in resin transfer molding as a method to enhance product quality. *Siam Rev.* **39** (1997), 714–727.

[13] E. Comparini and M. Ughi. Shock propagation in a one-dimensional flow through deformable porous media. To appear.

[14] E. Comparini and M. Ughi. On the existence of shock propagation in a flow through deformable porous media. To appear.

[15] H. Darcy. *Les Fontaines Publiques de la Ville de Dijon*, Dalmont, Paris 1856.

[16] A. Fasano. Some non-standard one-dimensional filtration problems. *Bull. Fac. Ed. Chiba Univ.* **44**, (1996), 5–29.

[17] A. Fasano. Nonlinear processes of injection of liquids through porous materials. *Rend. Sem. Mat. Fis. Milano.* To appear.

[18] A. Fasano, A. Mikelić. On the filtration through porous media with partially soluble permeable grains. to appear on NoDEA.

[19] A. Fasano, A. Mikelić, and M. Primicerio. Homogenization of flows through porous media with permeable grains. *Adv. Math. Sci. Appl.* **8** (1998), 1–31.

[20] W.H. Green and G.A. Ampt. Studies on soil physics. The flow of air and water through soils. *J. Agric. Sci.* **4**, (1911), 1–24.

[21] U. Hornung. Homogenization and Porous Media.. *Interdisciplinary Applied Mathematics* **6**, Springer Verlag (1996).

[22] M. Muskat. Two-fluid systems in porous media. The encroachment of water into an oil sand, Physics **5** (1934), 250–264.

[23] L. Preziosi. The theory of deformable porous media and its applications to composite material manufacturing. *Surveys Math. Ind.* **6**, (1996), 167–214.

[24] M. Primicerio. Analisi fisico-matematica di problemi di filtrazione. *Rend. Sem. Mat. Fis. Milano* **60** (1990), 133–155.

[25] M. Primicerio and R. Gianni. La Filtracion en Medios Porosos. *III Seminario sobre Problemas de Frontera Libre y sus Aplicaciones.* Cuadernos del Inst. de Matematica "Beppo Levi" **18** (1989).

[26] K.R. Rajagopal and L. Tao. *Mechanics Mixtures.* World Scientific (1995).

[27] L.A. Rapoport and W.J. Leas. Properties of linear waterfloods. *Trans. AIME* **196** (1953), 139.

[28] I. Rubinstein and L. Rubinstein, *Partial Differential Equations in Classical Mathematical Physics*, Cambridge University Press (1993).

[29] F. Talamucci. Analysis of the coupled heat-mass transport in freezing porous media. *Survey in Mathematics for Industry* **7**, (1997), 93–139.

[30] N.N. Verigin On the pressured forcing of binder solutions into rocks in order to increasing the solidity and imperviousness to water of the foundations of hydrotechnical constructions (in Russian). *Izv. Akad. Nauk.* SSSR Otd. Techn. Nauk. **5** (1952), 674–687.

NUMERICAL SOLUTION OF RESERVOIR FLOW MODELS BASED ON LARGE TIME STEP OPERATOR SPLITTING ALGORITHMS

— C.I.M.E. LECTURE NOTES —

MAGNE S. ESPEDAL AND KENNETH HVISTENDAHL KARLSEN

ABSTRACT. During recent years the authors and collaborators have been involved in an activity related to the construction and analysis of large time step operator splitting algorithms for the numerical simulation of multi-phase flow in heterogeneous porous media. The purpose of these lecture lecture notes is to review some of this activity. We illustrate the main ideas behind these novel operator splitting algorithms for a basic two-phase flow model. Special focus is posed on the numerical solution algorithms for the saturation equation, which is a convection dominated, degenerate convection-diffusion equation. Both theory and applications are discussed. The general background for the reservoir flow model is reviewed, and the main features of the numerical algorithms are presented. The basic mathematical results supporting the numerical algorithms are also given. In addition, we present some results from the BV (entropy) solution theory for quasilinear degenerate parabolic equations, which provides the correct mathematical framework in which to analyse our numerical algorithms. Two- and three-dimensional numerical test cases are presented and discussed. The main conclusion drawn from the numerical experiments is that the operator splitting algorithms indeed exhibit the property of resolving accurately internal layers with steep gradients, give very little numerical diffusion, and, at the same time, permit the use of large time steps. In addition, these algorithms seem to capture all potential combinations of convection and diffusion forces, ranging from convection dominated problems (including the pure hyperbolic case) to more diffusion dominated problems.

CONTENTS

1. INTRODUCTION

The flow in subsurface hydrology and the exploitation of hydrocarbons in a reservoir represent very complex processes of controlling the interaction between several fluids and rock [49, 75, 127]. Capillary and gravity forces are important for the dynamics, and the presence of heterogeneities in natural formations have a large effect on the flow. Accurate modelling of the rock lithology, including the modelling of scale dependent physical parameters and relations, is crucial for obtaining reliable results. The dynamics of the porous media itself add to the complexity. Change in the pressure balance between the overburden pressure and the fluid pressure in the reservoir will change porosity and permeability. Temperature gradients created by injecting fluids having lower temperature than the reservoir will create cracking which can effect the flow properties strongly. Models for flow dynamics should therefore be coupled to a geodynamical model describing the dynamics of the rock. Often a reservoir has horizontal wells with several branches, and the coupling between the wells and the reservoir represents a very difficult modelling task.

A series of models [9, 40, 137] have been constructed to describe the complex dynamics of reservoir flow processes. A black-oil model is commonly used to describe water injection. This model works well in simulating water flooding processes because reservoir hydrocarbons are not soluble in the water phase. But it may not be well suited when the composition of the reservoir fluid is dramatically changed because of chemical reactions. Consequently, fluid properties and phase behaviour should in many cases be functions of reservoir pressure, temperature, and fluid composition. The geological model together with the chemical and physical complexity of multi-phase, multi-component flow pose very large and difficult problems to solve. However, the vast expansion of computational capability has made a quantitative description of more realistic and complex models possible. Along with this development, there has also been a very active development of new numerical solution algorithms for such models.

The fluid flow models usually represent strongly coupled systems of nonlinear partial differential equations. A number of new algorithms [87, 102, 120, 129, 138, 152, 164] for the solution of (systems of) nonlinear partial differential equations have been developed over the years. But still there is acute need for better solution algorithms as well as mathematical theory supporting them (and the models). The different scales of variation appearing in a reservoir flow model demand an adaptive adjustment of the solution algorithm to the problem at hand. For compressible fluid or rock models, numerical algorithms which provide local conservation properties are needed. Moreover, the transport part of these models needs accurate pressure and velocity calculations. To resolve these issues one may need efficient finite volume methods, see, e.g., [1, 2, 3].

There is a close relationship between the challenges in mathematical and numerical modelling and the development of mathematical and numerical tools. An example is provided by operator splitting algorithms which calculate an approximate transport in a first step. These algorithms have become an important part of many solution methods for fluid dynamics, see, e.g., [12, 86, 116, 145, 162], and reservoir dynamics, see, e.g., [63, 106, 174]. The main purpose of these lecture notes is to present some recently developed operator splitting algorithms for models describing reservoir flow [9, 40, 137] and certain sedimentation processes [25, 27, 35, 44], as well as other models that can be reduced to a sequence of nonlinear convection-diffusion equations.

The reservoir flow model considered here contains a flow equation which is a highly nonlinear parabolic partial differential equation with a strong transport term. One important characteristic of the model is the degeneracy of the second order diffusion operator in the flow equation. The diffusion operator may be zero pointwise, it can be small or zero in regions of the solution space, and fairly large for other values of the solution. Consequently, solutions of the flow equation will in general possess minimal smoothness. The correct mathematical framework in which to treat such nonlinear partial differential equations — as well as to analyse their numerical solution algorithms — is provided by the BV (entropy) solution theory (analysis in classes of discontinuous functions [169]). For completeness, we will recall some important results from the BV (entropy) solution theory for second order partial differential equations. Mathematical theory supporting the numerical algorithms have been developed within this discontinuous solution framework, and central elements of this theory will be presented in a simplified context.

Roughly speaking, one can say that the splitting algorithms simplify the original problem into a hyperbolic problem and an almost symmetric (degenerate) parabolic problem, each of which is solved separately by suitable numerical methods. The main feature of the operator splitting algorithms is the ability to use long time steps and, at the same time, keep the numerical diffusion at a minimum. Several numerical examples will be given to illustrate these and other properties of the splitting algorithms. The nonlinearity inherent in the models gives a rich variation in the solutions. In particular, the solutions may develop sharp or even discontinuous fronts that propagate throughout the reservoir. Also, fractured and faulted regions may create large variations in the solutions. Local grid refinement techniques may therefore be necessary in order to resolve the physical phenomena within the limits of a computational model. Combining operator splitting and domain decomposition algorithms, we get very flexible adaptive mesh refinement and coarsening techniques [53, 146].

The remaining part of these notes is organised as follows: In §2, we present and discuss the reservoir flow model. This basic model consists of an elliptic equation which is coupled to a nonlinear convection-diffusion equation. A sequential time marching strategy will be used to decouple the equations. Since the solution algorithm for the elliptic equation is well described in the literature, we focus in these lecture notes entirely on the convection-diffusion equation. In §3, we recall a few results from the BV (entropy) solution theory for first and second order nonlinear partial differential equations. In §4, we give a fairly thorough presentation of the novel operator splitting algorithms for nonlinear convection-diffusion equations. Both analytical and numerical results for the splitting algorithms are presented. In §5, two numerical methods for hyperbolic problems are presented.

In §6, numerical methods for parabolic problems are presented. Finally, two reservoir simulation examples are presented and discussed in §7.

2. The Reservoir Flow Model

A reservoir may consist of several different types of sediments, which in general have different porosity, absolute permeability, relative permeabilities, and capillary pressure. To efficiently compute the solution of the nonlinear system of reservoir flow equations, one should carefully choose the primary variables based on all aspects of the computations and the inherent physics [9]. The pressure gradient is the driving force that causes the flow of reservoir fluids. Thus pressure should be chosen as a primary variable to describe the flow process. Normally a pressure equation is obtained by making a summation over all components and phases in the model. Below we give a pressure equation for a two-component, two-phase model which is derived in this way. For compressible flow, a mass balance formulation of the pressure equation based on a finite volume discretization is a better choice, see Reme *et al.* [147, 148]. Since the temperature T can vary, it should also be chosen as a primary variable. The temperature is normally governed by a convection-diffusion equation, see [147, 148]. Moreover, $N + 1$ primary variables are needed to characterise the component-transport phenomena (one water component and N hydrocarbon components). If the moles of component i in phase j (for $i = w, 1, \ldots, N$ and $j = w, o, g$), which we denote by m_j^i, is chosen as the primary variable for the ith component, the component-transport equations become coupled with strongly nonlinear phase equilibrium constraints. Consequently, computing the solution of the overall system can be extremely expensive. Furthermore, m_j^i is less smooth than the total moles of the component, i.e.,

$$(1) \qquad\qquad m^i = \sum_{j=o,g,w} m_i^j.$$

Thus a numerical approximation of m^i will be more accurate than an approximation of m_j^i, which also affects the accuracy of the overall approximations. Based on these observations, we choose m^i as the primary variable for the ith component. In this way the solution of the component-transport equations is decoupled from the computation of the phase equilibrium, which in turn reduces the nonlinearity inherent in the reservoir flow model. Since mass is exchanged between phases, component mass is not conserved within each phase, but the total mass of each component is conserved. The corresponding mass conservation can be described by the system of equations

$$(2) \qquad \partial_t\left(\phi m^i\right) + \nabla \cdot \left(\sum_{j=o,g,w} c_j^i \xi^i v_j + d\nabla m^i \right) = q^i, \qquad i = w, 1, \ldots, N,$$

where m^i is defined in (1), ϕ is the porosity of the rock, q^i is the molar flow rate of component i, c_j^i is the mole fraction of component i in phase j, ξ^i is the molar density of component i, d represents diffusion, and the Darcy velocities are given as

(3)
$$v_j = -\lambda_j(s_j)K(x)(\nabla p_j - \rho_j g \nabla h), \qquad j = w, o, g.$$

In (3), h is the height, g is the gravity constant, and ρ_j is the density of phase j. Furthermore, $K(x)$ is the absolute permeability of the rock, which may be a full tensor for anisotropic and heterogeneous porous media, $k_{rj}(s_j)$ are the relative permeabilities, μ_j are the viscosities of the fluids, s_j is the saturation of phase j, $\lambda_j = k_{rj}/\mu_j$ are the phase mobilities. In addition, the nonlinear partial differential equations presented above are coupled with a complex phase package that describes the relative phase equilibrium for a given pressure-volume-temperature through the equation of state (e.g., the Peng-Robinson equation of state). The phase saturations s_j, mole fractions c_i^j, and other secondary variables are determined by the thermodynamic equations. The fluid dynamics in a porous media may be influenced by changes in the properties of the rock. This means that we must also couple the flow model to a geomechanical model.

The molar mass model presented above gives a nonlinear system of convection-diffusion equations. We will in these notes use a two-component, two-phase, immiscible flow model as our main test case. This simplified model contains many of the mathematical and numerical challenges present in the general model described above. For the two-phase model, (2) can be expressed in terms of the phase saturations. Let $s = s_w$ denote the water saturation (and $1 - s$ the oil saturation). Then the incompressible displacement of oil by water in a porous medium can be described by the following set of partial differential equations (given in dimensionless form):

(4)
$$\nabla \cdot v = q_1(x),$$

(5)
$$v = -K(x)\lambda(x, s)(\nabla p - \rho(s)\nabla h),$$

(6)
$$\phi(x)\partial_t s + \nabla \cdot (f(s)v + f_g(s)K\nabla h) - \varepsilon\nabla \cdot (d(x, s)\nabla s) = q_2(x),$$

where q_1 and q_2 denote the injection/production wells, ε is a dimensionless scaling parameter, $v = v_w + v_o$ is the total Darcy velocity, $\lambda(x, s)$ denotes the total mobility of the phases,

$$\lambda(x, s) = \lambda_w(x, s) + \lambda_o(x, s).$$

We assume that the immobile water and oil are the same in different types of sediments, so that the saturation can be normalised globally. Furthermore, since the mobility is assumed to be constant in each sediment, we will from now on drop the space dependence in λ and λ_j. In (5), p is the global pressure [40]

(7)
$$p = \frac{1}{2}(p_w + p_o) + \frac{1}{2}\int_{s_c}^{s}\left(\frac{\lambda_o - \lambda_w}{\lambda}\frac{\partial p_c}{\partial \xi}\right)d\xi,$$

where p_w and p_o denotes the pressure of water and oil, respectively, $p_c(x, \xi)$ ($= p_o - p_w$) is the capillary pressure function (see (11) below), s_c is chosen such that $p_c(x, s_c) = 0$. In equations (5)-(6), we have

$$\rho(s) = g(\lambda_w\rho_w + \lambda_o\rho_o),$$

and

$$f_g(s) = (\rho_w - \rho_o)f(s)\lambda_o.$$

Equation (6) is the fractional flow formulation of the mass balance equation for water. The fractional flow function $f(s)$, which is typically an S-shaped function of s, is given as

(8)
$$f(s) = \frac{\lambda_w(s)}{\lambda_w(s) + \lambda_o(s)}.$$

The diffusion function is given as

(9)
$$d(x, s) = -K(x)f(s)\lambda_o(s)\frac{\partial p_c}{\partial s},$$

where the derivative of the capillary pressure function $p_c(x, s)$ is assumed to be negative, see (11) below. We refer to [9, 40, 137] (see also [7]) for a complete survey and justification of the model.

For computational purposes, the analytical forms for the relative permeabilities are chosen as

(10)
$$k_{rw} = s^p, \qquad k_{ro} = (1 - s)^q, \qquad p, q = 2 \text{ or } 3.$$

The capillary pressure may depend on the porosity and the permeability of the rock. We use a capillary pressure function of the form [62]

(11)
$$p_c(x, s) = 0.9 \cdot \phi^{-0.9}K^{-\beta}\frac{1 - s}{\sqrt{s}}, \qquad \beta > 0.$$

Note that the form of (10) and (11) imply that the global pressure function (7) is well-defined. In Figure 1 a), the variation of the convection term in the saturation equation (6) is shown for different values of the gravity term with a constant Darcy velocity v. In Figure 1 b), the diffusion function is plotted for different permeabilities. Note that the diffusion function (9) vanishes for $s = 0, 1$, and that it can be very small in regions close to these endpoints, see Figure 1 b). This means that (6) is a degenerate parabolic equation and close to being strongly degenerate in regions of the solution space. Mathematical and numerical theory for (strongly) degenerate parabolic equations will be given in §3 and §6.1, respectively.

The reservoir flow equations (4)-(5)-(6) are nonlinearly coupled. A sequential time marching strategy will be used to decouple the equations, see §7 for details. The pressure-velocity equations (4)-(5) are solved in a first step using standard finite element and domain decomposition methods, which are well documented in the literature [8, 156, 158]. We will in these lecture notes therefore limit our presentation to solution algorithms for the convection-diffusion equation (6).

3. Basic Mathematical Theory

The main purpose of this section is to recall some results from the theory of BV (entropy) solutions for second order nonlinear (or more precisely quasilinear) partial differential equations. For completeness, we also recall a few basic results from the theory of entropy solutions for hyperbolic partial differential

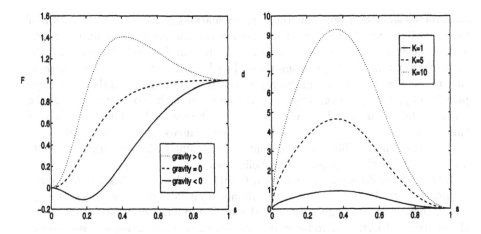

FIGURE 1. a) $F(x, s) = f(s)v + f_g(s)K\nabla h$ plotted as a function of saturation for a fixed Darcy velocity $v = 1$, permeability $K = 1$, and ∇h equal to -1, 0, and 1. b) $d(x, s)$ plotted as a function of saturation for $\beta = 0.1$ and permeabilities K equal to 1, 5, and 10.

equations. Although the main portion of our work is related to the design of numerical algorithms for the saturation equation (6), the mathematical theory presented below has been important for both the development and the analysis of these algorithms. We have tried to make this section (relatively) self-contained. Nevertheless, it presupposes that the reader has some basic knowledge about functional analysis, measure theory and BV theory, and thus no attempts have been made to discuss in detail the various function spaces, (semi) norms, and compactness theories that are used in these lecture notes. The reader who is primarily interested in numerical solution algorithms can skip this section.

We do not claim completeness or technical generality in this section. In fact, a significant limitation of our presentation is that only the Cauchy problem for homogeneous partial differential equations without source term is discussed. But we stress that the L^1/BV solution theory outlined below also applies (with necessary modifications) to various boundary value problems as well as partial differential equations with variable coefficients and source term.

Numerical algorithms and the techniques used for their analysis tend to be very different in the two limiting cases of hyperbolic and parabolic equations, see, e.g., [138]. However, independently of whether the problem is hyperbolic or parabolic, in this work *all* the analysis is carried out within the framework provided by the BV (entropy) solution theory. Since we try to approximate solutions which in general possess minimal smoothness (due to hyperbolic dominance and parabolic degeneracy), the BV framework is well suited for our purpose. Moreover, the BV framework is also consistent with the theory of entropy solutions for hyperbolic partial differential equations.

Later we show that our numerical approximations converge to the solution of the underlying problem as the discretization parameters tend to zero (see §4). The convergence proof is based on deriving uniform L^∞ and BV (space) estimates for the numerical approximations. The BV estimate is then used to show that the approximations are uniformly L^1 continuous in time, and therefore compact in L^1_{loc}. Compactness arguments are classical in the context of hyperbolic partial differential equations and go back to Oleĭnik [141]. The advantage of this "hyperbolic approach" is that the derived estimates are independent of the lower bound on the diffusion coefficient. Consequently, the convergence results are also valid in the degenerate parabolic case.

If Ω is a domain in $\mathbb{R}^n$, $n \geq 1$, then $C^p(\Omega)$, $p = 1, \ldots, \infty$, denotes the space of functions $z : \Omega \to \mathbb{R}$ possessing continuous partial derivatives of order $\leq p$. When there is no ambiguity, we omit the set Ω and write only C^p (similarly for the spaces and norms introduced below). The space consisting of functions in $C^p(\Omega)$ with compact support is denoted by $C_0^p(\Omega)$. We denote by $\mathrm{Lip}(\Omega)$ the Lipschitz space consisting of bounded functions $z : \Omega \to \mathbb{R}$ that satisfy

$$(12) \qquad \|z\|_{\mathrm{Lip}(\Omega)} = \sup\left\{ \left| \frac{z(y_2) - z(y_1)}{y_2 - y_1} \right| : y_1, y_2 \in \Omega, y_1 \neq y_2 \right\} < \infty.$$

If $\left| \frac{z(y_2)-z(y_1)}{y_2-y_1} \right|$ in (12) is replaced by $\max\left(0, \frac{z(y_2)-z(y_1)}{y_2-y_1}\right)$, we write $\|z\|_{\mathrm{Lip}^+(\Omega)}$. The space of locally Lipschitz continuous functions $z : \Omega \to \mathbb{R}$ is denoted by $\mathrm{Lip}_{\mathrm{loc}}(\Omega)$, i.e., functions whose restriction to any compact subset $\mathcal{K} \subset \Omega$ is Lipschitz continuous. The classical L^p spaces of real-valued functions on Ω are denoted by $L^p(\Omega)$, and the norms on $L^p(\Omega)$ are denoted by $\|\cdot\|_{L^p(\Omega)}$. We denote by $L^p_{\mathrm{loc}}(\Omega)$ the space of functions on Ω that are in $L^p(\mathcal{K})$ for any compact subset $\mathcal{K} \subset \Omega$.

As we have tried to indicate above, the space $BV(\Omega)$ consisting of functions $z : \Omega \to \mathbb{R}$ of bounded variation is of fundamental importance to us. A function $z \in L^1_{\mathrm{loc}}(\Omega)$ is an element of $BV(\Omega)$ if and only if its first order distributional derivatives are represented by locally finite Borel measures on Ω, i.e., if there exist Borel measures μ_j, $j = 1, \ldots, n$, such that

$$-\int_\Omega z \frac{\partial \phi}{\partial y_j}\, dy = \int_\Omega \phi\, d\mu_j, \qquad \forall \phi \in C_0^\infty(\Omega),$$

and

$$|\mu_j|(\mathcal{K}) < \infty \text{ for each compact subset } \mathcal{K} \subset \Omega,$$

where $|\mu_j|$ denotes the total variation of the measure μ_j. Let the total variation of $z \in L^1_{\mathrm{loc}}(\Omega)$, which we denote by $|z|_{BV(\Omega)}$, be defined as

$$|z|_{BV(\Omega)} = \sup\left\{ \int_\Omega z\, \nabla \cdot \phi\, dy : \phi \in C_0^\infty(\Omega; \mathbb{R}^n), |\phi(y)| \leq 1, \forall y \in \Omega \right\}.$$

Then using Riesz's theorem on functionals in the space of continuous functions, we obtain that $BV(\Omega)$ can be equivalently defined as

$$BV(\Omega) = \left\{ z \in L^1_{loc}(\Omega) : |z|_{BV(\Omega)} < \infty \right\}.$$

We say that $z(y)$ has bounded Tonelli variation if for any $j = 1, \ldots, n$, $u(y)$ considered as a function of y_j has essential (up to the set of one-dimensional measure zero) variation which is integrable with respect to the other variables $y_1, \ldots, y_{j-1}, y_{j+1}, \ldots, y_n$. A function $z(y)$ is in $BV(\Omega)$ if and only if the Tonelli variation of $z(y)$ is bounded. Furthermore, $z \in BV(\Omega)$ if and only if there is a constant C such that for any $h \in \mathbb{R}^n$ with $y + h \in \Omega$,

$$\int |z(y+h) - z(y)| \, dy \le C|h|.$$

The space $BV(\Omega)$ is a Banach space when equipped with the norm

$$\|z\|_{BV(\Omega)} = \|z\|_{L^1(\Omega)} + |z|_{BV(\Omega)}.$$

It is well known that the following inclusions hold:

$$BV(\Omega) \subset L^{\frac{n}{n-1}}(\Omega) \text{ for } n > 1 \text{ and } BV(\Omega) \subset L^\infty(\Omega) \text{ for } n = 1.$$

Furthermore,

$$BV(\Omega) \text{ is compactly imbedded into } L^p(\Omega) \text{ for } 1 \le p < \frac{n}{n-1}.$$

After a possible modification of $z(y)$ on a set of zero n - dimensional Lebesgue measure, the domain Ω of definition of $z(y)$ is the disjoint union of a set A_z of points of approximate continuity, a set Γ_z of points of approximate jump discontinuity, and a set I_z of irregular points;

$$\Omega = A_z \cup \Gamma_z \cup I_z.$$

The set of irregular points has zero $(n-1)$ - dimensional Hausdorff measure. Hence, with the exception of a small set of irregular points, an arbitrary BV function $z : \Omega \to \mathbb{R}$ is either approximately continuous or has an approximate jump discontinuity. If $z(y)$ is a BV solution of a nonlinear partial differential equation, the set Γ_z can be though of as representing the discontinuities (shocks) in $z(y)$, the set A_z as representing the discontinuity free (smooth) region, and I_z as representing the points of shock formation and shock collision. We refer to [60] (and the references cited therein) for a more complete discussion and interpretation of the sets A_z, Γ_z, I_z in the context of BV solutions of hyperbolic partial differential equations. For an introduction to BV (solution) theory, we refer to [66, 166, 167, 169, 185].

For completeness, we recall a few results from the Kružkov theory of entropy solutions for scalar hyperbolic partial differential equations. For an introduction to general theory for hyperbolic problems, we refer to [102, 136, 159]. We consider nonlinear hyperbolic problems of the form

$$(13) \quad \begin{cases} \partial_t u + \nabla \cdot f(u) = 0, & (x,t) \in \Pi_T = \mathbb{R}^m \times (0,T), \ m \ge 1, \\ u(x,0) = u_0(x), & x \in \mathbb{R}^m, \end{cases}$$

where $u : \Pi_T \to \mathbb{R}$ is the unknown function. We always assume that the vector valued flux function $f = (f_1, \ldots, f_m)$ is sufficiently smooth (e.g., in Lip_{loc}) and that the initial function u_0 belongs to $L^1 \cap L^\infty \cap BV$. It is well known that non-linearity has dramatic effects on hyperbolic waves, most notably in the formation of shock waves (discontinuous solutions), a feature that can reflect the physical phenomenon of breaking of waves. Due to this loss of regularity it is necessary to work with weak solutions. For hyperbolic equations, due to neglected physical (e.g., dissipative) mechanisms, weak solutions are not uniquely determined by their initial data. However, weak solutions satisfying an entropy condition (see below) are uniquely determined by their data.

Recall that a weak solution of the Cauchy problem (13) is a bounded measurable function $u(x, t)$ that satisfies the integral identity

$$(14) \qquad \iint \left(u \partial_t \phi + f(u) \cdot \nabla \phi \right) dt dx + \int u_0(x) \phi(x, 0) \, dx = 0,$$

for all test functions $\phi \in C_0^\infty$ such that $\phi|_{t=T} = 0$. Let $\eta : \mathbb{R} \to \mathbb{R}$ be a convex C^2 function, often referred to as the entropy function, and $q_1, \ldots, q_m$ the associated entropy fluxes satisfying the compatibility conditions

$$(15) \qquad q_j'(u) = \eta'(u) f_j'(u), \qquad j = 1, \ldots, m.$$

We let q denote the vector $(q_1, \ldots, q_m)$. Let u denote the L_{loc}^1 limit of classical solutions u_ε of the parabolic regularisation

$$(16) \qquad \partial_t u_\varepsilon + \nabla \cdot f(u_\varepsilon) = \varepsilon \Delta u_\varepsilon, \qquad u_\varepsilon(x, 0) = u_0(x),$$

as $\varepsilon \to 0$. Equipped with the compatibility conditions (15), it is straightforward to show that the "vanishing viscosity" limit u satisfies the entropy inequality

$$\iint \left(\eta(u) \partial_t \phi + q(u) \cdot \nabla \phi \right) dt dx \geq 0,$$

for all non-negative $\phi \in C_0^\infty(\Pi_T)$, see Kružkov [123]. By a limiting argument, we can let

$$\eta(u) \to |u - k|,$$

for any given $k \in \mathbb{R}$, and use

$$q(u) = \text{sign}(u - k)(f(u) - f(k)).$$

We thus end up with the following definition due to Kružkov [123] (see also Vol'pert [166]):

Definition 3.1. *A weak solution $u(x, t)$ of (13) is called an entropy weak solution if, for all non-negative test functions $\phi \in C_0^\infty(\Pi_T)$ and $k \in \mathbb{R}$,*

$$(17) \qquad \iint \left(|u - k| \partial_t \phi + \text{sign}(u - k)(f(u) - f(k)) \cdot \nabla \phi \right) dt dx \geq 0.$$

Kružkov [123] (see also [14, 124]) proved that an entropy weak solution of (13) is uniquely determined by its initial data, and thus coincides with the vanishing viscosity solution. More precisely, he proved the following result: If u and v are entropy weak solutions of (13) with initial data u_0 and v_0, respectively, then

$$(18) \qquad \|u(\cdot,t) - v(\cdot,t)\|_{L^1} \leq \|u_0 - v_0\|_{L^1}.$$

We note that Vol'pert [166] only proved uniqueness of entropy weak solutions in the BV class. The Kružkov theory for the pure initial value problem (13) can be extended to the initial-boundary value problem, see Bardos, LeRoux, and Nédélec [10] and Otto [143] for details.

Soon after the penetrating work of Kružkov, it is was realized by Kuznetsov [126] that the method of "doubling of the variables" could be used to establish a general approximation theory for entropy solutions of hyperbolic partial differential equations. In the subsequent years, the theory of Kuznetsov was used (by himself and others) to derive error estimates for the method of vanishing viscosity as well as many numerical methods. We refer to [102] for an introduction to this theory. For recent contributions to the approximation theory of entropy solutions, we refer to Cockburn and Gremaud [41], Bouchout and Perthame [20], Perthame [144], and Tadmor [160].

For completeness, we recall that the solution of (13) satisfies following estimates:

Theorem 3.1. *Let u, v be entropy weak solutions of the hyperbolic problems*

$$\partial_t u + \nabla \cdot f(u) = 0, \qquad u|_{t=0} = u_0,$$
$$\partial_t v + \nabla \cdot g(v) = 0, \qquad v|_{t=0} = v_0.$$

Then we have
$$(19)$$
$$\|u(\cdot,T) - v(\cdot,T)\|_{L^1} \leq \|u_0 - v_0\|_{L^1} + T \min\big(|u_0|_{BV}, |v_0|_{BV}\big) \max_j \|f_j - g_j\|_{\text{Lip}}.$$

Furthermore, the unique entropy weak solution of (13) satisfies the following estimates:

$$(20) \quad \begin{cases} \text{(a)} & \|u(\cdot,t)\|_{L^\infty} \leq \|u_0\|_{L^\infty}, \\ \text{(b)} & |u(\cdot,t)|_{BV} \leq |u_0|_{BV}, \\ \text{(c)} & \|u(\cdot,t_2) - u(\cdot,t_1)\|_{L^1} \leq \text{Const} \cdot |t_2 - t_1|, \qquad \forall t_1, t_2 \geq 0. \end{cases}$$

The first part of this theorem was first proved by Lucier [135] in the one-dimensional case, and in [105] and by Bouchut and Perthame [20] in the multi-dimensional case. It can also be obtained by using Theorem 3.3 below and then passing to the limit in the parabolic regularisation (16). The second part of the theorem follows, e.g., from the estimates derived by Kružkov [123].

For later use, let us recall a few additional facts about entropy weak solutions of the one-dimensional conservation law

$$(21) \quad \begin{cases} \partial_t u + \partial_x f(u) = 0, & (x,t) \in \mathbb{R} \times (0,T), \\ u(x,0) = u_0(x), & x \in \mathbb{R}. \end{cases}$$

Let us assume that the entropy weak solution of (21) has discontinuities of the simplest kind; namely, we suppose that there is a smooth curve Γ_u, given by

$x = x(t)$, such that u is smooth on either side of Γ_u, with a simple jump across Γ_u. Letting u^l and u^r denote the left and right limits of $u(\cdot, t)$, it then follows from the integral identity (14), via Green's formula, that the Rankine-Hugoniot condition holds across the curve Γ_u;

$$(22) \qquad s := x'(t) = \frac{f(u^l) - f(u^r)}{u^l - u^r}.$$

The following geometric interpretation of the Kružkov entropy condition (17) is known as Oleĭnik's theorem or entropy condition [141].

Theorem 3.2 ([141]). *Let $u(x,t)$ be a piecewise smooth solution to (21) with jumps across Γ_u satisfying (22). Then the entropy condition (17) holds if and only if:*

- *($u^r < u^l$): The graph $y = f(u)$ restricted to $[u^r, u^l]$ lies below or equals the chord connecting the point $(u^r, f(u^r))$ to the point $(u^l, f(u^l))$, i.e.,*

$$(23) \qquad \frac{f(u^l) - f(k)}{u^l - k} \geq s, \qquad \forall k \in (u_r, u_l).$$

- *($u^r > u^l$): The graph $y = f(u)$ restricted to $[u^l, u^r]$ lies above or equals the chord connecting the point $(u^l, f(u^l))$ to the point $(u^r, f(u^r))$, i.e.,*

$$(24) \qquad \frac{f(u^r) - f(k)}{u^r - k} \leq s, \qquad \forall k \in (u_l, u_r).$$

Remark 3.1. *Oleĭnik's entropy condition states that the entropy loss associated with a shock in the solution of (21) manifests itself in the form of a local convexification of the flux function. This geometric interpretation of the entropy loss turns out to be the heart of the matter of the corrected operator splitting algorithms described §4.*

We next consider nonlinear convection-diffusion problems of the form

$$(25) \qquad \begin{cases} \partial_t u + \nabla \cdot f(u) = \nabla \cdot (d(u)\nabla u) = \Delta D(u), & (x,t) \in \Pi_T = \mathbb{R}^m \times (0,T), \; m \geq 1, \\ u(x,0) = u_0(x), & x \in \mathbb{R}^m, \end{cases}$$

where $f = (f_1, \ldots, f_m)$ is the convection flux and $D(u)$ is the diffusion function given as

$$(26) \qquad D(u) = \int_0^u d(\xi) \, d\xi.$$

If not otherwise stated, it is always understood that the functions f, D are sufficiently smooth and that u_0 (at least) belongs to $L^1 \cap L^\infty \cap BV$. The L^1/BV solution theory presented below can, of course, be extended to more general convection-diffusion equations with variable coefficients, a source term, and a second order diffusion term containing mixed partial derivatives.

Let us first assume that the convection-diffusion equation is uniformly parabolic, i.e.,

$$(27) \qquad d(\xi) \geq \gamma > 0, \qquad \forall \xi.$$

With a solution of (25) we then understand a function $u(x,t)$ which is at least twice continuously differentiable in x and at least once in t such that the partial differential equation is satisfied in the classical sense for $t > 0$. Furthermore, we require that $u(t) \to u_0$ in the weak sense, i.e.,

$$\int \big(u(x,t) - u_0(x)\big)\phi(x)\, dx \to 0 \text{ as } t \to 0+, \qquad \forall \phi \in C_0(\mathbb{R}^m).$$

It well known that (25) possesses such a solution and that it is unique. We refer to the survey paper of Oleĭnik and Kružkov [142] for a nice overview of the theory of parabolic equations.

For later use, let us collect some important L^1 type estimates not found in [142].

Theorem 3.3. *Let u, v be smooth solutions of the convection-diffusion problems*

$$(28) \qquad \partial_t u + \nabla \cdot f(u) = \Delta D(u), \qquad u|_{t=0} = u_0,$$

$$(29) \qquad \partial_t v + \nabla \cdot g(v) = \Delta D(v), \qquad v|_{t=0} = v_0.$$

Then we have

$$(30)$$
$$\|u(\cdot,T) - v(\cdot,T)\|_{L^1} \le \|u_0 - v_0\|_{L^1} + T \min\big(|u_0|_{BV}, |v_0|_{BV}\big) \max_j \|f_j - g_j\|_{\text{Lip}}.$$

Furthermore, a smooth solution of (25) satisfies the following estimates:

$$(31) \quad \begin{cases} \text{(a)} & \|u(\cdot,t)\|_{L^\infty} \le \|u_0\|_{L^\infty}, \\ \text{(b)} & |u(\cdot,t)|_{BV} \le |u_0|_{BV}, \\ \text{(c)} & \|u(\cdot,t_2) - u(\cdot,t_1)\|_{L^1} \le \text{Const} \cdot \sqrt{|t_2 - t_1|}, \qquad \forall t_1, t_2 \ge 0. \end{cases}$$

Remark 3.2. *We will later show that our approximate solutions satisfy estimates very similar to those in (31), which in turn will imply compactness of the numerical approximations. It is important to notice that estimates (a)-(c) in (31) continue to hold in the degenerate case. The fact that these estimates are independent of γ (the lower bound on $d(u)$) makes the L^1 space rather attractive from the point of view of numerical analysis for parabolic equations.*

To prove estimate (c) in (31), we need a version of an interpolation lemma due to Kružkov [122]. This lemma is also used in §4 for convergence analysis of operator splitting.

Lemma 3.1. *Let there be finite constants C_1, C_2, and C_3 such that $z : \bar{\Pi}_T \to \mathbb{R}$ satisfies the two estimates*

$$\|z(\cdot,t)\|_{L^\infty} \le C_1, \qquad \text{for all } t \in [0,T],$$
$$|z(\cdot,t)|_{BV} \le C_2, \qquad \text{for all } t \in [0,T],$$

and the weak time estimate

$$\left|\int \big(z(x,t_2) - z(x,t_1)\big)\phi(x)\, dx\right| \le C_3\big(\|\phi\|_{L^\infty} + \max_j \|\phi_{x_j}\|_{L^\infty}\big)|t_2 - t_1|,$$

for all $\phi \in C_0^1$ and $0 \le t_1, t_2 \le T$. Then there is a constant C, depending in particular on C_1 and C_2, such that the following interpolation result is valid:

$$\|z(\cdot, t_2) - z(\cdot, t_1)\|_{L^1} \le C\sqrt{|t_2 - t_1|}, \qquad 0 \le t_1, t_2 \le T.$$

See [112, 114] for a proof of this lemma.

Proof (of Theorem 3.3). The proof of (30) uses a classical dual (error) argument. The quantity $e = v - u$ solves the (error) equation

$$(32) \qquad \partial_t e + \nabla \cdot (a(x,t)e) - \Delta(b(x,t)e) = \mathcal{T}(x,t),$$

where the vector $a(x,t) = (a_1(x,t), \ldots, a_m(x,t))$ and the scalar $b(x,t)$ are given as

$$a_j(x,t) = \frac{f_j(v) - f_j(u)}{v - u}, \qquad j = 1, \ldots, m,$$

$$b(x,t) = \frac{D(v) - D(u)}{v - u},$$

and $\mathcal{T}(x,t)$ denotes the truncation error,

$$\mathcal{T}(x,t) = \partial_t v + \nabla \cdot f(v) - \Delta D(v).$$

Now, for given $T > 0$, let ψ solve the backward problem

$$(33) \qquad \begin{cases} \partial_t \psi + a(x,t) \cdot \nabla \psi + b(x,t) \Delta \psi = 0, & t < T, \\ \psi|_{t=T} = \phi \in C^\infty(\mathbb{R}^m). \end{cases}$$

Then $\psi(t)$ is well defined for $t \le T$ and the maximum principle yields

$$\|\psi(t)\|_{L^\infty} \le \|\phi\|_{L^\infty}.$$

By integrating the error equation (32) against ψ over Π_T, and noting that (33) is just the adjoint problem of (32), we obtain (assuming that e decreases rapidly to zero as $|x| \to \infty$)

$$(34) \qquad \int e(x,T)\phi(x)\,dx = \int e(x,0)\psi(x,0)\,dx + \iint \mathcal{T}(x,t)\psi(x,t)\,dtdx.$$

We are now equipped with the tool (34) needed to prove (30). Assume first $f = g$ so that $\mathcal{T} \equiv 0$. Then by choosing $\phi = \mathrm{sign}(e)$ (omitting a standard approximation argument) and using the maximum principle for $\psi(t)$, we obtain

$$(35) \qquad \begin{aligned} \int |v(x,T) - u(x,T)|\,dx &\le \|\psi(\cdot,0)\|_{L^\infty} \int |v_0(x) - u_0(x)|\,dx \\ &\le \int |v_0(x) - u_0(x)|\,dx. \end{aligned}$$

Observe now that estimate (b) is a direct consequence of the L^1 contraction property (35) since the convection-diffusion equation (25) is translation invariant.

Equipped with (34) and (35), it remains to estimate the truncation error in (34). Using (b) and again the maximum principle for ψ, we can readily calculate

$$\left| \iint \mathcal{T}(x,t)\psi(x,t)\,dt\,dx \right| = \left| \iint \nabla \cdot (f(v) - g(v))\psi(x,t)\,dt\,dx \right|$$

$$\leq \max_j \|f_j - g_j\|_{\mathrm{Lip}} \sum_j \iint |\partial_{x_j} v||\psi(x,t)|\,dt\,dx$$

$$\leq T \max_j \|f_j - g_j\|_{\mathrm{Lip}} |v_0|_{BV} \|\phi\|_{L^\infty}.$$

Choosing $\phi = \mathrm{sign}(e)$ (again omitting a standard approximation argument) in (34), we now obtain

$$\|v(\cdot,T) - u(\cdot,T)\|_{L^1} \leq \|v_0 - u_0\|_{L^1} + T|v_0|_{BV} \max_j \|f_j - g_j\|_{\mathrm{Lip}}.$$

Finally, using symmetry we derive the desired stability result (30).

It remains to prove (a) and (c). The first claim follows from the maximum principle. We are going to use (b) and Lemma 3.1 to derive the time estimate (c). To this end, we shall employ a technique introduced by Kružkov [122]. Let $\phi(x)$ be a test function on $\mathbb{R}^m$. Multiplying (25) by ϕ, integrating the result in space, and subsequently integrating by parts, yields

$$\left| \int \phi(x)\partial_t u\,dx \right| \leq \left| \int f'(u) \cdot \nabla u \phi\,dx \right| + \left| \int D'(u)\nabla u \cdot \nabla\phi \right|$$

$$\leq \mathrm{Const} \cdot \left(\|\phi\|_{L^\infty} + \max_j \|\partial_{x_j}\phi\|_{L^\infty} \right),$$

where the BV estimate (b) has been used. From this estimate we get the following weak continuity result

(36)
$$\left| \int (u(x,t_2) - u(x,t_1))\phi(x)\,dx \right| \leq \mathrm{Const} \cdot \left(\|\phi\|_{L^\infty} + \max_j \|\partial_{x_j}\phi\|_{L^\infty} \right)|t_2 - t_1|.$$

Applying Lemma 3.1 to (36), we get estimate (c). $\qquad\square$

Remark 3.3. *The proof of Theorem 3.3 was based on a classical dual argument. It seems difficult to get a stability result with respect to the diffusion function using this technique. However, Cockburn and Gripenberg [42] have recently obtained such stability using an elegant extension of Kružkov's "doubling of the variables", see also Evje, Karlsen, and Risebro [74].*

We now turn our attention to degenerate convection-diffusion equations. To this end, we replace condition (27) by

(37) $$d(\xi) \geq 0, \qquad \forall \xi.$$

When (25) is non-degenerate, the equation admits classical solutions. This fact contrasts with the case where (25) is allowed to degenerate ($d(u) = 0$) for some values of u. In general, a striking manifestation of the point degeneracy is the finite speed of propagation of disturbances. Thus, if $d(0) = 0$ and at some fixed time the solution u has compact support, then it will continue to have compact support for all later times. The transition from a region where $u > 0$ to one where

$u = 0$ is not smooth and it is therefore necessary to deal with (continuous) weak solutions rather than classical solutions, see the book [157] (and the references cited therein) for further details.

Recall that the convection-diffusion equations arising in reservoir simulation typically degenerates at two points, see §2. A natural generalisation would be to consider equations for which the function $D(u)$ (see (26)) is strictly increasing in u. Note that a sufficient condition for $D(u)$ to be strictly increasing is that

$$\text{meas}\{\xi : d(\xi) = 0\} = 0,$$

which does not rule out the possibility that $d(u)$ has an infinite number of zero points. To deal with this kind of parabolic degeneracy, we introduce the following notion of a generalised solution:

Definition 3.2. *Suppose that $D(u)$ is strictly increasing. Then a function $u(x,t)$ is called a BV weak solution of the Cauchy problem (25) if:*

1. $u \in L^\infty(\Pi_T) \cap BV(\Pi_T)$ and $\nabla D(u) \in L^2_{\text{loc}}(\Pi_T; \mathbb{R}^n)$.

2. For all test functions $\phi \in C_0^\infty$ such that $\phi|_{t=T} = 0$,

$$(38) \qquad \iint \left(u\partial_t\phi + [f(u) - \nabla D(u)] \cdot \nabla\phi \right) dtdx + \int u_0(x)\phi(x,0) \, dx = 0.$$

Provided u_0 is sufficiently smooth, existence of a BV weak solution is established in the work of Vol'pert and Hudjaev [168] by sending ε to zero in the parabolic regularisation

$$(39) \qquad \partial_t u_\varepsilon + \nabla \cdot f(u_\varepsilon) = \Delta D(u_\varepsilon) + \varepsilon\Delta u_\varepsilon, \qquad u_\varepsilon(x,0) = u_0(x).$$

Yin [181] has proved that BV weak solutions are uniquely determined by their initial data. More precisely, he proved that if u and v are BV weak solutions of (25) with data u_0 and v_0, respectively, then the L^1 stability result (18) holds. Using the essential condition that $D(\cdot)$ is strictly increasing, the L^1 stability result is proved by showing that the set of discontinuity points of BV weak solutions is of m - dimensional measure zero. Zhao [182] has proved that if (x_0, t_0) is a point of approximate continuity of a BV solution u such that $d(u(x_0, t_0)) > 0$, then u is a classical solution in a neighbourhood of (x_0, t_0). Furthermore, in one space dimension ($m = 1$), the BV solution is continuous, see [182]. Under the hypothesis

$$D \in C^1(\mathbb{R}), \ D' > 0 \text{ a.e. in } \mathbb{R} \qquad \text{and} \qquad |f'|^2 \le \sigma D', \ \sigma \in C(\mathbb{R}),$$

Bénilan and Gariepy [13] have shown that the unique BV weak solution of (25) is actually a strong solution, i.e., $\partial_t u$, $\nabla \cdot f(u)$, and $\Delta D(u)$ are functions in $L^1_{\text{loc}}(\Pi_T)$ (and not merely locally finite measures on Π_T). Zhao [184] has shown recently that the requirement $u \in BV(\Pi_T)$ in Definition 3.2 can be replaced by the weaker requirement $u \in BV_x(\Pi_T)$, where $BV_x(\Pi_T) \supset BV(\Pi_T)$ denotes the space consisting of locally integrable functions $z(x,t)$ for which $\partial_x z$ is a locally finite measure on Π_T. Note that if $z \in BV_x(\Pi_T)$, then $z(\cdot, t) \in BV(\mathbb{R})$ for a.e. $t \in (0, T)$. It is possible to go a step further and replace $u \in BV_x(\Pi_T)$ by $u \in L^1(\Pi_T) \cap C(0, T; L^1(\mathbb{R}^d))$, see Remark 3.4 below.

Bürger *et al.* [25, 26, 27, 31, 33, 44] have in a series of papers (see also the recent book [35]) proposed and analysed a sedimentation model which contains a certain nonlinear partial differential equation. This partial differential equation is similar to the saturation equation in the reservoir flow model. The main difference is, however, that the diffusion coefficient in the sedimentation model is degenerate (zero) on intervals of solution values, and not only at isolated points as is the case with the saturation equation (6). The splitting algorithms described in §4 can also be applied to the sedimentation model, see [28, 30, 97] for details.

We end this section with a discussion of the strongly degenerate case, i.e., the case where $D(u)$ is merely non-decreasing. In this case there exists at least one interval $[\alpha, \beta]$ such that

$$d(\xi) = 0, \qquad \forall \xi \in [\alpha, \beta].$$

A simple example of a strongly degenerate equation is a hyperbolic equation. We can therefore conclude that strongly degenerate equations will in general possess discontinuous (weak) solutions. Moreover, discontinuous solutions are not uniquely determined by their initial data. In fact, an additional condition — the entropy condition — is needed to single out the physically relevant weak solution of the problem. For strongly degenerate equations that possess discontinuous solutions, a natural generalisation of Definition 3.2 is the following:

Definition 3.3. *Suppose that $D(u)$ is non-decreasing. Then a BV weak solution $u(x, t)$ of (25) is called a BV entropy weak solution if, for all non-negative test functions $\phi \in C_0^\infty(\Pi_T)$ and $k \in \mathbb{R}$,*

$$(40) \qquad \iint \left(|u - k| \partial_t \phi + \text{sign}(u - k) [f(u) - f(k) - \nabla D(u)] \cdot \nabla \phi \right) dt dx \geq 0.$$

Remark 3.4. *In the one-dimensional case, the condition $\partial_x D(u) \in L^2_{\text{loc}}(\Pi_T)$ can be replaced by the stronger condition $D(u) \in C^{1,\frac{1}{2}}(\bar{\Pi}_T)$ in the sense that it is still possible to prove existence of an BV entropy weak solution. Here, $C^{1,\frac{1}{2}}(\bar{\Pi}_T)$ denotes the space of functions that are Hölder continuous with exponent 1 in the space variable and 1/2 in the time variable. Furthermore, the requirement $u \in BV(\Pi_T)$ in Definition 3.2 can be replaced by the much weaker requirement*

$$u \in L^1(\Pi_T) \cap C(0, T; L^1(\mathbb{R}^d))$$

(or even $u \in L^1(\Pi_T)$) and uniqueness still holds (see [36, 37, 115] and discussion below). In this case, we call u an entropy weak solution.

Note that the entropy condition (40) reduces to Kružkov's entropy condition (17) for hyperbolic problems when $d \equiv 0$, and thus Definition 3.3 contains the hyperbolic problem (13) as a special case. This condition was first proposed by Vol'pert and Hudjaev [168], who also showed, provided u_0 is sufficiently smooth, existence of a BV entropy weak solution by passing to the limit in the parabolic regularisation (39). It is easy to see that the entropy condition (40) implies that the partial differential equation in (25) holds in the distributional sense. Consequently, the generalised derivative $\Delta D(u)$ is a locally finite measure on Π_T, since $\partial_t u$ and $\nabla \cdot f(u)$ are locally finite measures, and the partial differential equation in (25) holds in the sense of equality of measures.

In the one-dimensional case, Wu and Yin [178] have proved uniqueness of BV entropy weak solutions. More precisely, they have proven that if u and v are BV entropy weak solutions with initial data u_0 and v_0, respectively, then (18) holds. For further results — via nonlinear semigroup theory — on existence, uniqueness, and continuous dependence on the data of entropy weak solutions in the one-dimensional case, we refer to Bénilan and Touré [15, 16, 17].

Uniqueness results for multi-dimensional problems are obtained in Brézis and Crandall [23] (when $f \equiv 0$) and in the recent work of Carrillo [36, 37]. In particular, Carrillo's work represents an important step forward in developing an entropy solution theory for multi-dimensional (strongly) degenerate parabolic equations, which contains the Kružkov theory for first hyperbolic equations as a special case! In [36, 37], the author showed uniqueness of entropy weak solutions for a particular boundary value problem with the boundary condition "$A(u) = 0$". His method of proof is an elegant extension of the by now famous "doubling of variables" device introduced by Kružkov [123]. The author of [36, 37] also showed existence of an entropy weak solution using the semigroup method. In Bürger, Evje, and Karlsen [29] (see also Rouvre and Gagneux [153]), the uniqueness proof of Carrillo was adopted to several initial-boundary value problems arising the theory of sedimentation-consolidation processes, which in some cases call for the notion of an entropy boundary condition (see also [32, 175, 176, 177, 179, 183] for related existence and uniqueness results for initial-boundary value problems). Karlsen and Risebro [115] generalised Carrillo's uniqueness result [36, 37] by showing that it holds for entropy weak solutions of the Cauchy problem with a flux function of the form $f = f(x, t, u)$ where the spatial dependence is non-smooth. Only the case $f = f(u)$ was studied in [37]. In the spirit of Theorem 3.4 below, the authors also proved (using the "doubling of variables" device) continuous dependence of the entropy weak solution on the flux function in the case $f(x, t, u) = k(x) f(u)$, where $k(x)$ is a vector-valued function and $f(u)$ is a scalar function. Weakly coupled systems of (strongly) degenerate parabolic equations are treated in Holden, Karlsen, and Risebro [99].

Equipped with Theorem 3.3 and the uniqueness theorem for entropy weak solutions [36, 37, 115], we can pass to the limit in the parabolic regularisation (39) (following [168]) and conclude that the following continuous dependence theorem holds:

Theorem 3.4. *Let u and v be entropy weak solutions of* (28) *and* (29), *respectively. If u and v belong to $BV_x(\Pi_T)$, then* (30) *holds.*

This theorem shows that the unique entropy weak solution u of (28) depends continuously on the initial function u_0 and the flux function f. For results regarding continuous dependence on the diffusion function A, see Cockburn and Gripenberg [42] and Evje, Karlsen, and Risebro [74].

The uniqueness proof due to Wu and Yin [178] is based on a characterisation of the set of discontinuity points of u (the jump conditions). In [178], the authors corrected a previous result by Vol'pert and Hudjaev [168] stating that the jump conditions for strongly degenerate parabolic equations coincide with the jump conditions for purely hyperbolic equations. Before stating the (correct) jump conditions, we must introduce some notation. Let Γ_u be the set of jumps, that is, $(x_0, t_0) \in \Gamma_u$ if and only if there exists a unit vector $\nu = (\nu_t, \nu_x)$ such that the approximate limits of u at (x_0, t_0) from the sides of the half-planes $(t - t_0)\nu_t + (x - x_0)\nu_x < 0$ and $(t - t_0)\nu_t + (x - x_0)\nu_x > 0$, denoted by $u^-(x_0, t_0)$ and $u^+(x_0, t_0)$, respectively, exist and are not equal. Similarly, let $u^l(x, t)$ and $u^r(x, t)$ denote the left and right approximate limits of $u(\cdot, t)$ respectively. One should note the difference between u^-, u^+ and u^l, u^r. The approximate limits u^-, u^+ are well-defined for $u \in BV(\Pi_T)$ while the limits u^l, u^r exist under the weaker assumption $u \in BV_x(\Pi_T)$. Introduce the notations $\text{sign}^+ := \text{sign}$ and $\text{sign}^- := \text{sign}^+ - 1$, and let $\text{int}(a, b)$ denote the closed interval bounded by a and b. Finally, let H_1 denote the one-dimensional Hausdorff measure.

Theorem 3.5 ([178]). *Let $u(x, t)$ be a BV entropy weak solution (see Definition 3.3) of (25) with $m = 1$. Then H_1 - almost everywhere on Γ_u, we have:*

$$(41) \qquad d(k) = 0, \qquad \forall k \in \text{int}(u^-, u^+), \qquad \nu_x \neq 0.$$

$$(42) \qquad (u^+ - u^-)\nu_t + (f(u^+) - f(u^-))\nu_x - (\partial_x D(u)^r - \partial_x D(u)^l)|\nu_x| = 0.$$

For all $k \in \mathbb{R}$,

(43)
$$|u^+ - k|\nu_t + \text{sign}(u^+ - k)[f(u^+) - f(k) - (\partial_x D(u)^r \text{sign}^+ \nu_x - \partial_x D(u)^l \text{sign}^- \nu_x)]\nu_x$$
$$\leq |u^- - k|\nu_t + \text{sign}(u^- - k)[f(u^-) - f(k) - (\partial_x D(u)^l \text{sign}^+ \nu_x - \partial_x D(u)^r \text{sign}^- \nu_x)]\nu_x.$$

Conditions (42) and (43) are, respectively, generalisations of the Rankine-Hugoniot condition (22) and Oleĭnik's entropy condition (see Theorem 3.2) for conservation laws. One should observe that for the multi-dimensional case, it is not possible to conclude from the Definition 3.3 that each $\partial_{x_j} D_j(u)$ is a finite measure on Π_T, although $\Delta D(u)$ is. This fact prevents one from deriving the analogue of the jump conditions in the multi-dimensional case, although the BV_x theory of Wu and Yin [178] can be extended to the multi-dimensional case.

We next note that the jump conditions (42) and (43) can be more instructively stated as follows (see [71] for a proof):

Corollary 3.1. *Let $u(x, t)$ be a piecewise smooth BV entropy weak solution (see Definition 3.3) of (25) with $m = 1$. Assume that $d(k) = 0$ for all $k \in [u_*, u^*]$ for some $u_*, u^* \in \mathbb{R}$. A jump between two values u^l and u^r of the solution $u(x, t)$, which is referred to as a shock, can occur only for $u^l, u^r \in [u_*, u^*]$. This shock must satisfy:*

1. The shock speed s is given by

$$(44) \qquad s = \frac{f(u^r) - f(u^l) - (\partial_x D(u)^r - \partial_x D(u)^l)}{u^r - u^l}.$$

2. *For all $k \in \text{int}(u^l, u^r)$, the following entropy condition holds:*

(45)
$$\frac{f(u^r) - f(k) - \partial_x D(u)^r}{u^r - k} \leq s \leq \frac{f(u^l) - f(k) - \partial_x D(u)^l}{u^l - k}.$$

It is important to realize that solutions of strongly degenerate parabolic equations in general have a more complex structure than solutions of hyperbolic equations. The following example demonstrates this.

Example 3.1 (Structure of Solutions). *This example is taken from Evje and Karlsen [71]. We consider the Burgers type equation*

(46)
$$\partial_t u + \partial_x(u^2) = \partial_x(d(u)\partial_x u),$$

where

(47)
$$d(u) = \begin{cases} 0, & \text{for } x \in [0, 0.5], \\ 2.5u - 1.25, & \text{for } x \in [0.5, 0.6], \\ 0.25, & \text{for } x \in [0.6, 1]. \end{cases}$$

Note that $d(\cdot)$ is continuous and degenerates on the interval $[0, 0.5]$. In Figure 2 we have plotted the initial function, the solution of the corresponding conservation law, i.e., $d \equiv 0$ in (46), and the solution of (46) at time $T = 0.15$. A finite difference method (with very fine discretization parameters) is used to compute the solutions. An interesting observation is that the solution of (46) has a "new" increasing jump, despite of the fact that f is convex. Thus the solution is not bounded in the Lip^+ norm, as opposed to the solution of the conservation law (see [160]). Moreover, while the speed of a jump in the conservation law solution is determined solely by f (see Theorem 3.2), the speed of a jump in the solution of (46) is in general determined by the jumps in both $f(u)$ and $\partial_x D(u)$ (see Corollary 3.1). In Figure 3, we have given a geometric

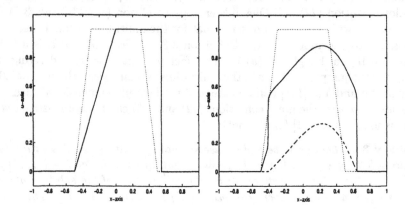

FIGURE 2. Left: The solution (solid) of the inviscid Burgers type equation. Right: The solution (solid) of the Burgers type equation with a strongly degenerate diffusion term and the corresponding diffusion function $D(u(\cdot, t))$ (dashed). The initial function is shown as dotted.

interpretation of the entropy condition (45) for the solution shown in Figure 2 (right), which possesses two shocks. (i) Left shock: note that $\partial_x D(u)^l = 0$ and $\partial_x D(u)^r > 0$, see Figure 2 (right). Condition (45) states that the graph of f restricted to the interval $[u^l, u^r]$ lies above or equals the straight line between $(u^l, f(u^l))$ and $(u^r, f(u^r) - \partial_x D(u)^r)$, see Figure 3 (left). (ii) Right shock: note that $\partial_x D(u)^l < 0$ and $\partial_x D(u)^r = 0$. Condition (45) now states that the graph of f restricted to $[u^r, u^l]$ lies below or equals the straight line between $(u^r, f(u^r))$ and $(u^l, f(u^l) - \partial_x D(u)^l)$, see Figure 3 (right).

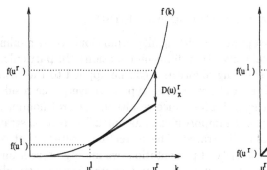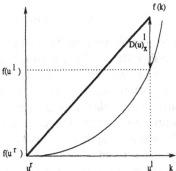

FIGURE 3. Geometric interpretation of the entropy condition (45) for the solution shown in Figure 2 (right).

The BV (entropy) solution theory can be generalised to doubly nonlinear degenerate parabolic equations of the form

$$(48)\quad \partial_t u + \partial_x f(u) = \partial_x \mathcal{A}(d(u)\partial_x u), \qquad \mathcal{A}(\pm\infty) = \pm\infty, \ \mathcal{A}'(s) \geq 0, \ d(s) \geq 0.$$

The functions $\mathcal{A}'(s)$ and $d(s)$ are allowed to have an infinite number of degenerate intervals in $\mathbb{R}$. Included in (48) are equations arising from the theory of non-Newtonian fluids. We say that $u(x,t)$ is a BV entropy weak solution of the Cauchy problem for (48) if:

1. $u \in L^\infty(\Pi_T) \cap BV(\Pi_T)$ and $D(u) \in C^{1,\frac{1}{2}}(\bar{\Pi}_T)$.
2. For all non-negative $\phi \in C_0^\infty$ such that $\phi|_{t=T} = 0$ and $k \in \mathbb{R}$,

$$\iint \left(|u - k|\partial_t \phi + \text{sign}(u - k)[f(u) - f(k) - \mathcal{A}(\partial_x D(u))]\partial_x \phi \right) dt\,dx$$

$$+ \int |u_0(x) - k|\phi(x,0)\, dx \geq 0.$$

We refer to Yin [180] for a treatment of the initial-boundary value problem for (48). Yin uses the method of parabolic regularisation for the existence proof. Evje and Karlsen [70] have treated the Cauchy problem for (48) using finite difference methods. Yin [180] has used the BV_x theory [178] to derive the jump conditions — and thus L^1 stability — for discontinuous solutions of doubly nonlinear problems. Without going into details, we only mention that for doubly nonlinear problems the shock speed (44) is replaced by

$$s = \frac{f(u^r) - f(u^l) - \left(\mathcal{A}(\partial_x D(u))^r - \mathcal{A}(\partial_x D(u))^l\right)}{u^r - u^l},$$

while the entropy condition (45) is replaced by

(49) $$\frac{f(u^r) - f(k) - \mathcal{A}(\partial_x D(u))^r}{u^r - k} \leq s \leq \frac{f(u^l) - f(k) - \mathcal{A}(\partial_x D(u))^l}{u^l - k},$$

see also [70]. Finally, we refer to Evje [67] for a more detailed review of BV (entropy) solution theory for strongly degenerate parabolic equations.

4. THE OPERATOR SPLITTING ALGORITHMS

In this section we describe the operator splitting algorithms for solving nonlinear, possibly strongly degenerate, convection-diffusion equations. In particular, as we demonstrate in §7, these splitting algorithms can be applied to the saturation equation (6). We also state and prove some typical convergence results for some of the splitting algorithms. Performance results for several nonlinear model problems are presented, and the importance of having a "correction strategy" for reducing splitting errors is illustrated. The correction strategy that we employ is described in detail. For clarity of presentation, we mainly present and analyse the operator splitting algorithms in their semi-discrete form for a simplified convection-diffusion problem. For numerical purposes, we use fully discrete splitting algorithms based on the (large time step) hyperbolic solvers described in §5 and the parabolic solvers described in §6.

It is well known that accurate modelling of convective and diffusive processes is one of the most challenging tasks in the numerical approximation of partial differential equations. Particularly difficult is the case where convection dominates diffusion. This is often the case in models of two-phase flow in oil reservoirs. Accurate numerical simulation of such models is consequently often complicated by unphysical oscillations and/or numerical diffusion [138].

In the last two decades we have seen an enormous activity on developing sophisticated numerical methods for hyperbolic equations. We refer to [87, 102, 120, 129, 164] for a general introduction to modern numerical methods for hyperbolic partial differential equations. It seems reasonable to employ some of these hyperbolic solvers as "building blocks" in numerical algorithms for convection-diffusion problems. Indeed, our numerical strategy is based on splitting the convection-diffusion equation (6) into a hyperbolic equation for convection and a parabolic equation for diffusion. We then try to reproduce the solution of (6) using numerical methods for these simpler equations as building blocks. Variations on this operator splitting approach have been taken in various contexts by many authors. The novelty of our work lies in the use of large time step methods for the convection step and a correction strategy (or an appropriate flux splitting) for reducing potential splitting errors. Applying large time step algorithms for the convection step has some advantages. In particular, equipped with an implicit diffusion solver, the resulting operator splitting algorithms are unconditionally stable in the sense that there is no CFL condition (see below) restricting the time step. However, it turns out that a reasonable choice of time step is highly dictated by the degree of (nonlinear) interplay between convective and diffusive forces. In particular, large time steps can lead to fronts that are too wide. However, as we will see later, it is possible to identify and correct (or counterbalance) this nonlinear splitting error. We refer to the books [138, 152] for an introduction to more standard numerical algorithms for convection-diffusion problems.

For simplicity of presentation, we restrict our attention to convection-diffusion problems of the form

$$(50) \quad \begin{cases} \partial_t u + \nabla \cdot f(u) = \varepsilon \Delta u, & (x,t) \in \Pi_T = \mathbb{R}^m \times (0,T), \ m \geq 1, \\ u(x,0) = u_0(x), & x \in \mathbb{R}^m, \end{cases}$$

where the flux vector $f = (f_1, \ldots, f_m)$ is sufficiently smooth and the initial function u_0 belongs to $L^1 \cap L^\infty \cap BV$. We emphasise that the numerical solution algorithms and their convergence analysis presented below carry over to more general convection-diffusion equations with variable coefficients, a source term, and a second order diffusion term containing mixed partial derivatives. Also, due to space limitation, we only treat the Cauchy problem. Details on the extension of the operator splitting algorithms to various boundary value problems can be found in [28, 110].

As was discussed above, an underlying design principle for many successful solution algorithms for problems such as (50) is operator splitting. Operator splitting means here that we split the time evolution in (50) into two partial steps in order to separate out the effects of convection and diffusion. To describe this operator splitting more precisely, we need the solution operator taking the initial data $v_0(x)$ to the entropy weak solution at time t of the hyperbolic problem

$$(51) \quad \begin{cases} \partial_t v + \nabla \cdot f(v) = 0, & (x,t) \in \mathbb{R}^m \times \{t > 0\}, \\ v(x,0) = v_0(x), & x \in \mathbb{R}^m. \end{cases}$$

This solution operator we denote by $\mathcal{S}^f(t)$. Similarly, let $\mathcal{H}(t)$ be the solution operator (at time t) associated with the parabolic problem

$$(52) \quad \begin{cases} \partial_t w = \varepsilon \Delta w, & (x,t) \in \mathbb{R}^m \times \{t > 0\}, \\ w(x,0) = w_0(x), & x \in \mathbb{R}^m. \end{cases}$$

Now choose a time step $\Delta t > 0$ and an integer N such that $N\Delta t = T$. Furthermore, let $t_n = n\Delta t$ for $n = 0, \ldots, N$ and $t_{n+1/2} = (n+\frac{1}{2})\Delta t$ for $n = 0, \ldots, N-1$. We then let the operator splitting solution $u_{\Delta t}$ be defined at the discrete times $t = t_n$ by the product formula

$$(53) \quad u_{\Delta t}(x, n\Delta t) = \left[\mathcal{H}(\Delta t) \circ \mathcal{S}^f(\Delta t) \right]^n u_0(x).$$

In applications, the exact solution operators $\mathcal{S}^f(t)$ and $\mathcal{H}(t)$ are replaced by appropriate numerical approximations. We use the modified method of characteristics (together with a suitable flux splitting) or front tracking methods for the hyperbolic part, see §5 for details. For the parabolic part we use finite difference methods or a Petrov-Galerkin method, see §6 for details.

For later reference, let us define the CFL (Courant-Friedrichs-Lewy) number. If Δx is the grid spacing and Δt is the time step associated with a numerical method for (51), the CFL number is defined as

$$(54) \qquad \text{CFL} := \max_u |f'(u)| \frac{\Delta t}{\Delta x},$$

where the maximum is taken over all u in $\left[-\|u_0\|_{L^\infty}, \|u_0\|_{L^\infty}\right]$. The classical CFL condition for finite difference methods for (51) states that (54) should be bounded by one.

Here we mention that Dawson, Wheeler, and collaborators [56, 57, 55, 58, 174] are using operator splitting algorithms similar to (53). In their splitting algorithm, the hyperbolic equation (51) is solved by $M \geq 1$ local time steps (for each global splitting step) with an explicit high resolution Godunov type method, while the diffusion equation (52) is solved implicitly. The point is that such splitting algorithms may be more efficient than standard numerical methods when $M > 1$, since the implicit equations need not be solved during each update of the conservation law. Since the explicit nature of the convection solver requires a CFL time step constraint, the basic splitting procedure may be expensive, especially if the CFL constraint is severe. Different from the view taken by Dawson *et al.*, we also insist on using large time step methods for the convection updates, i.e., methods that avoid a CFL constraint. We will come back to this point later when discussing the corrected operator splitting algorithms.

Note that we have only defined $u_{\Delta t}$ at the discrete times t_n. In between two consecutive discrete times, we use the following time interpolant:

$$(55) \qquad u_{\Delta t}(x,t) = \begin{cases} \mathcal{S}^f(2(t-t_n))u^n, & t \in (t_n, t_{n+1/2}], \\ \left[\mathcal{H}(2(t-t_{n+1/2})) \circ \mathcal{S}^f(\Delta t)\right]u^n, & t \in (t_{n+1/2}, t_{n+1}], \end{cases}$$

where $u^n = u_{\Delta t}(t_n)$. Regarding $u_{\Delta t}$ we have the following lemma:

Lemma 4.1 ([112]). *The following a priori estimates hold:*

$$(56) \quad \begin{cases} \text{(a)} & \|u_{\Delta t}(\cdot, t)\|_{L^\infty} \leq \|u_0\|_{L^\infty}, \\ \text{(b)} & |u_{\Delta t}(\cdot, t)|_{BV} \leq |u_0|_{BV}, \\ \text{(c)} & \|u_{\Delta t}(\cdot, t_2) - u_{\Delta t}(\cdot, t_1)\|_{L^1} \leq \text{Const} \cdot \sqrt{|t_2 - t_1|}, \qquad \forall t_1, t_2 \geq 0. \end{cases}$$

Proof. Claim (a) is true because the solution operators $\mathcal{S}^f(t)$ and $\mathcal{H}(t)$ do not introduce new minima or maxima. Similarly, claim (b) is true since the solution operators $\mathcal{S}^f(t)$ and $\mathcal{H}(t)$ do not increase the total variation of their initial data.

Fix a test function $\phi \in C_0^\infty(\mathbb{R}^m)$. Using estimate (c) in (20), we get

$$(57) \qquad \left|\int \left(\mathcal{S}^f(t_2)v_0 - \mathcal{S}^f(t_1)v_0\right)\phi\,dx\right| \leq \text{Const} \cdot \|\phi\|_{L^\infty}|t_2 - t_1|.$$

Using the differential equation for $w(x, t) = \mathcal{H}(t)w_0(x)$ and integration by parts, we get the bound

$$
\left| \int \left(\mathcal{H}(t_2)w_0 - \mathcal{H}(t_1)w_0 \right) \phi \, dx \right| = \left| \int \left(\int_{t_1}^{t_2} \varepsilon \Delta w(x, t) \, dt \right) \phi \, dx \right|
$$

(58)

$$
\leq \text{Const} \cdot \max_j \|\phi_{x_j}\|_{L^\infty} |t_2 - t_1|.
$$

Using (55), (57), and (58), we now readily compute

(59)

$$
\left| \int \left(u_{\Delta t}(x, t_2) - u_{\Delta t}(x, t_1) \right) \phi \, dx \right| \leq \text{Const} \cdot \left(\|\phi\|_{L^\infty} + \max_j \|\phi_{x_j}\|_{L^\infty} \right) |t_2 - t_1|.
$$

Finally, applying Lemma 3.1 to (59) yields estimate (c). $\qquad \square$

In view of estimates (a) and (b) in (56), a classical application of Helly's theorem yields the existence of a subsequence $\{u_{\Delta t_j}(\cdot, t)\}$ converging in L^1_{loc} to a function $u(\cdot, t)$ in $L^\infty \cap BV$ for each fixed t. By a diagonalization argument we obtain the existence of a further subsequence, still denoted by $\{u_{\Delta t_j}(\cdot, t)\}$, which converges for all t in some dense countable subset of $(0, T)$. By appealing to (c), we obtain convergence for all t in $(0, T)$. Summing up, for any given sequence $\{\Delta t\}$ tending to zero, there exists a subsequence $\{\Delta t_j\}$ and a limit function u such that

(60)
$$
u_{\Delta t_j} \to u \text{ in } L^1_{\text{loc}}(\Pi_T) \text{ as } j \to \infty.
$$

We now prove our main convergence theorem:

Theorem 4.1 ([112]). *Suppose $u_0 \in L^1 \cap L^\infty \cap BV$. Then the operator splitting solution $u_{\Delta t}$ converges in $L^1_{\text{loc}}(\Pi_T)$ to the unique classical solution of the Cauchy problem (50) as $\Delta t \to 0$.*

Proof. We will show that the limit in (60) is a weak solution of (50). To this end, fix a test function $\phi \in C_0^\infty(\mathbb{R} \times [0, T])$ such that $\phi|_{t=T} = 0$, and define a new test function by

$$
\varphi(x, t) = \phi\left(x, \frac{t}{2}\right).
$$

Let

$$
v^n(t) = \mathcal{S}^f(t)u^n, \qquad t \in (0, \Delta t).
$$

Since $v^n(t)$ satisfies the hyperbolic problem (51) in the sense of distributions on $\mathbb{R}^m \times (0, \Delta t)$ with initial data u^n, the following integral equality holds

(61)

$$
\int \int_{t_n}^{t_{n+1/2}} \left(\frac{1}{2} u_{\Delta t} \partial_t \phi + f(u_{\Delta t}) \cdot \nabla \phi \right) dt \, dx
$$

$$
= \frac{1}{2} \int \int_0^{\Delta t} \left(v^n(\tau) \partial_\tau \varphi(x, \tau + 2t_n) + f(v^n(\tau)) \cdot \nabla \varphi(x, \tau + 2t_n) \right) d\tau \, dx
$$

$$
= \frac{1}{2} \int u^{n+1/2} \phi(x, t_{n+1/2}) \, dx - \frac{1}{2} \int u^n \phi(x, t_n) \, dx,
$$

where we have used the substitution $\tau = 2(t - t_n)$ and introduced the short-hand notation $u^{n+1/2} = \mathcal{S}^f(\Delta t)u^n$. Similarly, let

$$w^n(t) = \mathcal{H}(t)u^{n+1/2}, \qquad t \in (0, \Delta t).$$

Since $w^n(t)$ satisfies the parabolic problem (52) in the sense of distributions on $\mathbb{R}^m \times (0, \Delta t)$ with initial data $u^{n+1/2}$, the following integral equality holds

(62)
$$\int \int_{t_{n+1/2}}^{t_{n+1}} \left(\frac{1}{2} u_{\Delta t}\partial_t\phi + \varepsilon u_{\Delta t}\Delta\phi \right) dtdx$$

$$= \frac{1}{2} \int \int_0^{\Delta t} \left(w^n(\tau)\partial_\tau\varphi(x, \tau + 2t_{n+1/2}) + \varepsilon w_n(\tau)\Delta\varphi(x, \tau + 2t_{n+1/2}) \right) d\tau dx$$

$$= \frac{1}{2} \int u^{n+1}\phi(x, t_{n+1}) \, dx - \frac{1}{2} \int u^{n+1/2}\phi(x, t_{n+1/2}) \, dx,$$

where we have used the substitution $\tau = 2(t - t_{n+1/2})$. Adding together (61) and (62), multiplying with 2, and summing the result over all $n = 0, \ldots, N-1$, where $N\Delta t = T$, yields

(63)
$$\sum_{n=0}^{N-1} \int \int_{t_n}^{t_{n+1}} \left(u_{\Delta t}\partial_t\phi + 2\chi_n f(u_{\Delta t}) \cdot \nabla\phi + \varepsilon 2(1 - \chi_n)u_{\Delta t}\Delta\phi \right) dtdx$$

$$+ \int u_0(x)\phi(x, 0) \, dx = 0,$$

where $\chi_n = \chi_n(x, t)$ is the characteristic function of the set $\mathbb{R}^m \times [t_n, t_{n+1/2}]$. Since $\chi_n(x, t) \rightharpoonup \frac{1}{2}$ in $L^2(\Pi_T)$, we can pass to the limit in (63) and obtain

(64) $\quad \mathcal{L}(u; \phi) := \int\int \left(u\partial_t\phi + f(u) \cdot \nabla\phi + \varepsilon u\Delta\phi \right) dtdx + \int u_0(x)\phi(x, 0) \, dx = 0.$

Since ϕ was arbitrary, it follows that the limit u is a weak solution of the Cauchy problem (50). Finally, following Oleĭnik [141], we can actually show that this weak solution is a classical solution possessing the necessary smoothness for $t > 0$. This concludes the proof. $\qquad\square$

The analysis presented above is due to Karlsen and Risebro [112]. A refined analysis can be found in [114, 69, 96]. The refined analysis shows that the weak truncation error $|\mathcal{L}(u; \phi)|$ (see (64)) is of order $\mathcal{O}(\sqrt{\Delta t})$ when $u_0 \in BV$. For a sufficiently smooth initial function, it is possible to improve this to $|\mathcal{L}(u; \phi)| = \mathcal{O}(\Delta t)$, see [28]. Karlsen and Lie [108] have analysed operator splitting for convection-diffusion equations with variable coefficients and source term. Similar convergence results hold for fully discrete operator splitting algorithms. For example, we can replace the hyperbolic solution operator $\mathcal{S}^f(t)$ by front tracking and the parabolic solution operator $\mathcal{H}(t)$ by an explicit or implicit central difference method, see, e.g., [108, 112] for details.

Remark 4.1. *It should be noted that if the numerical method for $\mathcal{H}(t)$ is chosen properly, our numerical solution algorithm is unconditionally stable in the sense that the time step Δt is not limited by the space discretization Δx, i.e., no* CFL *type condition is associated with the algorithm. Hence, whenever the "physics" of the problem allows for it, large time steps can be used in the numerical simulations.*

Let us for a while restrict our attention to the one-dimensional convection-diffusion problem

$$(65) \qquad \partial_t u + \partial_x f(u) = \varepsilon \partial_x(d(u)\partial_x u), \qquad u(x,0) = u_0(x),$$

where also a nonlinear diffusion function $d(\cdot) \geq 0$ has been included. The standard operator splitting algorithm for (65) takes the form

$$(66) \qquad u_{\Delta t}(x, n\Delta t) = \left[\mathcal{H}(\Delta t) \circ \mathcal{S}^f(\Delta t)\right]^n u_0(x)$$

where $\mathcal{S}^f(t)$ is the solution operator of the hyperbolic equation

$$(67) \qquad \partial_t v + \partial_x f(v) = 0,$$

and $\mathcal{H}(t)$ is the solution operator of the parabolic equation

$$(68) \qquad \partial_t w = \varepsilon \partial_x(d(w)\partial_x w).$$

Evje and Karlsen [69] have analysed operator splitting for nonlinear, strongly degenerate convection-diffusion initial value problems in one space dimension. Holden, Karlsen and Lie [96] extended the results in [69] to multi-dimensional equations with variable coefficients. Bürger, Evje, and Karlsen [28] have treated various initial-boundary value problems. We refer to Holden, Karlsen, Lie, and Risebro [98] for an L^1 convergence theory for *general* operator splitting methods for weakly coupled systems of degenerate convection-diffusion equations.

The analysis in the degenerate case is similar to the one presented above. Of course, in the case of parabolic degeneracy, we have to show that the approximate solution $u_{\Delta t}$ defined in (66) converges to a limit function u that satisfies the conditions of Definition 3.2, or Definition 3.3 in the case of strong degeneracy. We will not go into all details about the analysis of (66) in the (strongly) degenerate case, see instead [69]. Here we only discuss the condition

$$(69) \qquad \varepsilon \partial_x D(u) \in L^2_{\text{loc}}(\Pi_T),$$

where $D(\cdot)$ is defined in (26). Suppose $u_{\Delta t} \to u$ in $L^1_{\text{loc}}(\Pi_T)$ as $\Delta t \to 0$. Then we would like to show that u satisfies condition (69). Below we sketch an argument which leads to (69) under the assumption that all functions are sufficiently regular. For details in the general (non-smooth) case, we refer to [69] (see also [96, 98]).

Introduce the two sequences $\{\tilde{u}_{\Delta t}\}$ and $\{g_{\Delta t}\}$,

$$\begin{cases} \tilde{u}_{\Delta t}(x,t) = \left[\mathcal{H}(t - t_n) \circ \mathcal{S}^f(\Delta t)\right] u^n, & (x,t) \in \mathbb{R} \times (t_n, t_{n+1}), \\ g_{\Delta t}(x,t) = \sqrt{\varepsilon d(\tilde{u}_{\Delta t})}\partial_x \tilde{u}_{\Delta t}, & (x,t) \in \mathbb{R} \times (t_n, t_{n+1}). \end{cases}$$

Since obviously

$$\|u_{\Delta t}(\cdot,t) - \tilde{u}_{\Delta t}(\cdot,t)\|_{L^1} = \mathcal{O}(\sqrt{\Delta t}),$$

we conclude that

$$\tilde{u}_{\Delta t} \to u \text{ in } L^1_{\text{loc}}(\Pi_T) \text{ as } \Delta t \to 0.$$

Multiplying the equation for $\tilde{u}_{\Delta t}$ on $\mathbb{R} \times (t_n, t_{n+1})$ by $\tilde{u}_{\Delta t}$, integrating over Π_T, and then integrating by parts in space, we get (recall that $u^{n+1/2} = \mathcal{S}^f(\Delta t)u^n$)

$$\|g_{\Delta t}\|^2_{L^2} = \iint \varepsilon d(\tilde{u}_{\Delta t})(\partial_x \tilde{u}_{\Delta t})^2 \, dt dx = -\iint \varepsilon \partial_x \left(d(\tilde{u}_{\Delta t}) \partial_x \tilde{u}_{\Delta t} \right) \tilde{u}_{\Delta t} \, dt dx$$

$$= -\iint \frac{1}{2} \partial_t (\tilde{u}_{\Delta t})^2 \, dt dx = -\frac{1}{2} \sum_{n=0}^{N-1} \int \left((\tilde{u}_{\Delta t}|_{t=t_{n+1}})^2 - (\tilde{u}_{\Delta t}|_{t=t_n})^2 \right) dx$$

$$= -\frac{1}{2} \sum_{n=0}^{N-1} \int \left([(u^{n+1})^2 - (u^n)^2] + [(u^n)^2 - (u^{n+1/2})^2] \right) dx$$

$$= -\frac{1}{2} \int_{\mathbb{R}} [(u^N)^2 - (u^0)^2] \, dx + \frac{1}{2} \sum_{n=0}^{N-1} \int_{\mathbb{R}} [(u^{n+1/2})^2 - (u^n)^2] \, dx =: I_1 + I_2,$$

where we have, without loss of generality, assumed that $(d(\tilde{u}_{\Delta t})\partial_x \tilde{u}_{\Delta t})\tilde{u}_{\Delta t} \to 0$ as $|x| \to \infty$.

Since the operators $\mathcal{S}^f(t)$ and $\mathcal{H}(t)$ both are L^1 contractive, we know that $\|\tilde{u}_{\Delta t}(\cdot, t)\|_{L^1} \leq \|\tilde{u}_{\Delta t}(\cdot, 0)\|_{L^1}$. Thus since $\tilde{u}_{\Delta t}$ is uniformly bounded and the initial function is integrable, the first term is clearly bounded independent of Δt,

(70) $$|I_1| \leq 2\|u^0\|_{L^\infty} \|u^0\|_{L^1} = \mathcal{O}(1).$$

Exploiting the L^1 Lipschitz continuity of $\mathcal{S}^f(t)$ and again that $\tilde{u}_{\Delta t}$ is bounded, we obtain for the second term that

(71) $$|I_2| \leq \|u_0\|_{L^\infty} \sum_{n=0}^{N-1} \|\mathcal{S}^f(\Delta t)u^n - u^n\|_{L^1} = \mathcal{O}(1)T.$$

From (70) and (71) we conclude that the following $L^2(\Pi_T)$ bound is valid

(72) $$\|g_{\Delta t}\|_{L^2} \leq M(T),$$

where $M(T)$ is a finite constant independent of η. By virtue of (72) we conclude that $\{g_{\Delta t}\}$ is weakly compact in $L^2_{\text{loc}}(\Pi_T)$. Without loss of generality, we may assume that the entire sequence $\{g_{\Delta t}\}$ converges weakly in $L^2_{\text{loc}}(\Pi_T)$ to a function g. Let G be defined such that $\partial G(u)/\partial u = \sqrt{\varepsilon d(u)}$ and let ϕ be a test function. We can then calculate

$$\iint g(x,t)\phi(x,t) \, dt dx = \lim_{\Delta t \to 0} \iint \partial_x G(\tilde{u}_{\Delta t})\phi \, dt dx = \lim_{\Delta t \to 0} \iint (-G(\tilde{u}_{\Delta t})\partial_x \phi) \, dt dx$$

$$= \iint (-G(u)\partial_x \phi) \, dt dx = \iint \frac{\partial G(u)}{\partial u} \partial_x u \phi \, dt dx = \iint r(u)\partial_x u \phi \, dt dx, \, r(u) = \sqrt{\varepsilon d(u)}.$$

Consequently, we have shown that $r(u)\partial_x u$ exists in the sense of distributions in $L^2_{\text{loc}}(\Pi_T)$. Finally, from the facts that $u \in L^\infty(\Pi_T)$, $r(u)\partial_x u \in L^2_{\text{loc}}(\Pi_T)$, and (41), we get $\varepsilon d(u)\partial_x u \in L^2_{\text{loc}}(\Pi_T)$, see [168] for details. Thus (69) holds.

Although the operator splitting algorithms converge to the solution of the underlying problem as various discretization parameters tend to zero, it turns out that splitting approximations are too diffusive near self-sharpening fronts, at least when the splitting step Δt is large. We next present a numerical example that confirms this claim.

Example 4.1 (Operator Splitting). *This example is taken from Karlsen and Risebro [112]. We consider Burgers' equation; that is, $f(u) = \frac{1}{2}u^2$ and $d(u) = u$ in (65). This equation, introduced by Burgers [34] in 1940, represents a simplified model of the more complicated Navier-Stokes equations and captures some of the essential features of incompressible fluid dynamics; namely, a nonlinear convection term and a viscous diffusion term. The Burgers equation has a time independent solution given by*

$$(73) \qquad u(x,t) = -\tanh\left(\frac{x}{2\varepsilon}\right),$$

so that it is well suited as a test case. The solution (73) corresponds to the case where the hyperbolic equation (67) would have a shock solution. We can also find an explicit solution in the case where the hyperbolic equation has a rarefaction wave solution, which corresponds to initial data of the form

$$(74) \qquad u_0(x) = \begin{cases} -1, & \text{for } x \leq 0, \\ 1, & \text{for } x > 0. \end{cases}$$

By applying the Hopf-Cole transform, one finds that the solution is

$$u(x,t) = \frac{g(-x,t) - g(x,t)}{g(-x,t) + g(x,t)}, \qquad g(x,t) = e^{\frac{t+2x}{4\varepsilon}} \, erfc\left(\frac{t+x}{\sqrt{4\varepsilon t}}\right).$$

In Figure 4, we show the operator splitting solutions at time $T = 1$ for both the "shock case" and the "rarefaction case". In this example, we use front tracking (see §5.2) to solve (67) and the Galerkin method (see §6.2) to solve (68). The most notable feature of these computations is the poor performance of the splitting algorithm in the shock case. The error turns out to be largely independent of the size of Δt in the rarefaction case, but it is very sensitive to the choice of Δt in the shock case. One should note that the error contribution in the shock case is due to the temporal splitting, and not the spatial discretization.

In view of Example 4.1, it can be tempting to conclude that operator splitting is a technique that is not particularly well suited to use with hyperbolic solvers that allow for large time steps. However, as we are about to learn, this is not the case! To better understand the (nonlinear) mechanisms behind the splitting error, one should bear in mind Oleĭnik's entropy condition, see Theorem 3.2. As we saw in Example 4.1, when Δt is larger than the diffusion scale ε, the standard splitting algorithm is too diffusive near the (self-sharpening) front. In view of Oleĭnik's theorem, this splitting error is simply a manifestation of the entropy condition being taken into account in the convection step. The entropy condition introduces a local convexification of $f(\cdot)$ representing the entropy loss associated with the shock in the hyperbolic solution. In other words, the operator splitting

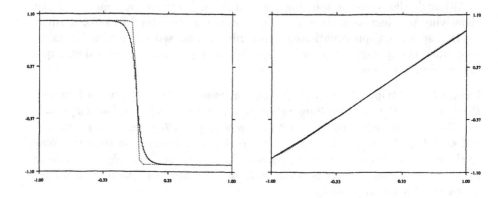

FIGURE 4. Exact solutions (dotted line) versus OS solutions (piecewise constant). The shock case (left) and the rarefaction case (right) are both calculated with $\Delta x = 0.01$, $\Delta t = 0.5$, and $\epsilon = 0.01$. Observe that there is a significant amount of splitting error in the shock case due to the large time step.

solution does not take into account the convex shape of the flux function, which in turn determines the self-sharpening nature of the (parabolic) front.

Luckily, the part of the flux function that is neglected (the entropy loss) can be identified as a residual flux term. For instance, assume that the solution of Burgers' equation is a moving steep front (as in Example 4.1) and that the hyperbolic solver produces a discontinuity with left and right limits v^l and v^r respectively. We can then identify the entropy loss associated with this shock as a residual flux term of the form

$$(75) \qquad f_{\text{res}} = f - f_c.$$

Here, $f_c = f_c(v; v^l, v^r)$ denotes the correct envelope (dictated by the entropy condition) of f in the interval bounded by v^l and v^r, i.e.,

(76)

$$f_c(v; v^l, v^r) = \begin{cases} \text{the lower convex envelope of } f \text{ between } v^l \text{ and } v^r, \text{ if } v^l < v^r, \\ \text{the upper concave envelope of } f \text{ between } v^r \text{ and } v^l, \text{ if } v^l > v^r. \end{cases}$$

There are two ways to take the residual flux term f_{res} into account; *that is, there are two ways to correct or counterbalance the splitting error.* We can, for instance, perform a separate correction step after the diffusion step. Correction is then realized by solving the "residual" equation

$$(77) \qquad \partial_t v + \partial_x f_{\text{res}}(v) = 0$$

over a time interval $(0, \tau)$, where $\tau > 0$ is some parameter that has to be chosen. Hence, instead of (66), we rely on an algorithm of the form (see Karlsen and

Risebro [114])

(78) $\qquad u_{\Delta t}(x, n\Delta t) = \left[\mathcal{C}(\tau) \circ \mathcal{H}(\Delta t) \circ \mathcal{S}^f(\Delta t) \right]^n u_0(x), \qquad \tau > 0,$

where $\mathcal{C}(\tau)$ is the solution operator of (77), also called the correction operator. Note that when $\tau \equiv 0$, (78) reduces to the standard splitting defined in (66). The residual equation (77) represents the entropy loss generated in the convection step. The purpose of the correction step in (78) is to counterbalance the entropy loss so that the correct width of the self-sharpening front is ensured.

Another approach is to include the residual term in the equation modelling diffusion; that is, instead of solving the equation (68), we solve

(79) $\qquad \partial_t w + \partial_x f_{\mathrm{res}}(w)_x = \varepsilon \partial_x(d(w) \partial_x w),$

thereby yielding a splitting algorithm of the form (see Karlsen *et al.* [107])

(80) $\qquad u_{\Delta t}(x, n\Delta t) = \left[\mathcal{P}^{f_{\mathrm{res}}}(\Delta t) \circ \mathcal{S}^f(\Delta t) \right]^n u_0(x),$

where $\mathcal{P}^{f_{\mathrm{res}}}(t)$ is the solution operator associated with (79). The point is that equation (79) contains the information needed to produce the correct width of the self-sharpening front. Since (80) does not involve the undetermined parameter τ that has to be "tuned", algorithm (80) is the most important one for applications and is the one that we put foremost in these notes.

Note that the residual term f_{res} is "small". In fact, following [114], it is not difficult to prove the estimate

$$\int \left| f_{\mathrm{res}}(u_{\Delta t}(x, t_n)) \right| dx = \mathcal{O}(\sqrt{\Delta t}),$$

where $u_{\Delta t}$ is defined in (78) or (80). Thus, (79) is much closer to being self adjoint than the original equation (65). This means that any iterative procedure will be more efficient for (79) than (65), and the numerical approximation properties will be better [11, 59]. Equation (79) may also be seen as an optimal upwind form of the original equation (65), where the amount of upwinding is determined by the mathematical model. Standard upwinding depends on the mesh size chosen for the problem, which may give severe grid orientation effects. A convection term determined by f_c in (75) gives a mass conserving upwinding and produces virtually no grid orientation effects [53]. It also produces a very good first approximation for (79). Within an iterative splitting strategy, the efficiency of the algorithm depends on the form chosen for f_c. A Riemann solver will give very accurate splittings because they provide "exact" information about the hyperbolic structure of the problem, see Example 4.2 below. Less accurate choices of f_c may give a simpler hyperbolic problem, but give a less efficient solution algorithm for the parabolic equation (79), see [88].

We have explained how to identify the entropy loss, i.e., the residual flux term, when the hyperbolic solution consists of a single shock wave. To describe the construction of a residual flux term in the general case, we assume that $f \in \mathrm{Lip}_{\mathrm{loc}}$ is piecewise linear with a finite number of breakpoints and that u^n is piecewise constant with a finite number of jumps. The reason for making these assumptions is that the exact solution $v(\cdot, t) = \mathcal{S}^f(t)u^n$ also will be piecewise constant with a finite number of jumps, see Theorem 5.1 in §5.2. Furthermore, the exact solution $v(\cdot, t)$ can be constructed by the front tracking method described in §5.2. In what follows, we suppress the time level dependency. The residual flux terms (associated with time level $t_n = n\Delta t$) can be constructed as follows: Observe that each jump in the solution $v(\cdot, t)$ is a shock obeying Oleinik's entropy condition, (see Theorem 3.2). Suppose that the function $v(\cdot, t)$ is piecewise constant on a finite number of intervals with entropy satisfying discontinuities located at $\{x^k\}$. Let v^{k+1} denote the value of v in $[x^k, x^{k+1})$, and let $\{(y_1^k, y_2^k)\}$ be a sequence of pairs of spatial positions chosen so that $x^k \in \langle y_1^k, y_2^k \rangle$ and $y_2^k \leq y_1^{k+1}$ for all k. Then we define the residual flux as

$$(81) \qquad f_{\text{res}}(x, v) = \begin{cases} f(v) - f_c(v; k), & \text{for } x \in [y_1^k, y_2^k) \text{ and } v \in [v^k, v^{k+1}], \\ 0, & \text{for } x \in [y_1^k, y_2^k) \text{ and } v \notin [v^k, v^{k+1}], \end{cases}$$

where $f_c(v; k)$ denotes the correct envelope of f restricted to $[v^k, v^{k+1}]$, see (76). In an actual implementation, all shocks with strength below a certain threshold are disregarded, i.e., we switch off all residual fluxes for which $|v^k - v^{k+1}|$ is less than some (small) constant. Furthermore, we enlarge the spatial support of the nonzero residual fluxes, see Figure 5. Of course, equipped with the residual flux (81) we are free to choose either one of the corrected splitting algorithms defined in (78) and (80). We refer to [114] for further details about the residual flux.

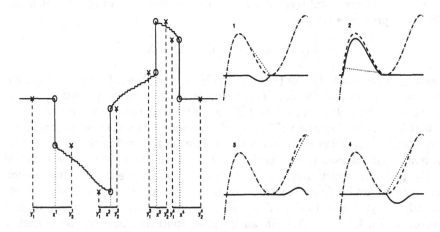

FIGURE 5. Left: Solution from a convection step where four shocks are identified in the spatial domain. Right: The corresponding residual flux functions: flux function (dash dotted), envelope function (dotted), and residual flux function (solid).

The idea of using a residual flux term in the diffusion step was introduced by Espedal and Ewing [63], and further developed and analysed by Dahle, Espedal, and their collaborators [50, 51, 54, 52, 53] in the context of reservoir simulation. In many reservoir flow problems, the residual flux term may change on a very slow time scale compared with the convection scale. This may be true even for heterogeneous models [65]. This means that just a few of the parabolic corrections may be needed in a given time interval. This can give a far more efficient computer code if an efficient hyperbolic solver is available. For such fairly stable flow problems, a reasonable residual flux term can often be derived a priori by solving a single Riemann problem. For instance, to solve the convection-diffusion problem in Example 4.1 these authors would use an operator splitting in which the convection step is to solve the (almost linear) hyperbolic equation

$$(82) \qquad \qquad \partial_t v + \partial_x f_c(v) = 0$$

and the nonlinear diffusion equation (79), see Dahle [50] for further details. Using operator notation, their algorithm can be stated as

$$(83) \qquad \qquad u_{\Delta t}(x, n\Delta t) = \left[\mathcal{P}^{f_{res}}(\Delta t) \circ \mathcal{S}^{f_c}(\Delta t) \right]^n u_0(x).$$

A fundamental difference between (83) and (80) is that (83) employs an a priori flux splitting; $f = f_c + f_{res}$, whereas (80) does not. The flux splitting is supposed be such that f_c represent most of the transport effects present in the original problem (65). If this is the case, then the modified method of characteristics [61] can be applied to (82) without severe time step restrictions, see §5.1 for further details. Moreover, the splitting algorithm defined in (83) has the advantage of giving the correct size of the shock layers, see Example 4.2 below. Following the method of proof that yielded Theorem 4.1, one can also show that the splitting algorithm defined in (83) converges to the solution of the underlying problem as $\Delta t \to 0$.

Of course, an a priori construction of a reasonable residual flux f_{res} is not possible for general problems, and new ideas were introduced by Karlsen and Risebro [114], and further developed and analysed by Karlsen et al. [107, 108, 111, 73, 24], which lead to the corrected operator algorithms defined in (78) and (80). To easily distinguish between the standard splitting algorithm defined in (66) and the more sophisticated splitting algorithms defined in (80) and (83), we refer to (66) as operator splitting (OS), whereas (80) and (83) are referred to as corrected operator splitting (COS) algorithms (we do not use (78) in these notes).

An extension of the corrected splitting algorithm defined in (80) to systems of convection-diffusion equations can be found in [109, 139]. Concerning algorithm (83), we only point out that it is not as easily extendible to systems of equations as (80). This is due to the fact that it is not easy to produce a reasonable flux splitting for a system of equations.

We next present an example that demonstrates the corrected splitting algorithm defined in (83), and how one can derive a reasonable splitting of the convective flux f into two parts.

Example 4.2 (Corrected Operator Splitting/Flux Splitting). *We consider the convection-diffusion problem* (65) *with fluxes*

$$(84) \qquad f(u) = \frac{u^3(1 - 10(1 - u)^3)}{u^3 + (1 - u)^3}, \qquad d(u) = 4u(1 - u),$$

and Riemann initial data

$$u(x,0) = \begin{cases} 0, & \text{for } x \leq 0.65, \\ 1, & \text{for } x > 0.65. \end{cases}$$

Note that the flux function $f(\cdot)$ is non-convex and that the diffusion function $d(\cdot)$ is nonlinear and has a two-point degeneracy; that is, $d(0) = d(1) = 0$. Consider a fluid in a one-dimensional homogeneous porous medium consisting of two immiscible phases; a wetting phase, say, water and a non-wetting phase, say, oil. Let u denote the water saturation (and thus $1 - u$ the oil saturation). Then the partial differential equation modelling the immiscible displacement of oil by water, under the influence of gravity, is of the form (65) with $f(\cdot)$ and $d(\cdot)$ given, e.g., as in (84).

To construct a convective flux splitting, we first note that the correpsonding hyperbolic problem,

$$\partial_t v + \partial_x f(v) = 0, \qquad v(x,0) = \begin{cases} 0, & \text{for } x \leq 0.65, \\ 1, & \text{for } x > 0.65, \end{cases}$$

admits a travelling wave solution $v(x,t) = v(x/t)$ of the form

$$(85) \qquad v\left(\frac{x}{t}\right) = \begin{cases} 0, & \text{for } \frac{x-0.65}{t} < f_c'(0), \\ (f_c')^{-1}\left(\frac{x}{t}\right), & \text{for } \frac{x-0.65}{t} \in [f_c'(0), f_c'(1)], \\ 1, & \text{for } \frac{x-0.65}{t} > f_c'(1), \end{cases}$$

where f_c denotes the lower convex envelope of f restricted to the interval $[0,1]$ and $(f_c')^{-1}$ the inverse of its derivative. Having the piecewise smooth solution (85) in mind, we define the flux splitting by letting $f_{\text{res}} = f - f_c$; consult Figure 6 (left).

We are going to compare OS and COS solutions constructed by the algorithms defined in (66) and (83), respectively. In(83), we use the modified method of characteristics to solve (82) and the Petrov-Galerkin method to solve (79). In (66), we use front tracking (see §5.2) to solve (67) and finite differences (see §6.1) to solve (68). Solutions are computed up to time $T = 0.5$ and the scaling parameter ε is set to 0.01. In Figure 6 (middle) we show the OS calculation on the interval $[0,2]$ using $\Delta x = 0.01$ and $\Delta t = 0.5$. In Figure 6 (right) we show the corresponding COS calculation. As was the case in Example 4.1, we see that OS produces too diffusive fronts when Δt is large. On the other hand, with the same Δt, we see that COS resolves the two fronts correctly.

We now turn our attention to corrected operator splitting for multi-dimensional problems. Provided that one is equipped with a reasonable flux splitting, algorithm (83) remains the same for multi-dimensional problems, see §5.1 and §6.2 for further details. On the other hand, algorithm (80) is genuinely one-dimensional. But there are two obvious ways of generalising it to several space dimensions; namely, the method of streamlines or the method of dimensional splitting. Here we will rely on the latter approach. The streamline approach will be considered elsewhere.

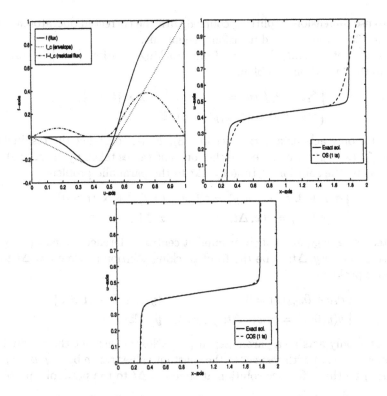

FIGURE 6. Left: The flux function (solid) and the flux splitting (dotted and dashdot). Middle: Exact solution versus OS using 1 time step and 200 mesh points. Right: The exact solution versus COS using 1 time step and 200 mesh points. We see that COS' temporal splitting error is negligible compared with OS.

For simplicity of notation, we only consider the two-dimensional problem

$$(86) \qquad \begin{cases} \partial_t u + \partial_x f(u) + \partial_y g(u) = \varepsilon \Delta u, & (x,y,t) \in \mathbb{R}^2 \times (0,T), \\ u(x,y,0) = u_0(x,y), & (x,y) \in \mathbb{R}^2. \end{cases}$$

The generalisation to higher dimensions (and more general diffusion functions) is straightforward. Consider a uniform Cartesian grid defined by the nodes $\{(i\Delta x, j\Delta y)\}$, where $\Delta x, \Delta y$ are given positive numbers and $i, j \in \mathbb{Z}$. Let π be the usual grid block averaging operator defined on this grid, that is,

$$(87) \qquad \pi u(x,y) = \frac{1}{\Delta x \Delta y} \int_{z_{i,j}} u(\tilde{x}, \tilde{y}) \, d\tilde{x} d\tilde{y}, \qquad \forall (x,y) \in z_{i,j},$$

where $z_{i,j}$ is grid block number (i,j) with lower left-hand corner in $(i\Delta x, j\Delta y)$. Let $f_\delta, g_\delta \in \text{Lip}_{\text{loc}}$ be piecewise linear approximations to f, g, respectively. Let u^n denote the fully discrete corrected splitting solution at some positive time $t = n\Delta t$, $u^0 = \pi u_0$. We next explain how to construct u^{n+1} from u^n. The idea is to use dimensional splitting coupled with the corrected operator splitting defined in (80) to solve one-dimensional equations.

x–sweep: Let $v(x, \Delta t; y)$ be the front tracking solution (see §5.2) at time $t = \Delta t$ to the hyperbolic problem

$$(88) \qquad \begin{cases} \partial_t v + \partial_x f_\delta(v) = 0, & (x, t) \in \mathbb{R} \times \{t > 0\}, \\ v(x, 0; y) = u^n(x; y), & x \in \mathbb{R}. \end{cases}$$

Note that y only acts as a parameter in (88). Next, construct the residual flux function $f_{\text{res}}(x, v; y)$ with respect to the constant values taken by $v(x, \Delta t; y)$. Let $w(x, \Delta t; y)$ be the solution at time $t = \Delta t$ to the parabolic problem

$$(89) \qquad \begin{cases} \partial_t w + \partial_x f_{\text{res}}(x, w; y) = \varepsilon \partial_x^2 w, & (x, t) \in \mathbb{R} \times \{t > 0\}, \\ w(x, 0; y) = v(x, \Delta t; y), & x \in \mathbb{R}, \end{cases}$$

computed using, e.g., an explicit or implicit central difference method (see §6.1).

y–sweep: Let $v(y, \Delta t; x)$ be the front tracking solution at time $t = \Delta t$ to the hyperbolic problem

$$(90) \qquad \begin{cases} \partial_t v + \partial_y g_\delta(v) = 0, & (y, t) \in \mathbb{R} \times \{t > 0\}, \\ v(y, 0; x) = (\pi w(\cdot, \Delta t; \cdot))(y; x), & y \in \mathbb{R}. \end{cases}$$

Note that x only acts as a parameter in (90). Next, construct the residual flux function $g_{\text{res}}(y, v; x)$ with respect to the constant values taken by $v(y, \Delta t; x)$. Let $w(y, \Delta t; x)$ be the difference solution at time $t = \Delta t$ to the parabolic problem

$$(91) \qquad \begin{cases} \partial_t w + \partial_y g_{\text{res}}(y, w; x) = \varepsilon \partial_y^2 w, & (y, t) \in \mathbb{R} \times \{t > 0\}, \\ w(y, 0; x) = v(y, \Delta t; x), & y \in \mathbb{R}. \end{cases}$$

The solution at time $t = (n + 1)\Delta t$ is defined as $u^{n+1} = \pi w(\cdot, \Delta t; \cdot)$.

In terms of approximate solution operators, the corrected operator splitting solution of (101) at time $t = T$ can be given by the composition

$$(92) \qquad u^N = \left[\mathcal{P}^{g_{\text{res}}, y}(\Delta t) \circ \mathcal{S}^{g_\delta, y}(\Delta t) \circ \mathcal{P}^{f_{\text{res}}, x}(\Delta t) \circ \mathcal{S}^{f_\delta, x}(\Delta t) \right]^N u^0,$$

where $\mathcal{P}^{g_{\text{res}}, y}$, $\mathcal{S}^{g_\delta, y}$, $\mathcal{P}^{f_{\text{res}}, x}$, and $\mathcal{S}^{f_\delta, x}$ denote the solution operators associated with the problems (91), (90), (89), and (88), respectively.

Remark 4.2. *Observe that by ignoring the residual terms in the diffusion steps (89) and (91), the resulting standard operator splitting algorithm is slightly different than (53). It can also be shown that this algorithm converges to the solution of (86).*

Remark 4.3. *The stability result in Theorem 3.3 provides us with an estimate of the error contribution coming from the flux approximation used in (92). Let u and u_δ denote the solutions of the parabolic problem (86) with flux functions f, g and f_δ, g_δ, respectively. Suppose $f, g \in \text{Lip}_{\text{loc}}$ are piecewise C^2, then the piecewise linear approximations f_δ, g_δ can be chosen so that*

$$\|f - f_\delta\|_{\text{Lip}}, \ \|g - g_\delta\|_{\text{Lip}} = \mathcal{O}(\delta).$$

Consequently, using (30), *we get*

$$\|u(\cdot, t) - u_\delta(\cdot, t)\|_{L^1} = \mathcal{O}(\delta).$$

Example 4.3 (Corrected Operator Splitting). *This example is taken from Karlsen et al.* [107]. *Consider* (86) *with flux functions of the form*

$$f(u) = \frac{u^2}{u^2 + (1 - u)^2}, \qquad g(u) = f(u)\big(1 - 5(1 - u)^2\big),$$

and initial data

$$u_0(x, y) = \begin{cases} 1, & \text{for } x^2 + y^2 < 0.5, \\ 0, & \text{otherwise.} \end{cases}$$

The solution is computed on the domain $[-1.5, 1.5] \times [-1.5, 1.5]$ *up to time* $T = 0.5$. *We use* $\delta = 0.01$ *for the flux approximations. The reference solution is computed with OS (as defined in* (92) *with the residual fluxes set to zero) using a* 1600×1600 *grid and 441 time steps* (CFL = 2.0). *Figure 7 a) shows a contour plot of the solution obtained by OS using 5 time steps. The shock layer, but also the rarefaction area, is too wide. Note also the presence of a small artificial, almost vertical shock layer on the left-hand side of the peak. This is a result of the dimensional splitting, which is not able to completely resolve the dynamics of the problem. In Figure 7 b) we have used 10 time steps. The artificial shock layer has now (nearly) disappeared, and the resolution of the physical shock layers is slightly improved. Figure 8 a) shows the solution obtained by COS (as defined in* (92)) *using 5 time steps. The shock layer is of correct size, but as in Figure 7 a) the artificial shock layer is present. In Figure 8 b) the number of time steps has been doubled, and the solution is now in good correspondence with the reference solution.*

5. Hyperbolic Solvers

Our numerical algorithms for solving the saturation equation (6) is based on splitting this convection-diffusion equation into a hyperbolic equation modelling convection and a (degenerate) parabolic equation modelling diffusion, see §4 for details. In the following we will describe two different methods for constructing the solution of the hyperbolic equation, see [87, 102, 120, 129, 164] for other alternatives. Numerical methods for the parabolic equation are presented in §6.

5.1. The Modified Method of Characteristics.
The modified method of characteristics for linear convection-diffusion problems was introduced by Douglas and Russell [61]. This method was then extended to nonlinear problems by Espedal and Ewing [63]. To describe the modified method of characteristics [63], we consider the nonlinear convection-diffusion problem

$$(93) \qquad \begin{cases} \partial_t u + \nabla \cdot F(x, u) = \varepsilon \nabla \cdot (d(x, u)\nabla u) + q(x), & (x, t) \in \Omega \times (0, T), \\ u(x, 0) = u_0(x), & x \in \Omega, \end{cases}$$

where $\Omega \subset \mathbb{R}^m$, $m = 1, 2, 3$, $q(x)$ is a source term, and the flux vector $F(x, u)$ is given as $f(u)v + f_g(u)K\nabla h$ (see §2). We assume that the initial function u_0

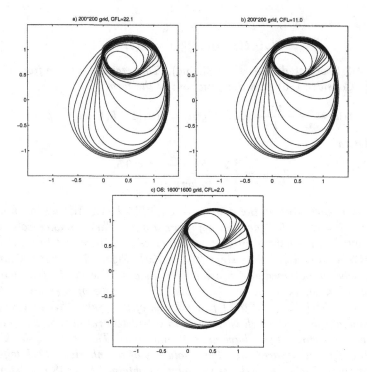

FIGURE 7. 2-D example. a) OS: 200*200 grid, CFL = 22.1. b) OS: 200*200 grid, CFL = 11.0. c) Reference solution on a 1600*1600 grid, CFL = 2.0.

is monotonically decreasing from 1 to 0 in each space direction. The solution of (93) is supposed to satisfy the boundary condition

$$(94) \qquad \big(F(x,u) - \varepsilon d(x,u)\nabla u\big) \cdot n = 0, \qquad x \in \partial\Omega,$$

where n is the outer normal vector to $\partial\Omega$. In view of the operator splitting methodology described in §4, the idea is to use a flux splitting to separate out the transport part of (93). We then apply the modified method of characteristics to the transport problem and a Petrov-Galerkin method to the parabolic residual problem (see §6.2). The resulting solution algorithm for (93), which is a variant of the corrected operator splitting algorithm defined in (83), is demonstrated in §7.

We have some freedom in the construction of the flux splitting. To simplify the presentation, we assume that the gravity term $f_g(u)K\nabla h$ is completely included in the residual flux, which gives a uniform splitting. A flux splitting based on the complete flux function $F(x,u)$, on the other hand, will give a non-uniform splitting which is space dependent. The uniform and non-uniform splittings are discussed by Hansen and Espedal [88] and Frøysa [82].

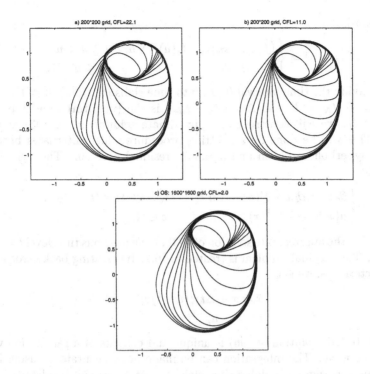

FIGURE 8. 2-D example. a) COS: 200*200 grid, CFL = 22.1. b) COS: 200*200 grid, CFL = 11.0. c) Reference solution on a 1600*1600 grid, CFL = 2.0.

Assuming that $f : \mathbb{R} \to \mathbb{R}$ is a S-shaped function, the entropy weak solution of the corresponding hyperbolic problem (with a decreasing u_0)

$$\begin{cases} \partial_t u + \nabla \cdot \big(f(u)v\big) = 0, & (x,t) \in \Omega \times (0,T), \\ u(x,0) = u_0(x), & x \in \Omega, \end{cases}$$

develops a discontinuity with top shock value $u = u^s$ and bottom shock value $u = 0$. Consistent with the entropy condition, u^s may be determined by the Buckley-Leverett condition

$$s := \frac{f(u^s)}{u^s} = f'(u^s).$$

We assume that the initial profile u_0 represents a shock solution. Equipped with this simplified initial condition, we introduce the residual flux function

$$f_{\text{res}}(x,u) := b(x,u)u = F(x,u) - f_c(u)v,$$

where

$$f_c(u) := \begin{cases} su, & 0 \le u < u^s, \\ f(u), & u^s \le u \le 1. \end{cases}$$

In other words,

$$(95) \qquad f_{\text{res}}(x, u) = \begin{cases} (f(u) - su)v + f_g(u)K\nabla h, & 0 \le u < u^s, \\ f_g(u)K\nabla h, & u^s \le u \le 1. \end{cases}$$

Let us divide the time interval $(0, T)$ into time slabs (t_{n-1}, t_n), $n = 1, \ldots, N$, where $t_0 = 0$, $t_N = T$, and $\Delta t = t_n - t_{n-1}$ is the (uniform) time step. For $n = 1, \ldots, N$, we will construct an approximate solution u_h^n of (93) at time $t = t_n$. This will be done by a splitting procedure which alternates between solving a hyperbolic problem and parabolic residual problem. The hyperbolic problem is

$$(96) \qquad \begin{cases} \partial_\tau u := \partial_t u + f_c'(u)v \cdot \nabla u = 0, & (x, t) \in \Omega \times (t_{n-1}, t_n), \\ u(x, 0) = u_h^{n-1}(x), & x \in \Omega, \end{cases}$$

where u_h^{n-1} is the approximate solution of (93) at the previous time level $t = t_{n-1}$ (see §6.2). The residual problem is treated in §6.2. Integrating backwards along the characteristics, we get

$$(97) \qquad \begin{cases} \bar{x} = x - \Delta t f_c'(\bar{u}^{n-1})v, \\ \bar{u}^{n-1} = u_h^{n-1}(\bar{x}). \end{cases}$$

The characteristic solution of (96) is unique and consists of a rarefaction wave and a shock wave. The integration can be made more accurate by using local time stepping within each global time slab (t_{n-1}, t_n), see [76] for details. This is especially important for heterogeneous models.

Finally, the approximate solution u_h^n of (93) at time level $t = t_n$ is taken as the solution of the parabolic residual problem with initial data $\bar{u}^{n-1}$. This solution is constructed with a Petrov-Galerkin finite element method, see §6.2 for details.

It is well known that the characteristic procedure described above will give a small mass error. This error may be removed by introducing an ELLAM discretization. The Eulerian-Lagrangian localised adjoint method (ELLAM) of Celia *et al.* [38] is a general characteristic-based numerical solution procedure that applies to a variety of convection-diffusion type equations. In particular, ELLAM provides a consistent framework for treating general boundary conditions and maintaining mass conservation. Several authors [38, 154, 77, 79, 165, 172, 173] have developed ELLAM methods for the solution of one-dimensional linear equations with general inflow and outflow boundary conditions, while nonlinear equations are addressed by Dahle *et al.* [54]. The method has also been extended to problems containing reactive terms [39, 78]. The asymptotic convergence analysis and optimal-order error estimates for ELLAM methods have been obtained by Ewing and Wang [77, 173]. While all of these works have been restricted to one spatial dimension, some research have been carried out on multidimensional problems. Russell and Trujillo [155] and Binning [18] have addressed various issues in multidimensional ELLAM methods. Wang [170] has developed an ELLAM simulator to solve two-dimensional linear equations with general inflow and outflow boundary conditions, see also [171]. Moreover, he has also proved optimal-order error estimates for the ELLAM scheme and performed different numerical experiments. Some of the results have been reported in [77, 79].

5.2. Front Tracking Methods.

We first describe a front tracking method for constructing approximate solutions to the one-dimensional hyperbolic problem

$$
(98) \qquad \begin{cases} \partial_t u + \partial_x f(u) = 0, & (x,t) \in \mathbb{R} \times (0,T), \\ u(x,0) = u_0(x), & x \in \mathbb{R}. \end{cases}
$$

The front tracking method was first introduced by Dafermos [48] and later used as a computational tool by many authors. In particular, Holden, Holden, and Høegh–Krohn [94, 95] proved that the method was well-defined and developed it into a numerical method. Front tracking was later extended to systems of equations by Risebro [149], see also Risebro and Tveito [150, 151]. Lie [130] recently extended the front tracking method to hyperbolic equations with a variable coefficient. Various implementation issues are discussed by Langseth [128]. Analysis of front tracking for hyperbolic equations with a flux function that depends discontinuously on u can be found in Gimse [84]. Front tracking for equations with a flux function that depends discontinuously on the space variable is analysed in Gimse and Risebro [85], Klingenberg and Risebro [118, 119], and Klausen and Risebro [117]. We refer to Holden and Risebro [102] for an excellent introduction to front tracking methods, see also Lie's thesis [131].

The front tracking method for (98) determines exact solutions within the class of step functions to a perturbed conservation law. For the moment, let us suppose that u_0 is piecewise constant and f is continuous and piecewise linear with breakpoints at $\{u^0, \ldots, u^N\}$. Observe that each jump in the initial data u_0 defines a Riemann problem; that is, (98) with data of the form

$$
u_0(x) = \begin{cases} u^l, & x < 0, \\ u^r, & x > 0. \end{cases}
$$

The solution of the Riemann problem generally consists of both rarefaction waves and shock waves. If the flux function is piecewise linear, the solution can be found by the technique outlined below.

Consider the Riemann problem with $u^l = u^0 \leq u^r = u^N$. Let f_c denote the lower convex envelope of f restricted to the interval $[u^l, u^r]$, see (76). Since f is piecewise linear then so is f_c. Let $\bar{u}^0 < \bar{u}^1 < \cdots < \bar{u}^M$ be such that

$$
\bar{u}^0 = u^0, \qquad \bar{u}^M = u^N, \qquad \{\bar{u}^0, \ldots, \bar{u}^M\} \subseteq \{u^0, \ldots, u^N\},
$$

and such that f_c is linear in each interval $[\bar{u}^i, \bar{u}^{i+1}]$. The solution of the Riemann problem with left state $u^l = u^0$ and right state $u^r = u^N$ is then given by

$$
(99) \qquad u(x,t) = \begin{cases} u^l, & x \leq \bar{s}^0 t, \\ \bar{u}^i, & \bar{s}^i t < x \leq \bar{s}^{i+1} t, \quad i = 0, \ldots, N-2, \\ u^r, & x > \bar{s}^{N-1} t, \end{cases}
$$

where $\bar{s}^i$ is the Rankine-Hugoniot shock speed,

$$\bar{s}^i = \frac{\bar{f}_{i+1} - \bar{f}_i}{\bar{u}^{i+1} - \bar{u}^i} = f_c'(u^i+), \qquad i = 0, \ldots, N-1,$$

and $\bar{f}_i = f(\bar{u}^i)$. When $u^l > u^r$ there is a similar formula involving the upper concave envelope. Note that since the flux function is piecewise linear, there are no rarefaction waves and each Riemann problem leads to a series of discontinuities propagating in the (x,t) - plane. The global solution of the perturbed problem (98) is obtained by connecting the solutions of the local Riemann problems defined by the piecewise constant initial data. This solution is well-defined until two or more discontinuities interact at some point. Then we have what is called a shock collision. A shock collision defines a new Riemann problem with left and right states given by the values immediately to the left and to the right of the colliding discontinuities. By solving this Riemann problem, the global solution is determined until the next shock collision occurs, and so on. We may continue in this fashion and thereby advance the (exact) solution up to any positive time. Holden, Holden, and Høegh-Krohn [94, 95] proved that this construction, which we call *front tracking*, is well-defined in the sense that there is a finite number of steps in the algorithm, even for infinite time. More precisely, the following theorem holds:

Theorem 5.1 ([94, 95, 102]). *Suppose that u_0 is a piecewise constant function with a finite number of discontinuities. Let $f \in \mathrm{Lip}_{\mathrm{loc}}$ be a piecewise linear function with a finite number of breakpoints. Then the problem (98) has an entropy weak solution $u(\cdot, t)$ which is piecewise constant for each fixed $t > 0$ and takes values in the set $\{u_0(x)\} \cup \{\text{the breakpoints of } f\}$. The solution $u(\cdot, t)$ can be constructed by front tracking in a finite number of steps for any $t > 0$.*

In the general case (arbitrary f and u_0), the front tracking method consists in replacing f by a suitable piecewise linear approximation $f_\delta \in \mathrm{Lip}_{\mathrm{loc}}$ and u_0 by a suitable piecewise constant approximation $u_{0,\Delta x}$. Here, $\delta > 0$ denotes the polygonal approximation parameter and Δx the spatial discretization parameter. Then this perturbed problem is solved according to the construction procedure outlined above.

Remark 5.1. *Note that there is no time step (or CFL condition) associated with the front tracking method, and it introduces no artificial diffusion since a grid is only used to specify the initial data.*

We denote the front tracking solution by u_Δ, where $\Delta = (\Delta x, \delta)$. We have the following estimates:

$$(100) \qquad \begin{cases} \text{(a)} & \|u_\Delta(\cdot, t)\|_{L^\infty} \le \|u_0\|_{L^\infty}, \\ \text{(b)} & |u_\Delta(\cdot, t)|_{BV} \le |u_0|_{BV}, \\ \text{(c)} & \|u_\Delta(\cdot, t_2) - u_\Delta(\cdot, t_1)\|_{L^1} \le C|t_2 - t_1|. \end{cases}$$

Thanks to estimates (a)-(c) in (100), $\{u_\Delta\}$ is bounded in $BV(\mathcal{K})$ for any compact set $\mathcal{K} \subset \Pi_T$. Since BV is compactly imbedded into L^1 on compact sets, it is possible, after a diagonalization procedure, to produce a subsequence that converges in $L^1_{loc}(\Pi_T)$ to some limit u,

$$u \in L^\infty(\Pi_T) \cap BV(\Pi_T).$$

Equipped with this strong convergence and the fact that each u_Δ is an entropy weak solution, it follows that the limit u is also an entropy weak solution. We can sum up as follows:

Theorem 5.2 ([48, 95, 94, 134]). *Suppose $u_0 \in L^1 \cap BV$ and $f \in \text{Lip}_{loc}$. Then the front tracking solution u_Δ converges in $L^1_{loc}(\Pi_T)$ to the unique entropy weak solution of (98) as $\Delta \to 0$. If, in addition, f is piecewise C^2, then the following error estimate holds:*

$$\|u(\cdot, T) - u_\Delta(\cdot, T)\|_{L^1} \leq \text{Const} \cdot \Delta x + \text{Const}_T \cdot \delta.$$

The first part of this theorem is proved by Dafermos [48] and also Holden, Holden, and Høegh-Krohn [95, 94]. If one chooses f_δ and $u_{0,\Delta x}$ properly, the error estimate is a direct consequence of the general stability result found in Theorem 3.1.

Remark 5.2. *Using an approximation theorem of Tadmor [160], one can obtain an improved error estimate for the front tracking method. To this end, we must restrict ourselves to Lip^+ bounded initial data u_0 and a strictly convex flux function f. We then have (see [106] for a proof)*

$$\|u(\cdot, T) - u_\Delta(\cdot, T)\|_{W^{-1,1}} \leq \text{Const} \cdot \Delta x^2 + \text{Const}_T \cdot \delta^2.$$

Recall that the norm $\|w(x, t)\|_{W^{-1,1}}$ can be defined when $\int w(x, t)\, dx = 0$ as

$$\|w(x, t)\|_{W^{-1,1}} = \left\| \int_{-\infty}^x w(\xi, t)\, d\xi \right\|_{L^1}.$$

One should note that front tracking is second order accurate in the weak $W^{-1,1}$ norm. This fact is in contrast to most difference methods, which are typically first order accurate in this norm.

The front tracking method for a hyperbolic equation with a variable coefficient, i.e., $\partial_x f(u)$ in (98) is replaced by $v(x)\partial_x f(u)$ for some smooth function $v(x)$, is more or less the same. The only difference is that the discontinuity lines $x(t)$ in the (x, t) plane are no longer straight lines, but given as solutions of the differential equation $x'(t) = v(x)s$, where s is the Rankine-Hugoniot shock speed $s = f_\delta(u^l) - f_\delta(u^r)/(u^l - u^r)$. By approximating the velocity $v(x)$ by a piecewise constant or piecewise linear function, the differential equation can be solved explicitly and the discontinuity lines $x(t)$ given in closed form, we refer to Lie [130] for details.

Holden and Risebro [101] proposed to extend the front tracking method to multi-dimensional equations by the means of dimensional splitting. For simplicity we consider the two-dimensional case. A generalisation to higher dimensions is straightforward. Let us start with the semi-discrete dimensional splitting method. Consider the two-dimensional hyperbolic problem

$$
(101) \qquad
\begin{cases}
\partial_t u + \partial_x f(u) + \partial_y g(u) = 0, & (x, y, t) \in \Pi_T = \mathbb{R}^2 \times (0, T), \\
u(x, y, 0) = u_0(x, y), & (x, y) \in \mathbb{R}^2,
\end{cases}
$$

whose entropy weak solution is denoted by $u(t) = S(t)u_0$. Let $v(t) = S^{f,x}(t)v_0$ and $w(t) = S^{g,y}(t)w_0$ denote the entropy weak solutions of the one-dimensional problems

$$
\partial_t v + \partial_x f(v) = 0, \qquad v|_{t=0} = v_0,
$$
$$
\partial_t w + \partial_y g(w) = 0, \qquad w|_{t=0} = w_0.
$$

Then the semi-discrete dimensional splitting solution is defined as

$$
(102) \qquad S(n\Delta t)u_0 \approx \left[S^{g,y}(\Delta t) \circ S^{f,x}(\Delta t) \right]^n u_0.
$$

The semi-discrete algorithm (102) was proved to be convergent by Crandall and Majda [46] using a compactness argument. A convergence rate estimate for (102) was later proved independently by Teng [163] and Karlsen [105] using Kuznetsov's approximation theory [126].

In applications, we replace the exact solutions operator by front tracking. More precisely, consider a uniform Cartesian grid defined by the nodes $\{(i\Delta x, j\Delta y)\}$, where $\Delta x, \Delta y$ are given positive numbers and $i, j \in \mathbb{Z}$. Let π be the usual grid block averaging operator defined on this grid, see (87). Furthermore, let f_δ and g_δ be piecewise linear approximations to f and g, respectively, and $S^{f_\delta,x}(t)$ and $S^{g_\delta,y}(t)$ the corresponding one-dimensional solution operators. Then the fully discrete dimensional splitting solution at time $t = t_n$ is defined as

$$
(103) \qquad u_\Delta(x, y, t_n) = \left[\pi \circ S^{g_\delta,y}(\Delta t) \circ \pi \circ S^{f_\delta,x}(\Delta t) \right]^n \pi u_0,
$$

where $\Delta = (\Delta x, \Delta y, \Delta t, \delta)$. The dimensional splitting method (103) consists in using front tracking in the x-direction for a small time step Δt. Note that the front tracking solution will not necessarily be piecewise constant on the original grid. The solution is therefore projected back onto this grid before we apply front tracking in the y-direction for a time step Δt, using the (projected) solution computed in the x-direction as initial data, and so on.

Remark 5.3. *It should be noted that no* CFL *condition is associated with the numerical method* (103). *Multi-dimensional computations using* CFL *numbers as high as* 10−20 *(with satisfactory results) have been reported, see Lie, Haugse, and Karlsen* [133]. *Computational results for multi-dimensional systems of equations can be found in Holden, Lie, and Risebro* [100] *for the Euler equations, Holdahl, Holden, and Lie* [93] *for the shallow water equations, and Haugse, Karlsen, Lie, and Natvig* [90] *for the polymer system.*

With $u^n = u_\Delta(t_n)$, we introduce the short-hand notations

$$
u^{n+1/2} = \pi \circ S^{f_\delta,x}(\Delta t)u^n, \qquad u^{n+1} = \pi \circ S^{g_\delta,y}(\Delta t)u^{n+1/2}.
$$

Note that we have only defined $u_{\Delta t}$ at the discrete times t_n. In between two consecutive discrete times, we use the following time interpolant:

$$(104) \qquad u_{\Delta}(x, y, t) = \begin{cases} S^{f_\delta, x}(2(t - t_n))u^n, & t \in (t_n, t_{n+1/2}), \\ u^{n+1/2}, & t = t_{n+1/2}, \\ S^{g_\delta, y}(2(t - t_{n+1/2}))u^{n+1/2}, & t \in (t_{n+1/2}, t_{n+1}), \\ u^{n+1}, & t = t_{n+1}. \end{cases}$$

We have the following convergence results for (104):

Theorem 5.3 ([101, 105]). *Suppose $u_0 \in L^1 \cap L^\infty \cap BV$ and $f, g \in \mathrm{Lip}_{\mathrm{loc}}$. Then the fully discrete dimensional splitting solution u_Δ converges in $L^1_{\mathrm{loc}}(\Pi_T)$ to the unique entropy weak solution of (101) as $\Delta \to 0$. If, in addition, f, g are piecewise C^2, then the following error estimate holds:*

$$\|u(\cdot, T) - u_\Delta(\cdot, T)\|_{L^1} \leq \mathrm{Const}_T \cdot \left(\sqrt{\Delta t} + \sqrt{\Delta x} + \delta \right).$$

The first part of this theorem was proved by Holden and Risebro [101]. The error estimate is due to Karlsen [105], see also Lie, Haugse, and Karlsen [133]. We refer to Lie [132] for the details concerning the extension of the dimensional splitting method (103) to equations with variable coefficients (velocity fields). Finally, we mention that a reservoir simulator based on front tracking methods is described in [21, 22].

6. PARABOLIC SOLVERS

To solve convection-diffusion problems we use the operator splitting algorithms described in §4. However, as we thoroughly explained in §4, it may become necessary to employ a correction strategy (or a suitable flux splitting) to reduce the splitting error when the time step is large. Hence the diffusion part of an operator splitting is not merely to solve a pure diffusion equation, but an equation containing both convection and diffusion terms. We thus need parabolic solvers that are capable of solving general convection-diffusion equations.

Below we describe some finite difference methods and a Petrov-Galerkin finite element method. The Petrov-Galerkin method is capable of solving convection-diffusion equations which may degenerate at some isolated points, and is thus well suited for the reservoir simulation problems we have in mind (see §2). The finite difference methods are capable of solving both degenerate and strongly degenerate convection-diffusion equations. Consequently, operator splitting algorithms based on finite differences for the parabolic updates can also be applied to mathematical models containing strongly degenerate parabolic equations, see [28, 30].

6.1. Finite Difference Methods. Let us first consider the one-dimensional, possibly strongly degenerate, convection-diffusion problem

$$(105) \qquad \begin{cases} \partial_t u + \partial_x f(u) = \partial_x(d(u)\partial_x u), & (x, t) \in \Pi_T = \mathbb{R} \times (0, T), \\ u(x, 0) = u_0(x), & x \in \mathbb{R}, \end{cases}$$

where $f, d \geq 0, u_0$ are given, sufficiently regular functions. As we mentioned in §3, solutions of (105) can become discontinuous in finite time. For a given initial condition a plenitude of weak solutions may exist. Consequently, we need a selection mechanism — an entropy condition — to single out the physical interesting weak solution, see §3 and Definition 3.3 for details. Evje and Karlsen [70, 71, 68, 72] and Karlsen and Risebro [113] have recently developed a convergence theory for a class of finite difference methods for problems such as (105). This theory, which roughly speaking states that any consistent, conservative-form, monotone difference method converges to the physically correct (entropy weak) solution of (105), can be viewed as a direct extension of the classical monotone difference theory developed by Harten, Hyman, and Lax [89] and Crandall and Majda [47] for conservation laws.

Selecting a mesh size $\Delta x > 0$, a time step $\Delta t > 0$, and an integer N so that $N\Delta t = T$, the value of our difference approximation at $(x, t) = (j\Delta x, n\Delta t)$ will be denoted by u_j^n. To simplify the notation, we introduce the difference operators

$$\Delta_- u_j = \frac{1}{\Delta x}(u_j - u_{j-1}), \qquad \Delta_+ u_j = \frac{1}{\Delta x}(u_{j+1} - u_j).$$

We consider consistent, conservative, monotone, $(2p+1)$ - point finite difference methods of the form

(106) $$\frac{u_j^{n+1} - u_j^n}{\Delta t} + \Delta_-\left(F(u^n; j) - \Delta_+ D(u_j^n)\right) = 0,$$

where $D(u)$ is defined in (26) and $F(u^n; j) = F(u_{j-p+1}^n, \ldots, u_{j+p}^n)$ is the numerical flux associated with the convection part of (105). The initial data for (106) is taken as

$$u_j^0 = \frac{1}{\Delta x}\int_{j\Delta x}^{(j+1)\Delta x} u_0(x)\, dx.$$

To make the methods (106) consistent with the convection-diffusion equation (105) it is sufficient to require that

$$F(u, \ldots, u) = f(u).$$

The assumption of monotonicity guarantees that (106), when viewed as an algorithm of the form

$$u_j^{n+1} = S(u_{j-p+1}^n, \ldots, u_{j+p}^n) =: S(u^n; j),$$

has the property that S is a non-decreasing function of all its arguments.

Let us give an example of a three-point $(p = 1)$ monotone scheme. For a monotone flux f, the upwind scheme is defined by

(107) $$F(u_j^n, u_{j+1}^n) = \begin{cases} f(u_j^n), & \text{if } f' > 0, \\ f(u_{j+1}^n), & \text{if } f' < 0. \end{cases}$$

More generally, for a non-monotone flux f, the generalised upwind scheme of Engquist and Osher is defined by

$$F(u_j^n, u_{j+1}^n) = f^+(u_j^n) + f^-(u_{j+1}^n),$$

where

$$f^+(u) = f(0) + \int_0^u \max(f'(s), 0)ds, \qquad f^-(u) = \int_0^u \min(f'(s), 0)ds.$$

A simple calculation reveals that the upwind method and the generalised upwind method both are monotone methods provided the following CFL type condition holds:

$$(108) \qquad \max|f'|\frac{\Delta t}{\Delta x} + 2\max|d|\frac{\Delta t}{\Delta x^2} \leq 1.$$

Let u_Δ, $\Delta = (\Delta x, \Delta t)$, be the interpolant of degree one associated with the discrete data points $\{u_j^n\}$; that is, u_Δ interpolates at the vertices of each rectangle

$$R_j^n = \left[j\Delta x, (j+1)\Delta x\right] \times \left[n\Delta t, (n+1)\Delta t\right].$$

Note that u_Δ is continuous everywhere and differentiable almost everywhere. Regarding the sequence $\{u_\Delta\}$, we have the following main convergence theorem:

Theorem 6.1 ([71]). *The sequence $\{u_\Delta\}$ built from (106) converges in $L^1_{\text{loc}}(\Pi_T)$ to the unique BV entropy weak solution (in the sense of Definition 3.3) of (105) as $\Delta \to 0$. Furthermore, $\{D(u_\Delta)\}$ converges uniformly on compact sets $\mathcal{K} \subset \Pi_T$ to $D(u) \in C^{1,\frac{1}{2}}(\bar{\Pi}_T)$ as $\Delta \to 0$.*

Here, $C^{1,\frac{1}{2}}(\bar{\Pi}_T)$ denotes the space of functions that are Hölder continuous with exponent 1 in the space variable and 1/2 in the time variable. An important part of the proof of this theorem is to establish the following three estimates for $\{u_j^n\}$:

$$\begin{cases} \text{(a) a uniform } L^\infty \text{ bound,} \\ \text{(b) a uniform total variation bound,} \\ \text{(c) } L^1 \text{ Lipschitz continuity in the time variable,} \end{cases}$$

and the following two estimates for the discrete total flux $F(u^n; j) - \Delta_+ D(u_j^n)$:

$$\begin{cases} \text{(d) a uniform } L^\infty \text{ bound,} \\ \text{(e) a uniform total variation bound,} \end{cases}$$

see [71] for details. Then, using the three estimates (a)-(c), it is not difficult to show that there is a finite constant $C = C(T) > 0$ (independent of Δ) such that

$$\|u_\Delta\|_{L^\infty(\Pi_T)} + |u_\Delta|_{BV(\Pi_T)} \leq C.$$

Hence, the sequence $\{u_\Delta\}$ is bounded in $BV(\mathcal{K})$ for any compact set $\mathcal{K} \subset \Pi_T$. Since $BV(\mathcal{K})$ is compactly imbedded into $L^1(\mathcal{K})$, it is possible to select a subsequence that converges in $L^1(\mathcal{K})$. Furthermore, using a standard diagonal process, we can construct a sequence that converges in $L^1_{\text{loc}}(\Pi_T)$ to a limit u,

$$u \in L^\infty(\Pi_T) \cap BV(\Pi_T).$$

For notational simplicity, let $w_\Delta = D(u_\Delta)$. It is possible to use estimates (d) and (e) to prove that w_Δ satisfies the following Hölder estimate (see [71]):

$$|w_\Delta(y,\tau) - w_\Delta(x,t)| \le C\left(|y-x| + \sqrt{|\tau - t|} + \Delta x + \sqrt{\Delta t}\right),$$

where $C > 0$ is a finite constant not depending on Δ, x, y, t, τ. By repeating the proof of the Ascoli-Arzela compactness theorem, we deduce the existence of a subsequence of $\{w_\Delta\}$ converging uniformly on each compactum $\mathcal{K} \subset \Pi_T$ to a limit w,

$$w \in C^{1,\frac{1}{2}}(\bar{\Pi}_T).$$

Let $\{\Delta_j\}$ be a sequence of discretization parameters tending to zero such that $u_{\Delta_j} \to u$ a.e. and $w_{\Delta_j} \to w$ uniformly on compacta as $j \to \infty$ (such a sequence can certainly be found). Since u_{Δ_j} converges to u a.e. and w is continuous, we conclude that

$$w = D(u).$$

Finally, convergence of $\{u_\Delta\}$ to the correct physical solution of (105) follows from the cell entropy inequality ($k \in \mathbb{R}$)

(109)
$$\frac{|u_j^{n+1} - k| - |u_j^n - k|}{\Delta t} + \Delta_-\left(F(u^n \vee k; j) - F(u^n \wedge k; j) - \Delta_+|D(u_j^n) - D(k)|\right) \le 0,$$

where $u \vee v = \max(u, v)$ and $u \wedge v = \min(u, v)$. This discrete entropy inequality is in turn an easy consequence of the monotonicity of $\mathcal{S}$. The reader is referred to [71] for further details on the convergence analysis.

Remark 6.1. *In many applications it is desirable to avoid the explicit stability restriction (108). One way to overcome (108) is of course to use an implicit version of (106). By combining the arguments in [71] with the Crandall and Liggett theory [45], it is possible to analyse implicit methods as well, see [68] for details.*

It is also possible to speed up explicit methods by using a so-called super time stepping procedure. Super time stepping (STS) is a simple and effective method that speeds up explicit methods for parabolic equations, rendering them as useful as any implicit method, while retaining its simplicity and better accuracy, see Alexiades et al. [5, 6] for details. In [73], the STS method has been used (with good results) as a part of our operator splitting methodology.

Remark 6.2. *A formally second order (in space) version of (106) can be obtained via a MUSCL type approach, which is by now a classical approach in the context of conservation laws. It uses a piecewise linear reconstruction, instead of piecewise constant, together with a limitation procedure, see Evje and Karlsen [72] for analytical and numerical results.*

Remark 6.3. *The finite difference theory can be generalised to doubly nonlinear degenerate parabolic equations of the form (48), see Evje and Karlsen [70] for details.*

A novel feature of our difference methods (106) is that they are based on differencing the conservative-form equation

$$\partial_t u + \partial_x \big(f(u) - \partial_x D(u)\big) = 0,$$

and not the equation in its original form. Of course, one can devise methods based on differencing (105) directly, yielding, for example, methods of the form

(110) $$\frac{u_j^{n+1} - u_j^n}{\Delta t} + \Delta_- \big(F(u^n; j) - d(u_{j+1/2}^n)\Delta_+ u_j^n\big) = 0,$$

where $u_{j+1/2}^n = \frac{1}{2}\big(u_j^n + u_{j+1}^n\big)$. Indeed, the diffusion discretization in (110) is commonly used in the case of uniformly parabolic equations, see, e.g., [138]. Although it is possible to prove that the non-conservative method (110) converges to a limit, this limit does not seem to satisfy the entropy condition; that is, the method (110) does not seem to converge to the physically correct solution of (105) in the case of strong degeneracy. The following numerical example demonstrates this.

Example 6.1 (Non-Conservative Methods). *This example is taken from Evje and Karlsen* [71]. *We consider* (105) *with fluxes*

$$\tilde{f}(u) = \frac{1}{4}u^2, \qquad \tilde{d}(u) = 4d(u),$$

where $d(\cdot)$ is given in (47). *We have computed solutions with* (106) *and* (110) *using very fine discretization parameters. In these calculations the upwind flux* (107) *was used in* (106) *and* (110). *The computed solutions are shown in Figure 9. Clearly, the non-conservative method* (110) *produces a* **wrong** *solution. Moreover, the difference between this solution and the (correct!) solution produced by* (106) *seems to increase with time. We are currently investigating this phenomenon. See Hou and LeFloch* [103] *for an analysis of difference methods for hyperbolic equations which use a non-conservative discretization of the flux function.*

Finally, let us very briefly discuss the multi-dimensional case. For simplicity of notation, we consider only the two-dimensional problem

(111) $$\begin{cases} \partial_t u + \partial_x f(u) + \partial_y g(u) = \partial_x(d(u)\partial_x u) + \partial_y(d(u)\partial_y u), \\ u(x, y, 0) = u_0(x, y), \end{cases}$$

where u_0 belongs to $L^1(\mathbb{R}^2) \cap L^\infty(\mathbb{R}^2)$. Let $u_{j,k}^n$ denote the finite difference approximation at the point $(x, y, t) = (j\Delta x, k\Delta y, n\Delta t)$. A consistent, conservative, monotone, finite difference method for (111) takes the form

(112) $$\frac{u_{j,k}^{n+1} - u_{j,k}^n}{\Delta t} + \Delta_{x,-}\big(F(u^n; j, k) - \Delta_{x,+}D(u_{j,k}^n)\big)$$
$$+ \Delta_{y,-}\big(G(u^n; j, k) - \Delta_{y,+}D(u_{j,k}^n)\big) = 0,$$

where $\Delta_{\ell,-}, \Delta_{\ell,+}$ are the backward and forward differences, respectively, in direction ℓ, for $\ell = x, y$, and F, G are convective numerical (e.g., Engquist and Osher) fluxes that are consistent with f, g, respectively. As usual, higher order

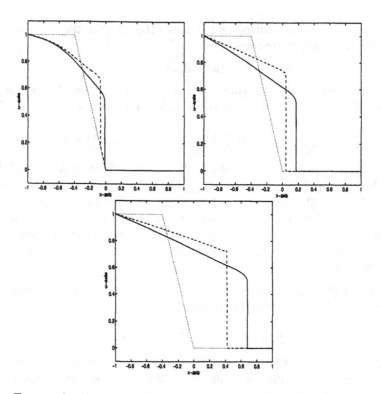

FIGURE 9. Plots of solutions produced by the difference methods defined
in (106) (solid) and (110) (dashed) at three different times; $T = 0.0625$
(left), $T = 0.25$ (middle), and $T = 1.0$ (right). The initial function is
shown as dotted.

methods can be built from the monotone ones via the MUSCL approach, see
[72] for details. The initial data for (112) is taken as

$$u_{j,k}^0 = \frac{1}{\Delta x \Delta y} \int_{z_{j,k}} u_0(x,y)\, dx dy,$$

$$z_{j,k} := \left[(j - \tfrac{1}{2})\Delta x, (j + \tfrac{1}{2})\Delta x\right) \times \left[(k - \tfrac{1}{2})\Delta y, (k + \tfrac{1}{2})\Delta y\right).$$

We assume, of course, that the CFL condition holds, which reads

$$\left(\max|f'| + \max|g'|\right)\frac{\Delta t}{h} + 4\max|d|\frac{\Delta t}{h^2} \leq 1$$

in the special case $\Delta x = \Delta y = h$. For the sake of convergence analysis, we need
to introduce the piecewise constant function

$$u_\Delta(x,y,t) = u_{j,k}^n \quad \text{for } (x,y) \in z_{j,k} \text{ and } t \in [n\Delta t, (n+1)\Delta t),$$

where $j, k \in \mathbb{Z}$, $n = 0, \ldots, N - 1$, and $\Delta = (\Delta x, \Delta t)$. We have the following
convergence theorem:

Theorem 6.2 ([72, 115]). *The sequence $\{u_\Delta\}$ of monotone difference approximations converges in $L^1_{\text{loc}}(\mathbb{R}^2 \times (0,T))$ to the unique entropy weak solution of*

(111) *as* $\Delta \to 0$. *Furthermore,* $\{D(u_\Delta)\}$ *converges in* $L^2_{\mathrm{loc}}(\mathbb{R}^2 \times (0,T))$ *to* $D(u)$ *as* $\Delta \to 0$ *and* $D(u) \in L^2(0,T;H^1(\mathbb{R}^2))$.

We continue with a few words about the proof of Theorem 6.2. The proof of the first of part of Theorem 6.2 is based on deriving uniform L^∞ and BV bounds on the approximate solution u_Δ. Equipped with the BV bound, we use the difference method (112) and the interpolation lemma of Kružkov (Lemma 3.1) to show that u_Δ is uniformly L^1 continuous in time. Kolmogorov's compactness criterion then immediately gives L^1_{loc} convergence (along a subsequence) of $\{u_\Delta\}$ to a function $u \in L^1 \cap L^\infty$ being L^1 continuous in time. Of course in the end, uniqueness of the entropy weak solution [36, 37, 115] will imply that the whole sequence $\{u_\Delta\}$ converges and not just some subsequence. To ensure that the limit u is the (unique) entropy weak solution of (111), we prove that the difference method satisfies a cell entropy inequality similar to (109) and hence it follows in the spirit of Lax and Wendroff that the limit u satisfies the entropy condition (40).

Finally, we show that the second part of Theorem 6.2 holds. Here the arguments are based on deriving a space estimate that is resemblant of the *weak BV* estimates employed by Eymard *et al.* [80] to prove convergence of finite volume methods on unstructured grids for first order hyperbolic equations, see also Eymard *et al.* [81] and Afif and Amaziane [4] for diffusion equations. The weak BV estimate implies a uniform (in Δ) space translation estimate of the form

$$(113) \qquad \left\| D(u_\Delta(\cdot + \xi, \cdot)) - D(u_\Delta(\cdot, \cdot)) \right\|_{L^2(\mathbb{R}^2 \times (0,T))} = \mathcal{O}(|\xi| + h), \qquad \forall \xi \in \mathbb{R}^2.$$

Equipped with (113) and the difference method (112), one can prove a uniform (in Δ) time translation estimate of the form

$$(114)$$
$$\left\| D(u_\Delta(\cdot, \cdot + \tau)) - D(u_\Delta(\cdot, \cdot)) \right\|_{L^2(\mathbb{R}^2 \times (0,T-\tau))} = \mathcal{O}(\sqrt{\tau + \Delta t}), \qquad \forall \tau \in (0,T).$$

In view of the space and time translation estimates (113) and (114), an application of Kolmogorov's compactness criterion implies strong L^2_{loc} convergence (along a subsequence) of $\{D(u_\Delta)\}$ to $D(u)$ and $D(u)$ possesses the regularity claimed in Theorem 6.2.

For further details on multi-dimensional difference methods, we refer to Evje and Karlsen [71] and Karlsen and Risebro [113]. Let us also mention that the difference methods described in this section apply equally well to equations with variable coefficients [113].

Numerical methods for degenerate convection-diffusion equations that are not discussed in these lecture notes include, among others, the kinetic BGK schemes in Bouchut, Guarguaglini, and Natalini [19], the local discontinuous Galerkin method in Cockburn and Shu [43], the central difference schemes in Kurganov and Tadmor [125], and the finite volume methods in Afif and Amaziane [4] and Ohlberger [140].

6.2. A Petrov-Galerkin Method.

In this section we describe a Petrov-Galerkin method for solving the parabolic residual problem resulting from the splitting algorithm described in §5.1. We refer to Morton [138] for a detailed introduction to Petrov-Galerkin methods. For $n = 1, \ldots, N$, the parabolic residual problem takes the form

(115) $$\begin{cases} \partial_t u + \nabla \cdot \big(b(x,u)u - d(x,u)\nabla u\big) = \bar{q}(x,t), & (x,t) \in \Omega \times (t_{n-1}, t_n), \\ u(x, t_{n-1}) = \bar{u}^{n-1}(x), & x \in \Omega, \end{cases}$$

where the residual fractional flow function $f_{\text{res}}(x,u) = b(x,u)u$ is defined in (95), $\bar{q}(x,t)$ is a source term, the initial condition $\bar{u}^{n-1}(x)$ is given in (97), and the time slab (t_{n-1}, t_n) is defined in §5.1. We note that with this definition of f_{res}, the numerical solution of (93) automatically satisfies the boundary condition (94) where $v(x) \cdot n = 0$.

Using the characteristic solution of (96) to approximate the time derivative and to linearise the nonlinear coefficients in (115), the Petrov-Galerkin method will introduce a symmetrization of (115), see [11, 104]. Let $H^1(\Omega)$ denote the usual Sobolev space formed by all functions in $L^2(\Omega)$ whose gradients belong to $L^2(\Omega)$. Let $H_0^1(\Omega)$ and V be subsets of $H^1(\Omega)$ such that [11, 50]

$$H_0^1 := \Big\{ w \in H^1(\Omega) : w(x) = 0 \text{ for } x \in \partial\Omega \Big\},$$

$$V := \Big\{ w \in H^1(\Omega) : w \text{ satisfies the given boundary conditions} \Big\}.$$

We now obtain the following weak formulation of (115): Find $u \in V$ such that

(116) $$\begin{cases} (\partial_t u, w) + B(u, w) = (\tilde{q}(x, t_{n-1}), w), & t \in (t_{n-1}, t_n), \ \forall w \in H_0^1, \\ u(x, t_{n-1}) = \bar{u}^{n-1}(x), & x \in \Omega, \end{cases}$$

where the bilinear form $B(\cdot, \cdot)$ is defined as

(117) $$B(u, w) = (\nabla \cdot (b(x,u)u), w) + \varepsilon(d(x,u)\nabla u, \nabla w)$$

and $(\cdot, \cdot)$ denotes the usual $L^2(\Omega)$ inner product. To deduce (116) from (115) we have used integration by parts. The boundary terms arising from this partial integration are included in the right-hand side $(\tilde{q}(x, t_{n-1}), w)$ of (116). Note that the boundary terms will in general depend nonlinearly on the unknown. We use, however, the characteristic solution $\bar{u}^{n-1}$ to linearise the boundary terms. One should also note that the the boundary condition for (115) will depend on the chosen flux splitting, see [82, 83, 88] for further details.

Equipped with the characteristic solution $\bar{u}^{n-1}$ of (96), we can replace the parabolic problem (116) by the following linearised elliptic problem: For $n = 1, \ldots, N$, find $u^n \in V$ such that

(118) $$(u^n, w) + \Delta t B(u^n, w) = (\bar{u}^{n-1}, w) + (\tilde{q}(x, t_{n-1}), w), \qquad \forall w \in H_0^1,$$

where the time step Δt has been introduced in §5.1. As a first step in an iterative procedure [88], the nonlinear coefficients are linearised as

$$b(x) := b(x, \bar{u}^{n-1}) \qquad \text{and} \qquad d(x) := d(x, \bar{u}^{n-1}).$$

With the components of $d(x)$ in $C^0(\bar{\Omega})$ and $b(x) \in H^1(\Omega)$, $B(\cdot, \cdot)$ defines a bilinear continuous form on $H_0^1 \times H_0^1$. Unfortunately, the sign of $B(\cdot, \cdot)$ is indefinite due to the transport term. However, we have [50, 53]

$$\left|\Delta t(\nabla \cdot (b(x)w), w)\right| < (w,w) + \varepsilon \Delta t(d(x)w, w),$$

which means that the complete bilinear form $A(\cdot, \cdot)$, where

$$A(u,w) = (u,w) + \Delta t B(u,w),$$

is coercive on $H^1(\Omega) \times H^1(\Omega)$. Hence the Lax-Milgram theorem ensures the existence of a unique element u^n satisfying the elliptic problem (118). In the following we restrict the presentation to $\Omega \subset \mathbb{R}^2$. The trial and optimal test spaces used within the Petrov-Galerkin formulation are given as follows: Let $\{x_{i,j}\}$ be the nodes generating a rectangular mesh covering Ω. We then introduce a trial space $S^h \subset H^1(\Omega)$ spanned by the trial functions $\{\theta_{i,j}\}$ and a test space $T^h \subset H^1(\Omega)$ spanned by the test functions $\{\psi_{i,j}\}$, where h denotes the grid spacing. Furthermore, we introduce the discrete subspaces

$$S_0^h = S^h \cap H_0^1, \qquad T_0^h = T^h \cap H_0^1, \qquad S_V^h = S^h \cap V.$$

Then the Petrov-Galerkin finite element formulation of (118) reads as follows: For $n = 1, \ldots, N$, find $u_h^n \in S_V^h$ such that

(119) $\qquad A(u_h^n, \psi) = (\bar{u}^{n-1}, \psi) + (\tilde{q}(x, t_{n-1}), \psi), \qquad \forall \psi \in T_0^h.$

It is well known that using $T^h = S^h$ (as in the usual Galerkin formulation) is a bad choice of test space when the transport term $b(x)$ dominates the diffusion term. This appears as unphysical oscillations in the numerical solution in the presence of a steep front. It may also be demonstrated that this problem is caused by the dominating transport term in the discretized bilinear form $B(\cdot, \cdot)$, i.e., the leading part of $A(\cdot, \cdot)$ in the steep front region. To handle such problems, Barrett and Morten [11] have developed a symmetrization technique in one space dimension that yields optimal approximation properties in suitable norms. The symmetrization technique used in one space dimension is in principle easily extendible to several space dimensions, but the extension may be technically involved and produces test functions that are difficult to use in practical computations. A procedure that resolves this problem has been developed by Demkowicz and Oden [59], who introduce the concept of "numerical optimal" test functions. Here, we use tensor products of one-dimensional test functions to define optimal test functions in several space dimensions.

Approximate optimal test functions in one space dimension, which yield an almost symmetric bilinear form when measured in a suitable norm [11, 91, 104], take the form

(120) $\qquad \psi_i(x) = \begin{cases} 0, & x < x_{i-1}, \\ \theta_i(x) + c_{i-1}\sigma_i(x), & x_{i-1} \leq x \leq x_i, \\ \theta_i(x) + c_i\sigma_i(x), & x_i < x \leq x_{i+1}, \\ 0, & x > x_{i+1}, \end{cases}$

where

$$c_i = 3\left(\frac{2}{\beta_i} - \coth\left(\frac{\beta_i}{2}\right)\right), \qquad \beta_i = \frac{b(x_i)h}{\varepsilon d(x_i)}.$$

In (120), $\theta_i(x)$ is the hat function

$$\theta_i(x) = \begin{cases} 0, & x < x_{i-1}, \\ \dfrac{x - x_{i-1}}{h}, & x_{i-1} \le x \le x_i, \\ \dfrac{x_{i+1} - x}{h}, & x_i < x \le x_{i+1}, \\ 0, & x > x_{i+1}, \end{cases}$$

and $\sigma_i(x)$ is the quadratic perturbation

$$\sigma_i(x) = \begin{cases} \dfrac{(x - x_{i-1})(x - x_i)}{h^2}, & x_{i-1} \le x \le x_i, \\ -\dfrac{(x - x_i)(x - x_{i+1})}{h^2}, & x_i < x \le x_{i+1}. \end{cases}$$

The test functions are depicted in Figure 10. The test functions in two dimensions are defined as

$$\psi_{i,j}(x,y) = \psi_i(x)\psi_j(y),$$

where $\psi_i(x)$ and $\psi_j(y)$ are defined above, see, e.g., [50] for further details.

As noted earlier, the porous media flow models may give solutions which vary on a wide range of scales in space and time. A proper resolution of the flow at wells, moving fronts, and the dynamics caused by large permeability variations at fractures and faults will require a very fine mesh. Such phenomena may be fairly local, separated by regions of slow variation. Thus, both the pressure-velocity equations and the Petrov-Galerkin formulation of the flow equation can represent very large elliptic problems. If a uniform mesh adequate for the fine scale variation is chosen, the problem may simply be too big for any computer. Therefore, an adaptive local grid refinement solution procedure is needed in order to reduce the problem to a solvable size. This can be achieved by using a preconditioned iterative solution procedure based on domain decomposition methods [158]. We will give the main steps in such a solution procedure for the flow equation (119), see [146, 53, 161] for further details. Assuming that we are given a coarse grid Ω_C on the computational domain Ω, we get the following algorithm:

1. Solve the hyperbolic problem (96) on the coarse grid Ω_C using the modified method of characteristics.
2. Identify coarse elements where the error is too large. This may be done simply by selecting elements which contain large gradients in the solution or by using an error estimator. Then activate refined overlapping/non overlapping sub grids Ω_k to each of these.
3. Solve (96) on the refined sub grids Ω_k.
4. Solve (119) on the refined coarse elements using domain decomposition methods. The characteristic solution of (96) is used as the boundary conditions for the sub domains Ω_k. Using an overlapping domain decomposition method will reduce the error introduced by the choice of boundary conditions, see [146, 161]

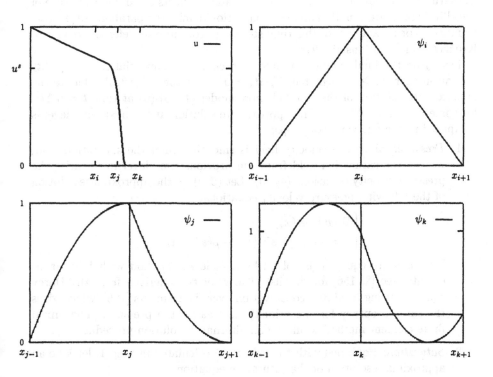

FIGURE 10. Typical test functions for the convection-diffusion problem problem (93). Upper left plot: A typical one-dimensional solution profile. The corresponding test functions are shown in the following three plots. Upper right plot: Test function for $u > u^s$. Lower left plot: Test function for $0 < u < u^s$. Lower right plot: Test function for $u = 0$.

It is well known that numerical algorithms based on domain decomposition methods have good parallel properties [158], which means that large complex problems may be solved [52].

Remark 6.4. *It is an open problem to prove L^1 convergence of an operator splitting algorithm based on, e.g., the modified method of characteristic and the Petrov-Galerkin method.*

7. RESERVOIR SIMULATION

We now apply the numerical algorithms developed in the previous sections to the reservoir flow model (4)-(5)-(6). The governing equations (4)-(5)-(6) constitute a coupled system of nonlinear partial differential equations. A sequential time stepping procedure is used to decouple the equations, which essentially consists of solving one equation at the time, starting with the pressure equation to generate a velocity field. Subsequently, this velocity field is used as input in the saturation equation, and so on. This strategy reflects the different nature of the elliptic pressure equation and the convection dominated parabolic saturation equation. For an analysis of this time stepping procedure we refer the reader to Kružkov and Sukorjanskiĭ [121].

Let T_s be the final computing time, and choose a sequential time step Δt_s and an integer $N_s \geq 1$ such that $N_s \Delta t_s = T_s$. Let the (p^n, v^n, s^n) denote the approximate solution of the reservoir flow model (4)-(5)-(6) at time $t = n\Delta t_s$, for some $n = 0, \ldots, N_s - 1$. The approximate solution at the next time level is computed in the following two steps:

1. *Pressure:* Since the velocity field is smoother than the saturation field, we use the saturation field from the previous time level to linearise the pressure-velocity equations (4)-(5). Let $(\bar{p}, \bar{v})$ be the approximate solution of the following pressure-velocity equations:

$$\begin{cases} \nabla \cdot \bar{v} = q_1(x), \\ \bar{v} = -K(x)\lambda(s^n)\big(\nabla\bar{p} - \rho(s^n)\nabla h\big). \end{cases}$$

   The pressure equation is solved by a Galerkin method with bilinear elements, see [8, 156] for details. The velocity is derived from the Darcy equation using local flux conservation over the elements [156], which gives the same accuracy for the velocity field as for the pressure. The mixed finite element method would be an alternative solution procedure.

2. *Saturation:* Equipped with the velocity $\bar{v}$ calculated in Step 1, let $\bar{s}$ be an approximate solution of the saturation equation

$$\begin{cases} \phi(x)\partial_t \bar{s} + \nabla \cdot \big(f(\bar{s})\bar{v} + f_g(\bar{s})K\nabla h\big) - \varepsilon\nabla \cdot \big(d(x, \bar{s})\nabla\bar{s}\big) = q_2(x), \\ \bar{s}(x, 0) = s^n(x). \end{cases}$$

   A good treatment of the saturation equation is essential for obtaining an accurate solution of the reservoir flow model (4)-(5)-(6). We use the corrected operator splitting algorithms described in §4 (see also §5 and §6) to solve the saturation equation, see the two examples presented below for further details. Finally, the approximate solution of (4)-(5)-(6) at the next time level is defined by

$$(p^{n+1}, v^{n+1}, s^{n+1}) = (\bar{p}, \bar{v}, \bar{s}\big|_{t=\Delta t_s}).$$

We now present two numerical examples. The first example is a two-dimensional, heterogeneous, quarter five-spot test case without a gravity term, while the second example is a three-dimensional homogeneous test case with a gravity term.

7.1. Two-Dimensional Test Case without Gravity.

This example is taken from Holden, Karlsen, and Lie [97]. The computations are based on the model obtained by neglecting the gravity term in equations (4)-(5)-(6). To simulate this model, we use the corrected splitting algorithm defined in (92) based on front tracking (see §5.2) for the convection updates and finite differences (see §6.1) for the diffusion updates. Similar simulations are presented in [65] using the algorithm defined in (83). These authors use the modified method of characteristics (see §5.1) for the convection updates and the Petrov-Galerkin method (see §6.2) for the diffusion updates.

The permeability field (see Figure 11) is generated as $K(x) = \exp(Z(x))$, where $Z(x)$ is a Gaussian field. Figure 12 shows saturation fields computed for viscosity ratios $\mu_o : \mu_w$ equal 1:1 and 5:1. The diffusion coefficient is $\varepsilon = 0.005$, the simulation grid has 129×129 blocks, and we use $N_s = 20$ sequential time steps to reach final time $T_s = 0.8$ with a CFL number 2.0 for the saturation solver (up to water breakthrough).

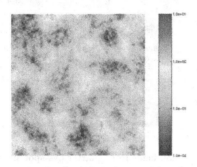

FIGURE 11. Permeability field plotted on a logarithmic colour scale.

Next, we consider a permeability field containing low-permeable blocks which are barriers to the flow (see Figure 13). Figure 14 shows the saturation fields computed for viscosity ratios $\mu_o : \mu_w$ equal 20:1 and 1:2. The diffusion coefficient is $\varepsilon = 0.005$ and we use a 257×257 grid with 80 sequential time steps to reach final time 0.8 and CFL number 2.0 in the saturation solver. As for the above case, the fingering effects are more pronounced at the adverse viscosity ratio. Notice also the improved areal sweep and penetration into low-permeable regions in the lower plot. Fine scale solutions such as those presented in this example are very accurate. Consequently, they can be used good reference solutions for upscaling problems, see, e.g., [64, 65, 92].

We refer to [97] for further details about the computations presented in this example.

7.2. Three-Dimensional Test Case with Gravity. This example is taken from Frøysa and Espedal [83], see also Frøysa [82]. The computations are based on the model described by equations (4)-(5)-(6). To simulate this model, we use the corrected splitting algorithm defined in (83) based on the modified method of characteristics (see §5.1) for the convection updates and the Petrov-Galerkin finite element method (see §6.2) for the diffusion updates. A flux splitting which is uniform in space has been applied (see §5.1), but local node based splittings

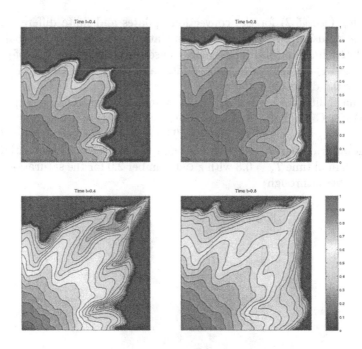

FIGURE 12. Saturation fields from quarter five-spot simulations for viscosity ratios $\mu_o : \mu_w$ equal 1:1 (top) and 5:1 (bottom).

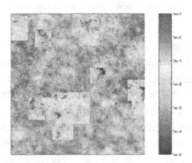

FIGURE 13. Permeability field with low-permeable regions plotted on a logarithmic colour scale.

have also been tested, see [83]. We use the following dimensionless data:

$$
\left\{
\begin{array}{lll}
\Omega & = & [0,1] \times [0,1] \times [0,0.5]. \\
\text{Grid} & = & 21 \times 21 \times 15. \\
\text{Rates} & = & 0.04 \text{ in injector, } -0.04 \text{ in producer.} \\
\text{Gravity} & = & g\nabla h = [0,0,-g]. \\
\text{Time step} & = & 0.01. \\
\text{Initial profile} & = & \text{if } (x+y) \leq 0.5 \\
& & \quad \text{if } z \geq 0.2 \\
& & \quad\quad s = 1.0 - (x+y) \\
& & \quad \text{else} \\
& & \quad\quad s = 5z(1.0 - (x+y)) \\
& & \text{else} \\
& & \quad s = 0
\end{array}
\right.
$$

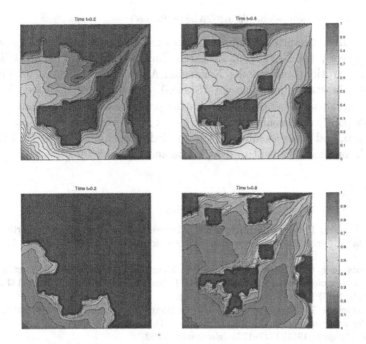

FIGURE 14. Saturation fields from quarter five-spot simulations (with low-permeable regions) for viscosity ratios $\mu_o : \mu_w$ equal 20:1 (top) and 1:2 (bottom).

The injection well is a line well located on the z-axis with $z \in [0.25, 0.5]$. A similar vertical production well is located at position $(x, y) = (1, 1)$ with $z \in [0, 0.25]$. In the test case, the z component of the velocity field mostly has the same sign as the gravity. The initial saturation gradient $\partial_z s$, however, is positive but changes sign as the saturation front moves along. The sign change is a function of both space and time. Thus we get a good indication of how well the methods handle a saturation field that "turns".

Figure 15 shows the saturation field with the uniform flux splitting described in §5.1. Similar results are obtained with a non-uniform splitting, see [82, 83]. After a few time steps the water has reached the bottom boundary and the saturation gradient $\partial_z s$ starts to change sign. At $t = 0.05$, there are regions with $\partial_z s < 0$ and regions with $\partial_z s > 0$, but no oscillations are visible. At $t = 0.3$, the gradient $\partial_z s$ is negative everywhere, except for a region in the vicinity of the injector. The present example gives a fairly difficult boundary condition, but the tests shows that the diffusion step in the solution algorithm handles these accurately. The results also indicate that the numerical algorithms are able to handle a "turning" field without introducing oscillations.

We refer to [83, 82] for further details about the computations presented in this example.

68

8. Acknowledgement

We gratefully acknowledge support from the University of Bergen (Computational Science), the Norwegian Research Council (PROPETRO), VISTA, a research cooperation between the Norwegian Academy of Science and Letters and Den norske stats oljeselskap a.s. (Statoil), and finally the Applied Mathematics for Industrial Flow problems (AMIF) program of the European Science Foundation (ESF). We would also like to thank H. K. Dahle, S. Evje, K.-A. Lie, N. H. Risebro, and other members of the applied mathematics groups in Bergen, Oslo, and Trondheim for fruitful discussions and contributions.

References

[1] I. Aavatsmark, T. Barkve, Ø. Bøe, and T. Mannseth. Discretization on non-orthogonal, quadrilateral grids for inhomogeneous, anisotropic media. *J. Comput. Phys.*, 127:2–14, 1986.

[2] I. Aavatsmark, T. Barkve, Ø. Bøe, and T. Mannseth. Discretization on unstructured grids for inhomogeneous, anisotropic media. I. Derivation of the methods. *SIAM J. Sci. Comput.*, 19(5):1700–1716 (electronic), 1998.

[3] I. Aavatsmark, T. Barkve, Ø. Bøe, and T. Mannseth. Discretization on unstructured grids for inhomogeneous, anisotropic media. II. Discussion and numerical results. *SIAM J. Sci. Comput.*, 19(5):1717–1736 (electronic), 1998.

[4] M. Afif and B. Amaziane. Convergence of finite volume schemes for a degenerate convection-diffusion equation arising in two-phase flow in porous media. Preprint, 1999.

[5] V. Alexiades. Overcoming the stability restriction of explicit schemes via super-time-stepping. In *Proceedings of Dynamic Systems and Applications, Vol. 2 (Atlanta, GA, 1995)*, pages 39–44, Atlanta, GA, 1996. Dynamic.

[6] V. Alexiades, G. Amiez, and P. Gremaud. Super time-stepping acceleration of explicit schemes. *Com. Num. Meth. Eng.*, 12:31–42, 1996.

[7] I. Allen, M. B., G. A. Behie, and J. A. Trangenstein. *Multiphase flow in porous media*. Springer-Verlag, Berlin, 1988. Mechanics, mathematics, and numerics.

[8] O. Axelsson and V. A. Barker. *Finite element solution of boundary value problems*. Academic Press Inc., Orlando, Fla., 1984. Theory and computation.

[9] K. Aziz and A. Settari. *Petroleum reservoir simulation*. Elsevier Applied Science Publishers, Essex, England, 1979.

[10] C. Bardos, A. Y. LeRoux, and J.-C. Nédélec. First order quasilinear equations with boundary conditions. *Comm. Partial Differential Equations*, 4(9):1017–1034, 1979.

[11] J. W. Barrett and K. W. Morton. Approximate symmetrization and Petrov-Galerkin methods for diffusion-convection problems. *Comput. Methods Appl. Mech. Engrg.*, 45(1-3):97–122, 1984.

[12] J. T. Beale, C. Greengard, and E. Thomann. Operator splitting for Navier-Stokes and Chorin-Marsden product formula. In *Vortex flows and related numerical methods (Grenoble, 1992)*, pages 27–38. Kluwer Acad. Publ., Dordrecht, 1993.

[13] P. Bénilan and R. Gariepy. Strong solutions in L^1 of degenerate parabolic equations. *J. Differential Equations*, 119(2):473–502, 1995.

[14] P. Bénilan and S. N. Kružkov. Conservation laws with continuous flux functions. *NoDEA Nonlinear Differential Equations Appl.*, 3(4):395–419, 1996.

[15] P. Bénilan and H. Touré. Sur l'équation générale $u_t = \varphi(u)_{xx} - \psi(u)_x + v$. *C. R. Acad. Sci. Paris Sér. I Math.*, 299(18):919–922, 1984.

[16] P. Bénilan and H. Touré. Sur l'équation générale $u_t = a(\cdot, u, \phi(\cdot, u)_x)_x + v$ dans L^1. I. Étude du problème stationnaire. In *Evolution equations (Baton Rouge, LA, 1992)*, pages 35–62. Dekker, New York, 1995.

[17] P. Bénilan and H. Touré. Sur l'équation générale $u_t = a(\cdot, u, \phi(\cdot, u)_x)_x + v$ dans L^1. II. Le problème d'évolution. *Ann. Inst. H. Poincaré Anal. Non Linéaire*, 12(6):727–761, 1995.

[18] P. J. Binning. *Modeling unsaturated zone flow and contaminant in the air and water phases*. PhD thesis, Department of Civil Engineering and Operational Research, Princeton University, 1994.

[19] F. Bouchut, F. R. Guarguaglini, and R. Natalini. Diffusive BGK approximations for nonlinear multidimensional parabolic equations. *Indiana Univ. Math. J.* To appear.

[20] F. Bouchut and B. Perthame. Kružkov's estimates for scalar conservation laws revisited. *Trans. Amer. Math. Soc.*, 350(7):2847–2870, 1998.

[21] F. Bratvedt, K. Bratvedt, C. F. Buchholz, T. Gimse, H. Holden, L. Holden, R. Olufsen, and N. H. Risebro. Three-dimensional reservoir simulation based on front tracking. In *North Sea Oil and Gas Reservoirs - III*, pages 247–257. Kluwer Academic Publishers, 1994.

[22] F. Bratvedt, K. Bratvedt, C. F. Buchholz, T. Gimse, H. Holden, L. Holden, and N. H. Risebro. FRONTLINE and FRONTSIM: two full scale, two-phase, black oil reservoir simulators based on front tracking. *Surveys Math. Indust.*, 3(3):185–215, 1993.

[23] H. Brézis and M. G. Crandall. Uniqueness of solutions of the initial-value problem for $u_t - \Delta\varphi(u) = 0$. *J. Math. Pures Appl. (9)*, 58(2):153–163, 1979.

[24] K. Brusdal, H. K. Dahle, K. H. Karlsen, and T. Mannseth. A study of the modelling error in two operator splitting algorithms for porous media flow. *Computational Geosciences*, 2(1):23–36, 1998.

[25] R. Bürger, M. C. Bustos, and F. Concha. Settling velocities of particulate systems: Phenomenological theory of sedimentation: 9. Numerical simulation of the transient behaviour of flocculated suspensions in an ideal batch or continuous thickener. *Int. J. Mineral Process.*, 55:267–282, 1999.

[26] R. Bürger and F. Concha. Simulation of the transient behaviour of flocculated suspensions in a continuous thickener. In H. Hoberg and H. v. Blottnitz, editors, *20th International Mineral Processing Congress (XX IMPC)*, volume 4, pages 91–101. GDMB, Clausthal-Zellerfeld, 1997.

[27] R. Bürger and F. Concha. Mathematical model and numerical simulation of the settling of flocculated suspensions. *Int. J. Multiphase Flow*, 24:1005–1023, 1998.

[28] R. Bürger, S. Evje, and K. H. Karlsen. Viscous splitting methods for degenerate convection-diffusion equations with boundary conditions. Preprint (in preparation).

[29] R. Bürger, S. Evje, and K. H. Karlsen. On strongly degenerate convection-diffusion problems modeling sedimentation-consolidation processes. *J. Math. Anal. Appl.*, 247(2):517–556, 2000.

[30] R. Bürger, S. Evje, K. H. Karlsen, and K.-A. Lie. Numerical methods for the simulation of the settling of flocculated suspensions. *Chemical Engineering Journal*, 80:119–137, 2000.

[31] R. Bürger and W. L. Wendland. Entropy boundary and jump conditions in the theory of sedimentation with compression. *Math. Methods Appl. Sci.*, 21(9):865–882, 1998.

[32] R. Bürger and W. L. Wendland. Existence, uniqueness, and stability of generalized solutions of an initial-boundary value problem for a degenerating quasilinear parabolic equation. *J. Math. Anal. Appl.*, 218(1):207–239, 1998.

[33] R. Bürger, W. L. Wendland, and F. Concha. Model equations for gravitational sedimentation-consolidation processes. *ZAMM Z. Angew. Math. Mech.*, 80(2):79–92, 2000.

[34] J. M. Burgers. Application of a model system to illustrate some points of the statistical theory of free turbulence. *Nederl. Akad. Wetensch., Proc.*, 43:2–12, 1940.

[35] M. C. Bustos, F. Concha, R. Bürger, and E. M. Tory. *Sedimentation and Thickening: Phenomenological Foundation and Mathematical Theory*. Kluwer Academic Publishers, Dordrecht, The Netherlands, 1999.

[36] J. Carrillo. Solutions entropiques de problèmes non linéaires dégénérés. *C. R. Acad. Sci. Paris Sér. I Math.*, 327(2):155–160, 1998.

[37] J. Carrillo. Entropy solutions for nonlinear degenerate problems. *Arch. Rational Mech. Anal.*, 147(4):269–361, 1999.

[38] M. A. Celia, T. F. Russell, I. Herrera, and R. E. Ewing. An Eulerian-Lagrangian localized adjoint method for the advection-diffusion equation. *Adv. Water Resources*, 13(4):187–206, 1990.

[39] M. A. Celia and S. Zisman. An Eulerian-Lagrangian localized adjoint method for reactive transport in groundwater. In Gambolati et al., editor, *Computational Methods in Water Resources VIII*, volume 1, pages 383–392. Springer, 1990.

[40] G. Chavent and J. Jaffre. *Mathematical models and finite elements for reservoir simulation*, volume 17 of *Studies in mathematics and its applications*. North Holland, Amsterdam, 1986.

[41] B. Cockburn and P.-A. Gremaud. A priori error estimates for numerical methods for scalar conservation laws. I. The general approach. *Math. Comp.*, 65(214):533–573, 1996.

[42] B. Cockburn and G. Gripenberg. Continuous dependence on the nonlinearities of solutions of degenerate parabolic equations. *J. Differential Equations*, 151(2):231–251, 1999.

[43] B. Cockburn and C.-W. Shu. The local discontinuous Galerkin method for time-dependent convection-diffusion systems. *SIAM J. Numer. Anal.*, 35(6):2440–2463 (electronic), 1998.

[44] F. Concha and R. Bürger. Wave propagation phenomena in the theory of sedimentation. In E. F. Toro and J. F. Clarke, editors, *Numerical Methods for Wave Propagation*, pages 173–196. Kluwer Academic Publishers, Dordrecht, The Netherlands, 1998.

[45] M. G. Crandall and T. M. Liggett. Generation of semi-groups of nonlinear transformations on general Banach spaces. *Amer. J. Math.*, 93:265–298, 1971.

[46] M. G. Crandall and A. Majda. The method of fractional steps for conservation laws. *Numer. Math.*, 34:285–314, 1980.

[47] M. G. Crandall and A. Majda. Monotone difference approximations for scalar conservation laws. *Math. Comp.*, 34(149):1–21, 1980.

[48] C. M. Dafermos. Polygonal approximation of solutions of the initial value problem for a conservation law. *J. Math. Anal. Appl.*, 38:33–41, 1972.

[49] G. Dagan. *Flow and Transport in Porous Formations*. Springer-Verlag, 1989.

[50] H. K. Dahle. *Adaptive characteristic operator splitting techniques for convection-dominated diffusion problems in one and two space dimensions*. PhD thesis, Department of Mathematics, University of Bergen, Norway, 1988.

[51] H. K. Dahle, M. S. Espedal, and R. E. Ewing. Characteristic Petrov-Galerkin subdomain methods for convection-diffusion problems. In *Numerical simulation in oil recovery (Minneapolis, Minn., 1986)*, pages 77–87. Springer, New York, 1988.

[52] H. K. Dahle, M. S. Espedal, R. E. Ewing, and O. Sævareid. Characteristic adaptive subdomain methods for reservoir flow problems. *Numer. Methods Partial Differential Equations*, 6(4):279–309, 1990.

[53] H. K. Dahle, M. S. Espedal, and O. Sævareid. Characteristic, local grid refinement techniques for reservoir flow problems. *International Journal for Numerical Methods in Enginering*, 34:1051–1069, 1992.

[54] H. K. Dahle, R. E. Ewing, and T. F. Russell. Eulerian-Lagrangian localized adjoint methods for a nonlinear advection-diffusion equation. *Comput. Methods Appl. Mech. Engrg.*, 122(3-4):223–250, 1995.

[55] C. N. Dawson. Godunov-mixed methods for advective flow problems in one space dimension. *SIAM J. Numer. Anal.*, 28(5):1282–1309, 1991.

[56] C. N. Dawson. Godunov-mixed methods for advection-diffusion equations in multidimensions. *SIAM J. Numer. Anal.*, 30(5):1315–1332, 1993.

[57] C. N. Dawson. High resolution upwind-mixed finite element methods for advection-diffusion equations with variable time-stepping. *Numer. Methods Partial Differential Equations*, 11(5):525–538, 1995.

[58] C. N. Dawson and M. F. Wheeler. Time-splitting methods for advection-diffusion-reaction equations arising in contaminant transport. In *ICIAM 91 (Washington, DC, 1991)*, pages 71–82. SIAM, Philadelphia, PA, 1992.

[59] L. Demkowicz and J. T. Oden. An adaptive characteristic Petrov-Galerkin finite element method for convection-dominated linear and nonlinear parabolic problems in two space variables. *Comput. Methods Appl. Mech. Engrg.*, 55(1-2):63–87, 1986.

[60] R. J. DiPerna. The structure of solutions to hyperbolic conservation laws. In *Nonlinear analysis and mechanics: Heriot-Watt Symposium, Vol. IV*, pages 1–16. Pitman, Boston, Mass., 1979.

[61] J. Douglas and T. F. Russell. Numerical methods for convection-dominated diffusion problems based on combining the method of characteristics with finite element or finite difference procedures. *SIAM J. Numer. Anal.*, 19(5):871–885, 1982.

[62] B. G. Ersland, M. S. Espedal, and R. Nybø. Numerical methods for flow in porous media with internal boundaries. *Computational Geosciences*. To appear.

[63] M. S. Espedal and R. E. Ewing. Characteristic Petrov-Galerkin subdomain methods for two-phase immiscible flow. *Comput. Methods Appl. Mech. Engrg.*, 64:113–135, 1987.

[64] M. S. Espedal, K. J. Hersvik, and B. G. Ersland. Domain decomposition methods for flow in heterogeneous porous media. *Contemporary Mathematics*, 218:104–120, 1998.

[65] M. S. Espedal and P. Langlo. Macrodispersion for two-phase immiscible flow in porous media. *Adv. Water Resources*, 17:297–316, 1994.

[66] L. C. Evans and R. F. Gariepy. *Measure theory and fine properties of functions*. CRC Press, Boca Raton, FL, 1992.

[67] S. Evje. *Numerical approximation of nonlinear degenerate convection-diffusion equations*. PhD thesis, Department of Mathematics, University of Bergen, Norway, 1998.

[68] S. Evje and K. H. Karlsen. Degenerate convection-diffusion equations and implicit monotone difference schemes. In *Hyperbolic problems: theory, numerics, applications, Vol. I (Zürich, 1998)*, pages 285–294. Birkhäuser, Basel, 1999.

[69] S. Evje and K. H. Karlsen. Viscous splitting approximation of mixed hyperbolic-parabolic convection-diffusion equations. *Numer. Math.*, 83(1):107–137, 1999.

[70] S. Evje and K. H. Karlsen. Discrete approximations of BV solutions to doubly nonlinear degenerate parabolic equations. *Numer. Math.*, 86(3):377–417, 2000.

[71] S. Evje and K. H. Karlsen. Monotone difference approximations of BV solutions to degenerate convection-diffusion equations. *SIAM J. Numer. Anal.*, 37(6):1838–1860 (electronic), 2000.

[72] S. Evje and K. H. Karlsen. Second order difference schemes for degenerate convection-diffusion equations. Preprint, Department of Mathematics, University of Bergen, 2000.

[73] S. Evje, K. H. Karlsen, K.-A. Lie, and N. H. Risebro. Front tracking and operator splitting for nonlinear degenerate convection–diffusion equations. In P. Bjørstad and M. Luskin, editors, *Parallel Solution of Partial Differential Equations*, volume 120 of *IMA Volumes in Mathematics and its Applications*, pages 209–228. Springer-Verlag, 2000.

[74] S. Evje, K. H. Karlsen, and N. H. Risebro. A continuous dependence result for nonlinear degenerate parabolic equations with spatially dependent flux function. Preprint, Department of Mathematics, University of Bergen, 2000.

[75] R. E. Ewing, editor. *The Mathematics of Reservoir Simulation*, volume 1 of *Frontiers in Applied Mathematics*. SIAM, 1983.

[76] R. E. Ewing and T. F. Russell. Efficient time-stepping methods for miscible displacement problems in porous media. *SIAM J. Numer. Anal.*, 19(1):1–67, 1982.

[77] R. E. Ewing and H. Wang. Eulerian-Lagrangian localized adjoint methods for linear advection or advection-reaction equations and their convergence analysis. *Comput. Mech.*, 12(1-2):97–121, 1993.

[78] R. E. Ewing and H. Wang. Eulerian-Lagrangian localized adjoint methods for variable-coefficient advective-diffusive-reactive equations in groundwater contaminant transport.

In *Advances in optimization and numerical analysis (Oaxaca, 1992)*, pages 185–205. Kluwer Acad. Publ., Dordrecht, 1994.

[79] R. E. Ewing and H. Wang. An optimal-order estimate for Eulerian-Lagrangian localized adjoint methods for variable-coefficient advection-reaction problems. *SIAM J. Numer. Anal.*, 33(1):318–348, 1996.

[80] R. Eymard, T. Gallouët, M. Ghilani, and R. Herbin. Error estimates for the approximate solutions of a nonlinear hyperbolic equation given by finite volume schemes. *IMA J. Numer. Anal.*, 18(4):563–594, 1998.

[81] R. Eymard, T. Gallouët, D. Hilhorst, and Y. Naït Slimane. Finite volumes and nonlinear diffusion equations. *RAIRO Modél. Math. Anal. Numér.*, 32(6):747–761, 1998.

[82] K. G. Frøysa. *Modelling and simulation of 3D flow with gravity forces in porous media.* PhD thesis, Department of Mathematics, University of Bergen, Norway, 1995.

[83] K. G. Frøysa and M. S. Espedal. Simulation of 3D flow with gravity forces in a porous media. *Computing and Visualization in Science.* Submitted.

[84] T. Gimse. Conservation laws with discontinuous flux functions. *SIAM J. Math. Anal.*, 24(2):279–289, 1993.

[85] T. Gimse and N. H. Risebro. Solution of the Cauchy problem for a conservation law with a discontinuous flux function. *SIAM J. Math. Anal.*, 23(3):635–648, 1992.

[86] R. Glowinski and O. Pironneau. Finite element methods for Navier-Stokes equations. In *Annual review of fluid mechanics, Vol. 24*, pages 167–204. Annual Reviews, Palo Alto, CA, 1992.

[87] E. Godlewski and P.-A. Raviart. *Numerical approximation of hyperbolic systems of conservation laws.* Springer-Verlag, New York, 1996.

[88] R. Hansen and M. S. Espedal. On the numerical solution of nonlinear reservoir flow models with gravity. *Int. J. for Num. Meth. in Engrg.*, 38:2017–2032, 1995.

[89] A. Harten, J. M. Hyman, and P. D. Lax. On finite-difference approximations and entropy conditions for shocks. *Comm. Pure Appl. Math.*, XXIX:297–322, 1976.

[90] V. Haugse, K. H. Karlsen, K.-A. Lie, and J. Natvig. Numerical solution of the polymer system by front tracking. *Transport in Porous Media.* To appear.

[91] P. W. Hemker. *A numerical study on stiff two-point boundary problems.* PhD thesis, Mathematiche Centrum, Amsterdam, 1977.

[92] K. J. Hersvik and M. S. Espedal. Adaptive hierarchical upscaling of flow in heterogeneous reservoirs based on an a posteriori error estimate. *Comput. Geosci.*, 2(4):311–336 (1999), 1998.

[93] R. Holdahl, H. Holden, and K.-A. Lie. Unconditionally stable splitting methods for the shallow water equations. *BIT*, 39(3):451–472, 1999.

[94] H. Holden and L. Holden. On scalar conservation laws in one-dimension. In S. Albeverio, J. E. Fenstad, H. Holden, and T. Lindstrøm, editors, *Ideas and Methods in Mathematics and Physics*, pages 480–509. Cambridge University Press, Cambridge, 1988.

[95] H. Holden, L. Holden, and R. Høegh-Krohn. A numerical method for first order nonlinear scalar conservation laws in one dimension. *Comput. Math. Appl.*, 15(6-8):595–602, 1988.

[96] H. Holden, K. H. Karlsen, and K.-A. Lie. Operator splitting methods for degenerate convection-diffusion equations I: convergence and entropy estimates. In *"Stochastic Processes, Physics and Geometry: New Interplays. A Volume in Honor of Sergio Albeverio"*. Amer. Math. Soc. To appear.

[97] H. Holden, K. H. Karlsen, and K.-A. Lie. Operator splitting methods for degenerate convection-diffusion equations II: numerical examples with emphasis on reservoir simulation and sedimentation. *Computational Geosciences.* To appear.

[98] H. Holden, K. H. Karlsen, K.-A. Lie, and N. H. Risebro. Operator splitting for nonlinear partial differential equations: An L^1 convergence theory. Preprint (in preparation).

[99] H. Holden, K. H. Karlsen, and N. H. Risebro. On the uniqueness and stability of entropy solutions of weakly coupled systems of nonlinear degenerate parabolic equations. Preprint, Department of Mathematics, University of Bergen, 2000.

[100] H. Holden, K.-A. Lie, and N. H. Risebro. An unconditionally stable method for the Euler equations. *J. Comput. Phys.*, 150(1):76–96, 1999.

[101] H. Holden and N. H. Risebro. A method of fractional steps for scalar conservation laws without the CFL condition. *Math. Comp.*, 60(201):221–232, 1993.

[102] H. Holden and N. H. Risebro. Front tracking for conservation laws. Lecture Notes, Department of Mathematics, Norwegian University of Science and Technology, 1997.

[103] T. Y. Hou and P. G. Le Floch. Why nonconservative schemes converge to wrong solutions: error analysis. *Math. Comp.*, 62(206):497–530, 1994.

[104] T. J. R. Hughes and A. Brooks. A theoretical framework for Petrov-Galerkin methods with discontinuous weighting functions: Application to the streamline-upwind procedure. In R. H. Gallagher et al., editor, *Finite Elements in Fluids*, volume 4, pages 47–65. John Wiley & Sons, 1982.

[105] K. H. Karlsen. On the accuracy of a dimensional splitting method for scalar conservation laws. Master's thesis, Department of Mathematics, University of Oslo, Norway, 1994.

[106] K. H. Karlsen. *Numerical solution of nonlinear convection-diffusion equations: operator splitting, front tracking and finite differences.* PhD thesis, Department of Mathematics, University of Bergen, Norway, 1998.

[107] K. H. Karlsen, K. Brusdal, H. K. Dahle, S. Evje, and K.-A. Lie. The corrected operator splitting approach applied to a nonlinear advection–diffusion problem. *Comput. Methods Appl. Mech. Engrg.*, 167:239–260, 1998.

[108] K. H. Karlsen and K.-A. Lie. An unconditionally stable splitting scheme for a class of nonlinear parabolic equations. *IMA J. Numer. Anal.*, 19(4):609–635, 1999.

[109] K. H. Karlsen, K.-A. Lie, J. Natvig, H. F. Nordhaug, and H. K. Dahle. Operator splitting methods for systems of convection-diffusion equations: nonlinear error mechanisms and correction strategies. Preprint, Department of Mathematics, University of Bergen, 2000.

[110] K. H. Karlsen, K.-A. Lie, and N. H. Risebro. A front tracking method for conservation laws with boundary conditions. In *Hyperbolic problems: theory, numerics, applications, Vol. I (Zürich, 1998)*, pages 493–502. Birkhäuser, Basel, 1999.

[111] K. H. Karlsen, K.-A. Lie, N. H. Risebro, and J. Frøyen. A front-tracking approach to a two-phase fluid-flow model with capillary forces. *In Situ (Special Issue on Reservoir Simulation)*, 22(1):59–89, 1998.

[112] K. H. Karlsen and N. H. Risebro. An operator splitting method for convection-diffusion equations. *Numer. Math.*, 77(3):365–382, 1997.

[113] K. H. Karlsen and N. H. Risebro. Convergence of finite difference schemes for viscous and inviscid conservation laws with rough coefficients. Preprint, Department of Mathematics, University of Bergen, 2000.

[114] K. H. Karlsen and N. H. Risebro. Corrected operator splitting for nonlinear parabolic equations. *SIAM J. Numer. Anal.*, 37(3):980–1003, 2000.

[115] K. H. Karlsen and N. H. Risebro. On the uniqueness and stability of entropy solutions of nonlinear degenerate parabolic equations with rough coefficients. Preprint, Department of Mathematics, University of Bergen, 2000.

[116] G. E. Karniadakis and R. D. Henderson. Spectral element methods for incompressible flows. In R. W. Johson, editor, *The Handbook of Fluid Dynamics*, volume 29, pages 1–42. 1998.

[117] R. A. Klausen and N. H. Risebro. Stability of conservation laws with discontinuous coefficients. *J. Differential Equations*, 157(1):41–60, 1999.

[118] C. Klingenberg and N. H. Risebro. Stability of a resonant system of conservation laws modeling polymer flow with gravitation. *J. Differential Equations*. To appear.

[119] C. Klingenberg and N. H. Risebro. Convex conservation laws with discontinuous coefficients. Existence, uniqueness and asymptotic behavior. *Comm. Partial Differential Equations*, 20(11-12):1959–1990, 1995.

[120] D. Kröner. *Numerical schemes for conservation laws.* John Wiley & Sons Ltd., Chichester, 1997.

[121] S. N. Kružkov and S. M. Sukorjanskiĭ. Boundary value problems for systems of equations of two-phase porous flow type; statment of the problems, questions of sovability, justification of approximate methods. *Math. USSR Sbornik*, 33(1):62–80, 1977.

[122] S. N. Kružkov. Results on the nature of the continuity of solutions of parabolic equations, and certain applications thereof. *Mat. Zametki*, 6:97–108, 1969.

[123] S. N. Kružkov. First order quasi-linear equations in several independent variables. *Math. USSR Sbornik*, 10(2):217–243, 1970.

[124] S. N. Kružkov and E. Y. Panov. Osgood's type conditions for uniqueness of entropy solutions to Cauchy problem for quasilinear conservation laws of the first order. *Ann. Univ. Ferrara Sez. VII (N.S.)*, 40:31–54 (1996), 1994.

[125] A. Kurganov and E. Tadmor. New high-resolution central schemes for nonlinear conservation laws and convection-diffusion equations. *J. Comput. Phys.*, 160:241–282, 2000.

[126] N. N. Kuznetsov. Accuracy of some approximative methods for computing the weak solutions of a first-order quasi-linear equation. *USSR Comput. Math. and Math. Phys. Dokl.*, 16(6):105–119, 1976.

[127] L. W. Lake. *Enhanced Oil Recovery*. Prentice-Hall, 1989.

[128] J. O. Langseth. On an implementation of a front tracking method for hyperbolic conservation laws. *Advances in Engineering Software*, 26(1):45–63, 1996.

[129] R. J. LeVeque. *Numerical methods for conservation laws*. Birkhäuser Verlag, Basel, second edition, 1992.

[130] K.-A. Lie. Front tracking for one-dimensional quasilinear hyperbolic equations with variable coefficients. *Numerical Algorithms*. To appear.

[131] K.-A. Lie. *Front tracking and operator splitting for convection dominated problems*. Dr. ing. thesis, Department of Mathematical Sciences, Norwegian University of Science and Technology, 1998.

[132] K.-A. Lie. A dimensional splitting method for quasilinear hyperbolic equations with variable coefficients. *BIT*, 39(4):683–700, 1999.

[133] K.-A. Lie, V. Haugse, and K. H. Karlsen. Dimensional splitting with front tracking and adaptive grid refinement. *Numer. Methods for Partial Differential Equations*, 14(5):627–648, 1998.

[134] B. J. Lucier. Error bounds for the methods of Glimm, Godunov and LeVeque. *SIAM J. Numer. Anal.*, 22:1074–1081, 1985.

[135] B. J. Lucier. A moving mesh numerical method for hyperbolic conservation laws. *Math. Comp.*, 46(173):59–69, 1986.

[136] J. Málek, J. Nečas, M. Rokyta, and M. Ružička. *Weak and measure-valued solutions to evolutionary PDEs*. Chapman & Hall, London, 1996.

[137] C. M. Marle. *Multiphase Flow in Porous Media*. Institut Francais du Petrole Publications, Editions Technip, 1981.

[138] K. W. Morton. *Numerical solution of convection-diffusion problems*. Chapman & Hall, London, 1996.

[139] J. Natvig. Operatorsplitting basert på frontfølging for polymersystemet. Diploma thesis, Department of Mathematical Sciences. Norwegian University of Science and Technology, 1998. In Norwegian.

[140] M. Ohlberger. A posteriori error estimates for vertex centered finite volume approximations of convection-diffusion-reaction equations. Preprint, Mathematische Fakultät, Albert-Ludwigs-Universität Freiburg, 2000.

[141] O. A. Oleĭnik. Discontinuous solutions of non-linear differential equations. *Amer. Math. Soc Transl. Ser. 2*, 26:95–172, 1963.

[142] O. A. Oleĭnik and S. N. Kružkov. Quasi-linear parabolic second-order equations with several independent variables. *Uspehi Mat. Nauk*, 16(5):115–155, 1961.

[143] F. Otto. Initial-boundary value problem for a scalar conservation law. *C. R. Acad. Sci. Paris Sér. I Math.*, 322(8):729–734, 1996.

[144] B. Perthame. Uniqueness and error estimates in first order quasilinear conservation laws via the kinetic entropy defect measure. *J. Math. Pures Appl.*, 77(10):1055–1064, 1998.

[145] O. Pironneau. On the transport-diffusion algorithm and its applications to the Navier-Stokes equations. *Numer. Math.*, 38(3):309–332, 1981/82.

[146] M. Rannacher and G. Zhou. Analysis of a domain-splitting method for nonstationary convection-diffusion problems. *East-West J. Numer. Math.*, 2(2):151–172, 1994.

[147] H. Reme and G. Å. Øye. Use of local grid refinement and a galerkin technique to study secondary migration in fractured and faulted regions. *Computing and Visualization in Science*. To appear.

[148] H. Reme, G. Å. Øye, M. S. Espedal, and G. E. Fladmark. Application of local grid refinements for simulation of faulted and fractured reservoirs. In *Proc. 6th European Conference on the Mathematics of Oil Recovery*. Peebles - Scotland, 8-11 September, 1998.

[149] N. H. Risebro. A front-tracking alternative to the random choice method. *Proc. Amer. Math. Soc.*, 117(4):1125–1139, 1993.

[150] N. H. Risebro and A. Tveito. Front tracking applied to a nonstrictly hyperbolic system of conservation laws. *SIAM J. Sci. Stat. Comput.*, 12(6):1401–1419, 1991.

[151] N. H. Risebro and A. Tveito. A front tracking method for conservation laws in one dimension. *J. Comp. Phys.*, 101(1):130–139, 1992.

[152] H.-G. Roos, M. Stynes, and L. Tobiska. *Numerical methods for singularly perturbed differential equations*. Springer-Verlag, Berlin, 1996. Convection-diffusion and flow problems.

[153] É. Rouvre and G. Gagneux. Solution forte entropique de lois scalaires hyperboliques-paraboliques dégénérées. *C. R. Acad. Sci. Paris Sér. I Math.*, 329(7):599–602, 1999.

[154] T. F. Russell. Eulerian-Lagrangian localized adjoint methods for advection-dominated problems. In *Numerical Analysis 1989 (Dundee, 1989)*, pages 206–228. Longman Sci. Tech., Harlow, 1990.

[155] T. F. Russell and R. V. Trujillo. Eulerian-Lagrangian localized adjoint methods with variable coefficients in multiple dimensions. In G. Gambolati et al., editor, *Proc. 8th Int. Conf. on Computational Methods in Water Resources*, pages 357–363. Computational Mechanics Publications, Southampton, UK, 1990.

[156] O. Sævareid. *On Local Grid Refinement Techniques for Reservoir Flow Problems*. PhD thesis, Department of Mathematics, University of Bergen, 1990.

[157] A. A. Samarskii, V. A. Galaktionov, S. P. Kurdyumov, and A. P. Mikhailov. *Blow-up in quasilinear parabolic equations*. Walter de Gruyter & Co., Berlin, 1995. Translated from the 1987 Russian original by Michael Grinfeld and revised by the authors.

[158] B. Smith, P. Bjørstad, and W. Gropp. *Domain decomposition*. Cambridge University Press, Cambridge, 1996. Parallel multilevel methods for elliptic partial differential equations.

[159] J. Smoller. *Shock waves and reaction-diffusion equations*. Springer-Verlag, New York, second edition, 1994.

[160] E. Tadmor. Approximate solutions of nonlinear conservation laws. In A. Quarteroni, editor, *Lecture notes of the 1997 C.I.M.E. course in Cetraro (Cosenza), Italy*. Springer Verlag, 1997.

[161] X.-C. Tai, T. O. W. Johansen, H. K. Dahle, and M. S. Espedal. A characteristic domain splitting method. In R. Glowinski et al., editor, *Domain decomposition methods in science and engineering, 8th international conference*, pages 308–317. John Wiley & Sons, England, 1997.

[162] X. C. Tai and P. Neittaanmäki. Parallel finite element splitting-up method for parabolic problems. *Numer. Methods Partial Differential Equations*, 7(3):209–225, 1991.

[163] Z.-H. Teng. On the accuracy of fractional step methods for conservation laws in two dimensions. *SIAM J. Numer. Anal.*, 31(1):43–63, 1994.

[164] E. F. Toro. *Riemann solvers and numerical methods for fluid dynamics*. Springer-Verlag, Berlin, 1997. A practical introduction.

[165] J. E. Våg, H. Wang, and H. K. Dahle. Eulerian-Lagrangian localized adjoint methods for systems of nonlinear advective-diffusive-reactive equations. *Advances in Water Resources*, 19:297–315, 1996.

[166] A. I. Vol'pert. The spaces BV and quasi-linear equations. *Math. USSR Sbornik*, 2(2):225–267, 1967.

[167] A. I. Volpert. Analysis in classes of discontinuous functions and partial differential equations. In *New results in operator theory and its applications*, pages 230–248. Birkhäuser, Basel, 1997.

[168] A. I. Vol'pert and S. I. Hudjaev. Cauchy's problem for degenerate second order quasilinear parabolic equations. *Math. USSR Sbornik*, 7(3):365–387, 1969.

[169] A. I. Vol'pert and S. I. Hudjaev. *Analysis in classes of discontinuous functions and equations of mathematical physics*. Martinus Nijhoff Publishers, Dordrecht, 1985.

[170] H. Wang. *Eulerian-Lagrangian localized adjoint methods: analyses, numerical implementations and their applications*. PhD thesis, Department of Mathematics, University of Wyoming, 1992.

[171] H. Wang, H. K. Dahle, M. S. Espedal, R. E. Ewing, R. C. Sharpley, and S. Man. An ELLAM scheme for advection-dispersion equations in two dimensions. *SIAM J. Sci. Comput.* To appear.

[172] H. Wang, R. E. Ewing, and T. F. Russell. Eulerian-Lagrangian localized adjoint methods for variable-coefficient convection-diffusion problems arising in groundwater applications. In T. F. Russell et al., editor, *Computational Methods in Water Resources IX. Numerical Methods in Water Resources*, volume 1, pages 25–31. Computational Mechanics Publications, Southampton, U.K., 1992.

[173] H. Wang, R. E. Ewing, and T. F. Russell. Eulerian-Lagrangian localized adjoint methods for convection-diffusion equations and their convergence analysis. *IMA J. Numer. Anal.*, 15(3):405–459, 1995.

[174] M. F. Wheeler, W. A. Kinton, and C. N. Dawson. Time-splitting for advection-dominated parabolic problems in one space variable. *Comm. Appl. Numer. Methods*, 4(3):413–423, 1988.

[175] Z. Wu. A note on the first boundary value problem for quasilinear degenerate parabolic equations. *Acta Math. Sci.*, 4(2):361–373, 1982.

[176] Z. Wu. A boundary value problem for quasilinear degenerate parabolic equation. MRC technical summary report #2484, University of Wisconsin, USA, 1983.

[177] Z. Wu and J. Wang. Some results on quasilinear degenerate parabolic equations of second order. In *Proceedings of the 1980 Beijing Symposium on Differential Geometry and Differential Equations*, volume 3, pages 1593–1609. Science Press, Beijing, Gordon & Breach, Science Publishers Inc., New York, 1982.

[178] Z. Wu and J. Yin. Some properties of functions in BV_x and their applications to the uniqueness of solutions for degenerate quasilinear parabolic equations. *Northeastern Math. J.*, 5(4):395–422, 1989.

[179] Z. Wu and J. Zhao. The first boundary value problem for quasilinear degenerate parabolic equations of second order in several space variables. *Chin. Ann. Math.*, 4B:57–76, 1983.

[180] J. Yin. On a class of quasilinear parabolic equations of second order with double-degeneracy. *J. Partial Differential Equations*, 3(4):49–64, 1990.

[181] J. Yin. On the uniqueness and stability of BV solutions for nonlinear diffusion equations. *Comm. in Partial Differential Equations*, 15(12):1671–1683, 1990.

[182] J. Zhao. Some properties of solutions of quasilinear degenerate parabolic equations and quasilinear degenerate elliptic equations. *Northeastern Math. J.*, 2(3):281–302, 1986.

[183] J. Zhao. Uniqueness of solutions for higher dimensional quasilinear degenerate parabolic equation. *Chinese Ann. Math*, 13B(2):129–136, 1992.

[184] J. Zhao and P. Lei. Uniqueness and stability of solutions for Cauchy problem of nonlinear diffusion equations. *Sci. China Ser. A*, 40(9):917–925, 1997.

[185] W. P. Ziemer. *Weakly differentiable functions*. Springer-Verlag, New York, 1989.

sat at time=0.05

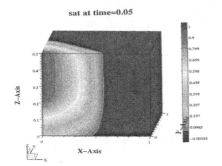

sat at time=0.3

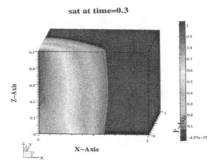

sat at time=0.5

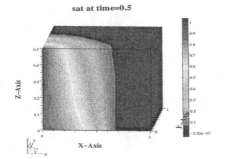

sat at time=1.5

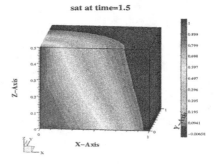

FILTRATION PROBLEMS IN VARIOUS INDUSTRIAL PROCESSES

Antonio Fasano

Dipartimento di Matematica "U. Dini"

Università di Firenze,

Viale Morgagni 67/A, 50130 Firenze, Italy

Contents

1 Introduction

In this series of lectures I wish to illustrate some processes of industrial relevance in which the leading phenomenon is the flow of a liquid through a porous material.

Although in all the cases we are going to deal with we assume that the simple Darcy's law is the basic equation governing the flow, the problems considered are much more complex than the well studied standard filtration, because of the presence of other effects of physical or chemical nature which deeply affect the rheological behaviour of the system.

It is well known that Darcy's law is a linear model which may become unsatisfactory under extreme circumstances and that various modifications have been proposed ([3], [4], [5] are fundamental references). However our aim is not to redefine Darcy's law in order to interpret nonlinear processes, but instead to provide convincing mathematical models describing a set of nonlinear phenomena which accompany and interact with the flow.

From the physical point of view the problems we want to discuss are quite far apart from each other and correspondingly we find substantial differences in their mathematical structure. However we can say that there is a common denominator: the presence of one or more free boundaries.

Our main scope is to show that not only there are quite interesting problems of industrial origin involving the flow of one or more substances through a porous medium, but also that this is a very rich and diversified field for mathematical research.

With this purpose in mind, our choice was not to insist on the details, but to summarize the main results, trying to put the emphasis on the crucial steps in the formulation of the mathematical models and on the techniques used in the analysis of the corresponding problems (mainly initial-boundary value problems for p.d.e.'s of various type and possibly o.d.e.'s).

We have selected the following classes of problems

(i) The so-called espresso-coffee problem, which is a filtration problem with strong mechanical and chemical interaction between the flow and the porous material (ground coffee).

(ii) The injection of polymerizing resins into porous materials (preforms) for the manufacturing of composite materials. Here we will focus on one of the most difficult aspects of the process, i.e. the fact that it is accompanied not only by the polymerization process (curing) but also by heat conduction and convection.

(iii) A series of other phenomena still in the area of composite material manufacturing, taking into account the deformation of the porous skeleton.

(iv) Liquid flows in the presence of absorbing granules: the device to be studied is a baby's diaper.

As we shall see in most cases there are numerous open questions of remarkable difficulty.

The class (i) has been largely studied during the last decade in cooperation with illy-caffé s.p.a. (Trieste, Italy). The problem is basically one-dimensional and both modelling and mathematical analysis have been developed to a large extent (see in particular the lecture notes [24] and the more recent papers [34], [27]). More would be desirable to know from the experimental point of view. This class proved to be a source of quite original research problems in various directions, so rich that here we will confine our attention to flows through completely saturated layers with undeformable porous matrix, leaving out the whole category of invasion problems (the first stage of water penetration in ground coffee) and problems with flow-induced deformation of the medium.

For the other classes of models, in most cases the physically interesting problem is multidimensional, while the only mathematical results available refer to schemes in one space dimension (still very complicated, as we shall see). On the other hand the effort in producing sufficiently general mathematical models has been intense. Basic references for flows through deformable porous media are the survey paper [45] and, in a more general frameworks the book [46]. We also also quote the very recent review [21] on the specific subject of composite materials, which contains an extensive bibliography.

The study of class (iv) is quite recent. We refer to the survey paper [25] which points out how several different boundary value problems can be appropriate for the description of the phenomenon, depending on the behaviour of the system at the inflow surface and at the wetting front. Even the one-dimensional case has received very partial answers.

Therefore I can say that the research problems we are going to illustrate are in full development. In particular, since the mathematical models are reasonably formulated even in the most complicated situations, there is much need of good numerical schemes.

It is also important to stress that several of the problems presented here could be effectively analyzed by means of homogenization techniques. This is particularly true for classes (i) and (iv) in which a mass exchange between the flowing liquid and the porous medium occurs at the pore scale. A first attempt in this direction is the paper [26], which will be followed by others taking a closer approach to the real problems.

Of course our selection of topics was complementary to the problems illustrated in the other lectures, but we must say that there are many more flow problems in different areas which are equally interesting and stimulating. For instance a large class of ground water flow problems with very peculiar characteristics and of relevant importance for geology and civil engineering is the one of ground freezing processes, including a number of impressive phenomena like heaving of the soil, formation of pure ice layers, etc. An interesting survey paper is [48].

In any case I hope that the material presented can give an idea of the variety of industrial applications and that some of the open problems can possibly attract the interest of young people.

2 The espresso-coffee problem

The Italian Company illy-caffé s.p.a (Trieste) has been investigating the physical and chemical aspects of coffee brewing for may years on a scientifical basis (see [39], [44]). The research performed jointly by illy-caffé and the industrial mathematics research group at the Mathematical Department U. Dini in Florence, under the aegis of ECMI, pointed out the existence of phenomena which make this area of research quite intriguing from the point of view of modelling as well as for the complexity of the related mathematical problems.

For a general overview we refer to the lecture notes [24] and to the very recent survey paper [27].

Here we concentrate on the illustration of the basic model ([28]) which allowed us to interpret the first series of laboratory experiments (performed with the use of cold water in order to eliminate most of the chemical processes) and of its generalization [34], which encompasses more phenomena and includes the chemistry of the process.

The physical system consists in a layer of pressed ground coffee placed between two sheets of paper permeable to water. The "cake" is placed in a percolation chamber and water is injected at some prescribed pressure (in the ordinary process the temperature of water is close to boiling point and the injection pressure is about 9 bar).

The basic experimental facts are described in [2], [31] and are summarized in fig. 2.1

Deviation from pure Darcyan flow is evident and occurs in two ways:

(a) in all cases the discharge is a decreasing function of time, tending to an asymptotic value (the initial peak is a transitory start phase)

(b) the asymptotic value is not proportional to the injection pressure (which would be an obvious consequence of Darcy's law), and is not even a monotone function of it.

These results, obtained at relatively low temperatures, were confirmed by a series of experiments performed at 4°C (fig. 2.2)

Clearly this counter intuitive behaviour must be attributed to the fact that the medium is not mechanically inert (like sand), but interacts in some way with the flow. The idea of using cold water was aimed at excluding the complex chemistry of the system, thus focussing on its mechanical behaviour.

Let us sketch the corresponding mathematical model and the consequent interpretation of (a), (b). We emphasize the fact that the water injection system is designed so that the flow is one-dimensional and all the relevant quantities are constant on planes normal to it.

In what follows we completely disregard the very early stage (raise of the applied pressure to its constant value and propagation of the wetting front) and we assume that the system is saturated from the very beginning. The wetting stage leads to considerably difficult problems (see [15], [16], [17], [35],). Filtration problems in geology accompained by diffusion-reaction phenomena affecting porosity, and therefore having some similarity with the process we are considering, have been studied in [12], [13], [14] [37].

A) Flow with removal and transport of fine particles.

A special percolation chamber which can be overtuaned shows that if after the stabilization of the discharge the coffee cake is placed upside down, then the discharge exhibits a new peak followed by the same decay detected in the first stage of the experiment.

Such a result provides the key evidence of the phenomenon responsible for the rheological peculiarities of the system, that we are going to illustrate in some detail.

Coffee in the cake is ground in a *bimodal* structure, i.e. with the grain size distribution having two maxima around two suitably chosen diameters. However the grinding process produces also much smaller particles (mainly cellulose), coming from the most fragile components of the roasted beaus. Such fine particles (in the micron range) adhere to the larger grains because of the oily substances present in the system and possibly also for bonds of electric type.

The flow may break such bonds and let the small particles be transported by the flowing liquid, reaching the terminal obstacle, which is permeable to liquid, but not to the solid component.

As a result the fine particles will build up a layer with a compact texture possessing a hydraulic resistivity much larger than the rest of the system. As the compact layer grows, the discharge decreases owing to the resistance in series and keeps decreasing as long as the layer becomes thicker.

Thus we have a clear explanation to (a).

Understanding (b) is far less easy. We have to relate somehow the rate at which the particles are released to the flow intensity. In this way we expect that a larger injection pressure can produce initially a more intense flow and therefore a larger number of particles can be set free, eventually producing a thicker compact layer. There may be cases in which the latter effect is so important to compensate for the larger applied pressure, thus producing a lower terminal discharge.

This is roughly speaking the basic idea of the model, but the real occurrence of (b) is linked, as we shall see, to a threshold phenomenon in the particle removal process.

Let x be the direction of the flow. At each time t the system is divided into two regions separated by an interface $x = s(t)$ (the free boundary): for $0 < x < s(t)$ the flow is accompanied by the removal and transport of fine particles, for $s(t) < x < L$ (total thickness of the cake) we have the compact layer.

The physical situation is such that during the whole process the relative thickness $\dfrac{L - s(t)}{L}$ is small.

Besides the function s, the unknowns of the problem are the pressure p, the volumetric velocity q and the concentrations of fine particles in the flow and in the solid matrix (denoted respectively by m and b) and measured in grams per unit volume of the whole system.

A first approximation we introduce in the model is that the fine particles can be considered to have the same size (we will consider a more general situation in the next section).

The relevant physical parameters are the hydraulic conductivity $K(b, m)$ in the removal-transport zone, the corresponding coefficient $K_c \ll K$ of the compact layer, and the overall particle concentration M in the compact layer.

We assume that the porosity ϕ is constant. This is not a bad approximation in the upper layer, since the solid particles occupy the same volume before and after they are released. We may expect to have a lower porosity in the compact layer, but the jump of porosity $[\phi]$ comes into play through the product $[\phi]\dot{s}$, defining the jump of volumetric velocity at the interface. Since however $(L-S)/L$ remains quite small during the process and the same is true for the ratio $|\dot{s}|/q$, we suppose that at all times $|[\phi]\dot{s}| \ll q$, which amounts to taking q continuous across the interface.

Let us introduce and discuss the governing equations.

(i) The removal process

$$\frac{\partial b}{\partial t} = -\gamma q (b - \beta(q))_+, \quad 0 < x < s(t), \ t > 0. \tag{2.1}$$

Here $\gamma > 0$ and the presence of the factor γq describes the flow efficiency for removal (it could be replaced by an increasing function of q vanishing for $q = 0$). The most important element is $\beta(q)$ that, through the positive part $(\cdot)_+$, acts as a *threshold concentration*.

We suppose that $\beta(q)$ is a decreasing function of q, having in mind that a more intense flow "sees" a larger population of removable particles. The effect of the presence of $\beta(q)$ is that, since tipically q decreases in time, β will increase and may switch off the removal process at some time. We shall see that this is indeed a crucial aspect of the model.

(ii) Particles transport equation

$$\frac{\partial m}{\partial t} + \frac{\partial}{\partial x}(\mu q m) = -\frac{\partial b}{\partial t}, \quad 0 < x < s(t), \ t > 0. \tag{2.2}$$

The water molecular velocity is $\dfrac{q}{\phi}$. The particle velocity will be smaller but still proportional to q (so $0 < \mu < 1$). This equation describes the evolution of m along the particles trajectories. Diffusion is neglected.

(iii) Flow equations We impose incompressibility, so that $\frac{\partial q}{\partial x} = 0$ and q is a function of time only. Darcy's law above and below the interface is

$$q(t) = -k(b,m)\frac{\partial p}{\partial x}, \quad 0 < x < s(t), \ t > 0 \tag{2.3}$$

$$q(t) = -k_c\frac{\partial p}{\partial x}, \quad s(t) < x < L, \ t > 0 \tag{2.4}$$

where gravity is neglected.

(iv) Interface conditions. As we said, we neglected the possible jump of q and we impose the continuity of pressure.

The growth of the compact layer is described by the following equation

$$[M - (m+b)]_{x=s(t)}\dot{s}(t) = -\mu q(t)m|_{x=s(t)}, \quad t > 0 \tag{2.5}$$

expressing the fact that the amount of fine particles needed for the interface advancement at the rate $|\dot{s}|$ is supplied by the particles mass current $\mu q m$.

We remark that the supposed continuity of p and q across $x = s(t)$ allows to deduce $q(t)$ from (2.3), (2.4) in the following form

$$q(t) = p_0\left\{\int_0^{s(t)} R(m,b)\,dx + R_c(L - s(t))\right\}^{-1}, \tag{2.6}$$

where p_0 is the pressure difference between the inflow $(x = 0)$ and the outflow $(x = L)$ surfaces.

(v) Initial and boundary conditions. For pressure

$$p(0,t) = p_0 > 0, \quad p(L,t) = 0, \quad t > 0. \tag{2.7}$$

For particles concentrations

$$b(x,0) = b_0 > 0, \quad 0 < x < L, \tag{2.8}$$

$$m(x,0) = m_0 \geq 0, \quad 0 < x < L \tag{2.9}$$

$$m(0,t) = 0, \quad t > 0. \tag{2.10}$$

Interface location

$$s(0) = L. \tag{2.11}$$

Remark 2.1 Once q, b, m, s are known the pressure is determined by (2.3), (2.4) with the conditions (2.7). Therefore the problem consists in solving (2.1), (2.2), (2.5), (2.6) (in the classical sense) with conditions (2.8)-(2.11).

Before sketching the derivation of the main results about the problem just stated, let us introduce the following non-dimensional variables

$$\bar{x} = x/L, \quad \bar{q} = q/q_0, \quad \bar{p} = p/p_0, \quad \bar{t} = t/t_0,$$
$$\bar{b} = b/b_0, \quad \bar{m} = m/b_0, \quad \bar{M} = M/b_0, \quad \bar{m}_0/b_0,$$
$$\bar{\beta}(\bar{q}) = \beta(\bar{q}q_0)/b_0, \quad \bar{s}(\bar{t}) = s(t)/L,$$

where $q_0 = k_0 \dfrac{p_0}{L}$, with $k_0 = k(b_0, m_0)$, is the initial value of q (at $t = 0$ the system behaves like an ordinary porous medium).

Moreover we put $\bar{\mu} = \mu \dfrac{L}{\gamma}, \bar{k} = k/k_0, \bar{k}_c = k_c/k_0$.

Now we rewrite (2.1), (2.2), (2.5), (2.6), (2.8)-(2.11) using the new variables and omitting the bars to simplify notation

$$\frac{\partial b}{\partial t} = -q(t)[b - \beta(q)]_+ , \qquad\qquad 0 < x < s(t), \ t > 0 \quad (2.12)$$

$$\frac{\partial m}{\partial t} + \mu q(t)\frac{\partial m}{\partial x} = -\frac{\partial b}{\partial t} , \qquad\qquad 0 < x < s(t), \ t > 0, \quad (2.13)$$

$$q(t) = \left\{ \int_0^{s(t)} R(b, m) \, dx + R_c(1 - s(t)) \right\}^{-1}, \qquad t > 0, \qquad (2.14)$$

$$\dot{s} = -\left.\frac{\mu q m}{M - (b + m)}\right|_{x=s(t)}, \qquad\qquad t > 0, \qquad (2.15)$$

$$b(x, 0) = 1, \qquad\qquad 0 < x < 1, \qquad (2.16)$$
$$m(x, 0) = m_0 \geq 0, \qquad\qquad 0 < x < 1, \qquad (2.17)$$
$$m(0, t) = 0, \qquad\qquad t > 0, \qquad (2.18)$$
$$s(0) = 1. \qquad\qquad (2.19)$$

The main assumptions are the following

$$R(b, m) \in C^1 \text{ for } b \in [0, 1], \ m \in [0, 1 + m_0], \qquad (2.20)$$
$$0 < R_m \leq R \leq R_M < R_c.$$

Define the constant

$$s_m = 1 - \frac{1 + m_0}{M} > 0 \qquad (2.21)$$

where we have used the hypothesis

$$M > 1 + m_0, \qquad (2.22)$$

which means that the cumulative initial concentration of fine particles is less than their concentration in the compact layer.

Defining the constants

$$q_m = \{R_M s_m + R_c(1 - s_m)\}^{-1} < q_M = R_M^{-1}, \qquad (2.23)$$

we can state our assumptions on the threshold concentration $\beta(q)$:

$$\beta \in C^1([q_m, q_M]), \quad \beta' \leq 0 \tag{2.24}$$
$$\beta(1) < 1, \tag{2.25}$$

meaning that the removal process does take place (otherwise the problem is trivial).

Now we remark that, having chosen a constant value for b at $t = 0$, the fact that q depends only on t implies, through equation (2.12), that

$$b = b(t), \quad \dot{b} \leq 0, \tag{2.26}$$

and therefore we can rewrite (2.18) as a conservation law for the sum $m + b$:

$$\frac{\partial}{\partial t}(m + b) + \mu q \frac{\partial}{\partial x}(m + b) = 0. \tag{2.27}$$

Thus $(m + b)$ is constant along the characteristics of (2.27). Since m_0 may be positive, the characteristic $x = \sigma(t)$ starting from the origin may carry a discontinuity for m.

Such a curve intersects the free boundary $x = s(t)$ at some finite time t^*, and in the half strip $\{(x, t) \mid 0 < x < 1, \ t > 0\}$ we can define the following regions:

- the compact layer $< x < 1, \ t > 0$,

- $Q_i = \{(x, t) \mid \sigma(t) < x <, \ 0 < t < t^*\}$,

- $Q_b = \{(x, t) \mid 0 < x < \sigma(t), \ 0 < t \leq t^*\} \cup \{(x, t) \mid 0 < x < s(t), \ t > t^*\}$.

The region Q_i carries the influence of the initial data, and we have

$$m(x, t) + b(x, t) = m_0 + 1, \quad (x, t) \in Q_i, \tag{2.28}$$

while the region Q_b is crossed by the characteristics with an endpoint on the fixed boundary $x = 0$. In Q_b we can define the function $\theta(x, t)$ as the time taken by the particles starting from $x = 0$ to reach the depth x at the time t:

$$x = \int_{t-\theta(x,t)}^{t} \mu q(\tau) \, d\tau. \tag{2.29}$$

Using (2.27), (2.28), we can write

$$m(x, b) = b(t - \theta(x, t)) - b(t), \quad (x, t) \in Q_b. \tag{2.30}$$

Another important consequence of (2.27) is the cumulative mass balance

$$\int_0^{s(t)} (m + b) \, dx + M(1 - s(t)) = 1 + m_0, \tag{2.31}$$

expressing the partition of available particles between the compact layer and its complementary region. Equation (2.21) is obtained integrating (2.27) separately in Q_i, Q_b, noting that the contributions on the two sides of $x = \sigma(t)$ cancel (due to the fact that $\sigma(t) = \mu q(t)$) and using (2.15) together with (2.16)-(2.19).

We can easily obtain bounds on θ and its first derivatives, which allow us to pass to the limit as $t \to +\infty$ in (2.31), getting the asymptotic value

$$s_\infty = \frac{s_m}{1 - b_\infty/M} \tag{2.32}$$

(where $b_\infty = \lim_{t \to +\infty} b(t)$), and from (2.14)

$$q_\infty = \{s_\infty R(b_\infty, 0) + R_c(1 - s_\infty)\}^{-1}, \tag{2.33}$$

where we have used the fact $\lim_{t \to \infty} m(x, t) = 0$ uniformly w.r.t. $x \in (0, s_\infty)$, that can be proved easily on the basis of (2.30).

The results obtained so far are of great help in the analysis of the qualitative behaviour of the solution. We omit the proof of the existence and uniqueness theorem, too long to be reported here, and we just sketch the main points of the qualitative analysis.

Since our objective is to check whether the fact (b) described in the introduction (i.e. the non-monotone dependence of q_∞ on the injection pressure p_0), can really occur, we introduce some assumption which simplify the analysis of the asymptotic behaviour of the system.

Let us suppose that besides (2.20) $R(b, m)$ satisfies

$$R(b, m) = R(m + b), \tag{2.34}$$

$$R'(\rho) < \frac{R_c - R_M}{M}, \qquad 0 \le \rho \le 1 + m_0. \tag{2.35}$$

Then we have

Proposition 2.1 $m(s(t), t) > 0 \Rightarrow \dot{q}(t) < 0$, $m(s(t), t) = 0 \Rightarrow \dot{q}(t) = 0$.

Proof. Differentiating (2.14) and using (2.27) we get

$$\dot{q}(t) = \mu q^3(t) m(s(t), t) \left\{ R'(\bar{\rho}) - \frac{R_c - R(\rho(s(t), t))}{M - \rho(s(t), t)} \right\} \tag{2.36}$$

for some $\bar{\rho} \in (0, 1 + m_0)$. The conclusion is now obvious.

An important consequence of the monotonicity of $q(t)$ is

Proposition 2.2 *The difference* $b(t) - \beta(q(t))$ *is non-increasing.*

We recall that initially $b(0) = 1 > \beta(q(0)) = \beta(1)$. Since $\dfrac{db}{dt}$ cannot remain less than a negative constant (remember that $b \to b_\infty$), then it must tend monotonically to zero and two cases are possible (fig. 2.4):

either

(i) $\dfrac{db}{dt}$ vanishes at some finite time t_E, the removal process becomes extinct and b remains constant for $t > t_E$,

or

(ii) $\dfrac{db}{dt} < 0$ for all t and removal never stops.

Remark 2.2 Assumption (2.34) looks physically acceptable. Of course one expects $R'(\rho) \geq 0$, so that (2.35) is a real limitation.

Remark 2.3 Clearly when case (ii) occurs the system never reaches a steady state, since the compact layer will continue being fed by a stream of particles. On the contrary, if (i) occurs, then the characteristic line originating at $x = 0$, $t = t_E$ bounds a subregion of Q_b where $m = 0$. The time t_S at which the same characteristic meets the free boundary is the time at which the steady state is reached (fig. 2.5).

Discriminating between (i) and (ii) is crucial in order to investigate the behaviour of the asymptotic discharge with respect to changes of the injection pressure p_0.

This study requires of course that we use the original dimensional variables, that we now denote by $\widetilde{q}, \widetilde{\beta}$, etc. . We recall that

$$\widetilde{q}_\infty = q_\infty k_0 p_0 / L = \widetilde{q}_0 q_\infty$$

and

$$\beta(q) = \widetilde{\beta}\left(\frac{k_0 p_0}{L} q\right) \Big/ b_0.$$

We further simplify our assumption on R, taking just $R =$ constant, and we prove the following

Proposition 2.3 *If* $R = \dfrac{1}{k_0} = $ *constant and the removal process never stops* *(case (ii)) then* $\dfrac{d\widetilde{q}_\infty}{dp_0} > 0$.

Proof. The proof starts from the fact that there is now a relationship between $\widetilde{b}_\infty$ and $\widetilde{q}_\infty$:

$$\widetilde{b}_\infty = \widetilde{\beta}(\widetilde{q}_\infty) \tag{2.37}$$

so that $\widetilde{s}_\infty$ is expressed by

$$\widetilde{s}_\infty = \frac{\widetilde{s}_m}{1 - \widetilde{\beta}(\widetilde{q}_\infty)/\widetilde{M}} \tag{2.38}$$

and we get an implicit formula for $\widetilde{q}_\infty$:

$$\widetilde{q}_\infty = \frac{k_c}{k_0}\left\{1 - \frac{\widetilde{s}_m}{1 - \widetilde{\beta}(\widetilde{q}_\infty)/\widetilde{M}}\left(1 - \frac{k_c}{k_0}\right)\right\}^{-1}\widetilde{q}_0, \tag{2.39}$$

where $\widetilde{q}_0$ is proportional to p_0. Now we calculate

$$\frac{d\widetilde{q}_\infty}{d\widetilde{q}_0} = \frac{\widetilde{q}_\infty}{\widetilde{q}_0} + \widetilde{q}_\infty^2 \left(\frac{k_c}{k_0}\widetilde{q}_0\right)^{-1}\left(1 - \frac{k_c}{k_0}\right)\widetilde{s}_m \frac{\widetilde{\beta}'(\widetilde{q}_\infty)/\widetilde{M}}{[1 - \widetilde{\beta}(\widetilde{q}_\infty)wM]^2}\frac{d\widetilde{q}_\infty}{d\widetilde{q}_0},\qquad (2.40)$$

which, owing to $\widetilde{\beta}' \le 0$, proves the desired result.

Remark 2.4 We have established that if R =constant, then (ii) is necessary for the occurrence of (b).

An obvious question is now whether both cases (i) and (ii) can indeed occur. The answer is positive:

Proposition 2.4 *We can choose $\beta(q)$ so that (i) or (ii) takes place.*

For the proof see [24].

The question of finding sufficient conditions for (b) is open, however specific examples have been constructed in which the experimental qualitative behaviour of the asymptotic discharge is reproduced. Numerical computations have been carried out in [32], [6] and by T. Susky (ICM, Warsow).

We may conclude that the model just illustrated is able to explain the principal facts observed, at least for the percolation process performed with cold water.

The process involving chemistry is accompanied also by mass removal by dissolution. The papers [29] and [49] studies a model in which particles removal and dissolution of several species takes place. Deformability of the medium is also introduced, but all species are allowed to leave the system, so that there is no compact layer.

It must be emphasized that letting the porosity depend on x, t (through b) implies that q and b no longer depend on t only. The mathematical implications of this fact is relevant. We will discuss briefly this aspect illustrating the more complete model of [34].

B) Removal, dissolution and transport of several species.

The model considered in [34] leaves aside the deformation of the medium (a possible result of the force exerted by the flow on the grains that in the specific case of the compressed coffee cake is of minor importance) and contains a careful analysis of all the phenomena involving mass removal and transport. It turns out that extending the formulation of the compact layer build up to the case of more than one species of transported solid particles is nontrivial.

We suppose that there are two families of substances that can be removed from the solid skeleton by the flow: a family of solid particles that are transported towards the compact layer and a family of substances which are dissolved in the flow and undergo a diffusion-convection process.

Now we have vectors of concentrations:

- $(b_1, \ldots, b_n)$ are the concentrations of the various (removable) species present in the porous medium (as before they are defined as grams p. unit volume of the total system)

- $(m_1, \ldots, m_n)$ the corresponding concentrations in the flow

- m_w the concentration of water.

We adopt the convention that indices up to $k < n$ refer to fine solid particles and the ones from $k + 1$ to n refer to soluble substances. We also introduce the specific volumes

$$\theta_i = b_i/\rho_0, \quad \eta_i = m_i/\rho_i, \quad i = 1, \ldots, n, \quad \eta_w = m_w/\rho_w,$$

ρ_j denoting the densities of the species.

In the following we will use the symbols

$$\eta_{(k)} = \sum_{i=1}^{k} \eta_i, \quad \eta^{(k)} = \sum_{i=k+1}^{n} \eta_i \eta_i, \quad \theta_{(k)} = \sum_{i=1}^{k} \theta_i, \quad \theta^{(k)} = \sum_{i=k+1}^{n} \theta_i.$$

A first basic equation is the definition of porosity:

$$\phi = \eta^{(k)} + \eta_w, \tag{2.41}$$

which accounts for saturation.

The complement of porosity, i.e. the solid volume fraction, is

$$1 - \phi = \eta_{(k)} + \theta_0 + \theta^{(k)} + \theta_{(k)}, \tag{2.42}$$

where θ_0 is the volume fraction of the rigid part of the skeleton.

The *volumetric compound velocity* is defined as

$$q = \eta_w V_w + \sum_{i=k+1}^{n} \eta_i V_i, \tag{2.43}$$

V_j being the molecular velocity of the species j.

The removal rates of solid particles can be expressed in a general way as

$$\frac{\partial b_i}{\partial t} = -F_i(q, \mathbf{b}) G_i(b_i - \beta_i(q, \mathbf{b})), \quad i = 1, \ldots, k \tag{2.44}$$

with $F_i, G_i \geq 0$ having Lipschitz continuous first derivatives and $G_i(\xi) = 0$ for $\xi < 0$, $F_i(0, \mathbf{b}) = 0$. The same regularity is required for the threshold concentrations β_i with $\dfrac{\partial \beta_i}{\partial q} \leq 0$.

The dissolution rates of soluble substances are given by

$$\frac{\partial b_i}{\partial t} = -H_i(q, \mathbf{b}), \quad i = k+1, \ldots, n \tag{2.45}$$

$H_i \geq 0$ having the same regularity as F_i and $H_i(q, \mathbf{b})_{b_i=0} = 0$.

For the two families of solid moving particles and solutes we write the following equations

$$\frac{\partial m_i}{\partial t} + \frac{\partial}{\partial x}\left(\alpha_i m_i \frac{q}{\phi}\right) = -\frac{\partial b_i}{\partial t}, \qquad i = 1, \ldots, k, \quad (2.46)$$

$$\frac{\partial m_i}{\partial t} + \frac{\partial}{\partial x}\left(-D_i \phi \frac{\partial}{\partial x}\frac{m_i}{\phi}\right) + \frac{\partial}{\partial x}\left(m_i \frac{q}{\phi}\right) = -\frac{\partial b_i}{\partial t}, \quad i = k+1, \ldots, n, \quad (2.47)$$

The first set of equations is the natural extension of the single transport equation we have seen in the previous model with $\alpha_i \in (0,1)$. The second set is the mass balance of solutes, based on the following form of their mass current:

$$m_i V_i = m_i V - \phi D_i \frac{\partial}{\partial x}\left(\frac{m_i}{\phi}\right), \qquad (2.48)$$

where

$$V = \frac{q}{\phi} \qquad (2.49)$$

is the compound molecular velocity. Here we have included diffusion (with diffusivity D_i) which is produced by the gradient of concentration within the pores, i.e. $\dfrac{m_i}{\phi}$, but we neglected interdiffusion (i.e. the mutual influence of various species on diffusion, see [18]).

Darcy's law is valid for the compound volumetric velocity

$$q(x,t) = -K(\mathbf{b}, \mathbf{m}, \phi)\frac{\partial p}{\partial x} \qquad (2.50)$$

(both in the compact layer and in its complement).

Next we have the global conservation laws describing the evolution of ϕ and q in the non-compact layer

$$\frac{\partial \phi}{\partial t} + \frac{\partial q}{\partial x} = -\frac{\partial}{\partial t}\theta^{(k)}, \qquad (2.51)$$

$$\frac{\partial q}{\partial x} + \frac{\partial}{\partial x}\sum_{i=1}^{k}\left(\alpha_i \eta_i \frac{q}{\phi}\right) = 0, \qquad (2.52)$$

that can be extended to the compact layer as $\dfrac{\partial \phi}{\partial t} = 0$, $\dfrac{\partial q}{\partial x} = 0$.

The first equation says that the medium remains saturated and that pore volume is produced because of dissolution of species with label $k < i \leq n$. It is derived by summing up equations (2.46) divided by the corresponding densities ρ_i, using the definition (2.43) of q, the definition (2.41) of ϕ, and the water mass balance

$$\frac{\partial \eta_w}{\partial t} + \frac{\partial}{\partial x}(\eta_w V_w) = 0, \qquad (2.53)$$

which is in fact the new information contained in (2.51).

Equation (2.52) expresses global volume conservation and follows from the sum of all equations (2.46), (2.47) (divided by the respective densities) and (2.53), introducing the information $\eta^{(k)} + \eta_{(k)} + \eta_w + \theta_0 + \theta^{(k)} + \theta_{(k)} = 1$, which makes the time derivatives disappear.

In conclusion, the differential system to be solved consists of (2.44)-(2.47), (2.50)-(2.52). Of these the set (2.44), (2.46), (2.51), (2.52) is to be considered for $0 < x < s(t)$, $0 < t$, because there is no removal nor transport of the solid particles in the compact layer $s(t) < x < L$, $t > 0$.

We still have to describe what happens at the interface $x = s(t)$. Here we have *continuity of pressure*

$$[p] = 0 \tag{2.54}$$

the symbol $[\ ]$ denotes jump across $x = s(t)$), the *continuity of pore concentration of solutes*

$$\left[\frac{m_i}{\phi}\right] = 0, \quad i = k+1, \ldots, n, \tag{2.55}$$

the continuity of *solutes diffusive flux*

$$\left[D_i \phi \frac{\partial}{\partial x} \frac{m_i}{\phi}\right] = 0, \quad i = k+1, \ldots, n. \tag{2.56}$$

Before discussing the law of growth of the compact layer, we write the initial conditions for the unknowns $b_i, m_i, \phi, s(t)$:

$$b_i(x,0) = b_{i,0}(x) > 0, \quad i = 1, \ldots, n, \; x \in [0,1], \tag{2.57}$$

$$m_i(x,0) = m_{i,0}(x) \geq 0, \quad i = 1, \ldots, n, \; x \in (0,1), \tag{2.58}$$

$$\phi(x,0) = \phi_0(x), \quad 1 < \phi_m \leq \phi_0 \leq \phi_M < 1, \; x \in [0,1] \tag{2.59}$$

$$s(0) = L. \tag{2.60}$$

At the inflow surface $x = 0$ we have

$$p(0,t) = p_0(t), \tag{2.61}$$

$$m_i(0,t) = 0, \quad\quad\quad i = 1, \ldots, k \tag{2.62}$$

$$D_i \phi(0,t) \frac{\partial}{\partial x}\left(\frac{m_i}{\phi}\right)_{x=0} = \frac{q(0,t) m_i(0,t)}{\phi(0,t)}, \quad i = k+1, \ldots, n, \tag{2.63}$$

meaning that there is no solutes flux.

At the outflow surface we have

$$p(L,t) = 0, \tag{2.64}$$

$$\frac{\partial}{\partial x}\left(\frac{m}{\phi}\right)_{x=L} = 0, \quad i = k+1, \ldots, n. \tag{2.65}$$

i.e. the absence of solutes diffusive flux.

Let us now come to the evolution of the compact layer. This is the result of several factors. First of all the mass balance of each species of moving particles

$$[m_i + b_i]\dot{s} = -\alpha_i m_i \frac{q}{\phi}\bigg|_{x=s(t)}, \quad i = 1, \ldots, k, \tag{2.66}$$

since the particles become immobile in the compact layer.

If to (2.66) we add the analogous law for the solutes

$$[m_i + b_i]\dot{s} - [m_i V_i] = 0, \quad i = k+1, \ldots, n \tag{2.67}$$

(where this time $[b_i] = 0$, while for m_i (2.55) holds), and for the water

$$[m_w]\dot{s} - [m_w V_w] = 0 \tag{2.68}$$

we obtain the cummulative volume balance at interface

$$[\phi]\dot{s} = [q], \tag{2.69}$$

having a very clear physical meaning.

The difficulty now comes from the fact that the concentrations $M_i(x)$, $i = 1, \ldots, k$, of the solid particles in the compact layer are not known a-priori, but depend on the history of the process. In (2.66) the sum $m_i + b_i$ evaluated for $x = s(t)^+$ is equal to $M_i(s(t))$. The latter quantities are not arbitrary, but they must satisfy a constraint $f(M_1, \ldots, M_\alpha) = 0$ that we call *packing configuration*. A very natural constraint is to require that the sum of the specific volumes $\dfrac{M_i}{\rho_i}$ is a given constant:

$$\sum_{i=1}^{k} \frac{M_i}{\rho_i} = \Theta \tag{2.70}$$

(note that if $k = 1$ then the concentration of the only species in the compact layer becomes known, as in the previous case).

Dividing (2.66) by ρ_i and summing up, we arrive at the desired growth law of the compact layer

$$(\Theta - (\eta_{(k)} + \theta_{(k)}))\dot{s} = -\frac{q}{\phi} \sum_{i=1}^{k} \alpha_i \eta_i \Big|_{x=s(t)}. \tag{2.71}$$

Remark 2.5 Equation (2.71) is the one which completes the problem (2.44)-(2.47), (2.50)-(2.52), (2.54)-(2.65) in the unknowns b_i, m_i, q, ϕ, p, s. Once the problem is solved, the equations (2.66), written explicitly as

$$(M_i(s(t)) - m_i(s(t)^-, t) - b_i(s(t)^-, t))\dot{s}(t) = -\alpha_i m_i \frac{q}{\phi} \Big|_{x=s(t)}, \quad i = 1, \ldots, k \tag{2.72}$$

provide the packing concentrations $M_1(x), \ldots, M_n(x)$.

The model has been stated. Now we list the assumptions required for existence. In the non-compact layer

$$0 < K_m \le K(\mathbf{b}, \mathbf{m}, \phi) \le K_M \tag{2.73}$$

and K having the same regularity as the functions F_i, G_i, H_i, i.e. C^1 with Lipschitz continuous derivatives. In the compact layer we take simply $K = K_0$, constant. The data $\phi_0, m_{i,0}, b_{i,0}$ and p_0 are continuously differentiable and

$$0 < p_0^{(m)} \le p_0(t) \le p_0^{(M)}. \tag{2.74}$$

The condition replacing (2.22) of the previous model is

$$\eta_{(k),0}(x) + \theta_{(k),0}(x) < \Theta < 1 - (\theta_0(x) + \theta_0^{(k)}(x)), \quad x \in [0,1]. \tag{2.75}$$

For the sake of simplicity we impose also the compatibility conditions

$$m_{i,0}(0) = 0, \quad i = 1, \ldots, k. \tag{2.76}$$

Existence and uniqueness in some time interval $[0, T]$ has been proved in [34]. It must be stressed that with respect to the one species model the real difficulty of this problem is not just the larger number of equations but its really different mathematical structure.

The proof is based on a fixed point argument and is quite long. We confine ourselves to pointing out the new difficulties which arise with respect to the previous model. One of the main differences is that the simple model with one moving species and constant porosity gives $q = q(t)$ as a function of $s(t)$ and a functional of $R(m, b)$. This is not possible in the more complete model, but we can obtain similar expressions for the two quantities $q(0, t)$, $q(x, 0)$.

Integrating (2.50) over the non-compact layer we get

$$\int_0^{s(t)} \frac{q(\xi, t)}{K(\mathbf{b}(\xi, t)\mathbf{m}(\xi, t), \phi(\xi, t))} \, d\xi = p_0(t) - p(s(t), t). \tag{2.77}$$

The volumetric velocity of the compact layer is a function of time only. If we call it $q_c(t)$ and take a constant hydraulic conductivity, we have

$$q_c(t) = K_0 \frac{p(s(t), t)}{1 - s(t)}. \tag{2.78}$$

We eliminate $p(s(t), t)$ obtaining

$$\int_0^{s(t)} \frac{q(\xi, t)}{K(\mathbf{b}(\xi, t)\mathbf{m}(\xi, t), \phi(\xi, t))} \, d\xi = p_0(t) - (1 - s(t)) \frac{q_c(t)}{K_0}. \tag{2.79}$$

Now we rewrite (2.52) in the form

$$q(x, t)(1 + l(x, t)) = f(t), \quad 0 \le x \le s(t), \tag{2.80}$$

where

$$l(x, t) = \frac{1}{\phi(x, t)} \sum_{i=1}^{k} \alpha_i \eta_i(x, t) \tag{2.81}$$

and $f(t)$ is unknown. We know however that $l(0,t) = 0$ because of (2.62), what leads to the identification of $f(t)$ with $q(0,t)$. Replacing $q(x,t)$ in (2.77) by the expression obtained from (2.80) and setting $t = 0$, we find

$$f(0) = q(0,0) = \frac{p_0(0)}{\int\limits_0^1 \frac{1}{1 + l_0(\xi)} \frac{1}{K(\mathbf{b}_0(\xi), \mathbf{m}_0(\xi), \phi_0(\xi))} d\xi}, \tag{2.82}$$

$l_0(x)$ being the initial value of $l(x,0)$.

Going back to (2.80) we find the initial volumetric velocity

$$q(x,0) = \frac{f(0)}{1 + l_0(x)}. \tag{2.83}$$

A relationship between $q_c(t)$ and $q(s(t)^-, t)$ is provided by the cumulative volume balance at interface (2.69). Since

$$[\phi] = \eta_{(k)} + \theta_{(k)}|_{s(t)^-,t)} - \Theta, \tag{2.84}$$

using (2.69) and the interface motion law (2.71) we eventually get

$$q_c(t) = q(1 + l)|_{x=s(t)} = q(1 + l)|_{x=0} = q(0,t) \tag{2.85}$$

(note the use of (2.80)). Inserting (2.80), (2.85) in (2.79), we arrive at the final expression

$$f(t) = q(0,t) = p_0(t) \left\{ \int\limits_0^{s(t)} \frac{1}{(1+l)K} d\xi + \frac{1 - s(t)}{K_0} \right\}^{-1}. \tag{2.86}$$

Formulas (2.83), (2.86) and the fact that once the pair (q, s) is given we can compute the functions b_i, m_i, ϕ, p suggest to set up a mapping from pairs (q, s) in a suitable space to $(\widetilde{q}, \widetilde{s})$ as follows. Impose that $s(0) = 1$ and $q(x,0)$ is given by (2.83), then after having calculated b_i, m_i, ϕ, p for a given selection of (q, s), define

$$\widetilde{q}(x,t) = \frac{f(t)}{1 + l(x,t)} \tag{2.87}$$

$$\widetilde{s}(t) = 1 - \int\limits_0^t \frac{l(s(\tau),\tau)q(s(\tau),\tau)}{\Theta - (\eta_{(k)} + \theta_k)_{x=s(\tau)}} d\tau \tag{2.88}$$

where $f(t)$ is just the function (2.86).

The proof of the existence and uniqueness of a fixed point of the mapping above is quite heavy, mainly because of topological questions. Consider for instance that the equations for m_i contain $\dfrac{\partial \phi}{\partial x}$, which does not necessarily exists, since ϕ must be calculated from (2.51). Therefore we have to overcome this difficulty by replacing terms involving $\dfrac{\partial \phi}{\partial x}$ by appropriate functions which are in turn solutions of differential systems.

An important detail of the existence proof is that one has to guarantee that the inequality $\eta_{(k)} + \theta_{(k)} < \Theta$ is satisfied everywhere. Indeed violating it would produce the onset of new compact layers.

This and other qualitative aspects are discussed at length in [34].

3 Non-isothermal resin injection

The manufacturing of composite materials is an inexhaustible source of very interesting mathematical problems. Composite materials are employed in a fast increasing way in aerospace and automotive industry for their exceptional mechanical properties. A typical manufacturing process consists in squeezing an array of alternate layers of fiber mats (preform) and liquid resin in an autoclave (fg. 1). Among the experimental works describing the process we quote the papers [41] [42] and the books [1], [40].

A cooperation on the mathematical modelling of injection processes has been promoted by ECMI (the European Consortion for Mathematics in Industry) among the University of Strathclyde (Glasgow), the Polytechnic of Turin and the Math. Department U. Dini of Florence

If one assumes as a first approximation that there are no deformations, the temperature is constant in space and time and that planar geometry is conserved, then each penetration front moves according to the so-called Green-Ampt model [38] and the whole process is described by the combination of the motions of each pair of layers (fig. 2). This is the point of view adopted in [11]. The above picture of a typical manufacturing process is by far too simple when we consider different geometries, particularly for the following reasons:

(i) even if the medium to be impregnated is a parallelepiped and we may expect a planar penetration front, the pressure gradients involved are so large that mechanical deformations take place,

(ii) the mechanical strenght of a composite material is due to the fact that while penetrating the resin undergoes a polymerization process (*curing*), which is largely exothermic, so that the role of temperature should be taken into account.

The consequence of (i) is that the flow is actually coupled to the mechanics of the fibers of the porous medium. This coupling is illustrated in detail in [45]. Moreover, if deformations produce a bending of the penetration front the flow problem is multidimensional.

The curing phenomenon is a peculiar aspect of the process and it produces various additional difficulties:

(a) the necessity of describing polymerization along the flow,

(b) the fact that the polymerization degree of the resin affects its viscosity in a quite significant way,

(c) the considerable amount of heat released during polymerization creates a non-uniform thermal field and therefore the heat conduction-convection problem becomes coupled to the flow and to the curing process in a by no means trivial way

If $\alpha \in [0, 1]$ is the *polymerization degree* (no curing for $\alpha = 0$, curing completed for $\alpha = 1$), the resin viscosity depends on temperature and on α in the following way

$$\eta = \eta_\infty \exp\left(\frac{\Delta E}{RT} + K\alpha\right), \tag{3.1}$$

where $\Delta E > 0$ is a kind of activation energy, R is the gas constant and $K > 0$.

In this section we focus our attention on the thermal aspect (c), studying the injection problem with curing in a one-dimensional setting with no deformation of the porous matrix. Even with this simplification the problem is remarkably difficult and some question is still open. A model including deformation has been studied in [19, 20].

We consider a layer $0 < x < 1$, in non-dimensional variables, occupied by a porous medium, in which non-polymerized resin is injected at a pressure $p_0 > 0$ ($p = 0$ being the pressure at the penetration front). The situation is such that capillarity and gravity can be neglected, so that the flow problem is governed by

$$R(\alpha, u)q = -\frac{\partial p}{\partial x}, \qquad 0 < x < s(t), \ t > 0, \tag{3.2}$$

where R is the hydraulic resistivity depending on the polymerization degree α and on the temperature u. As usual q denotes the volumetric velocity. Incompressibility implies $q = q(t)$. The penetration speed is given by

$$\dot{s}(t) = \frac{q(t)}{\phi}, \tag{3.3}$$

(ϕ =porosity) since the invasion front is a material surface, so that one easily deduces

$$\dot{s}(t) = \frac{p_0}{\phi}\left\{\int_0^{s(t)} R(\alpha(x, t), u(x, t)) \, dx\right\}^{-1}. \tag{3.4}$$

The thermal problem is complicated by the fact that we have a mixture in which the liquid component moves, changing its chemical nature, and an unknown boundary $x = s(t)$ across which all the thermal parameters are discontinuous.

Using the index s for the solid component, the index p for the polymer component and the index l for non-polymerized resin, we have the compound heat capacity

$$c = (1 - \xi)c_s + \phi[\alpha c_p + (1 - \alpha)c_l], \qquad 0 < x < s(t), \ t > 0 \tag{3.5}$$
$$c = (1 - \phi)c_s, \qquad s(t) < x < 1, \ t > 0 \tag{3.6}$$

(neglecting air heat capacity), and thermal conductivity

$$\lambda = (1 - \phi)\lambda_s + \phi[\alpha\lambda_p + (1 - \alpha)\lambda_l], \qquad 0 < x < s(t), \ t > 0, \qquad (3.7)$$
$$\lambda = (1 - \phi)\lambda_s, \qquad s(t) < x < 1, \ t > 0, \qquad (3.8)$$

(neglecting also the air conductivity).

So we can write the equation for heat conduction-convection in the impregnated region

$$c\frac{\partial u}{\partial t} - \frac{\partial}{\partial x}\left(\lambda\frac{\partial u}{\partial x}\right) + \dot{s}c_f\frac{\partial u}{\partial x} = h, \quad 0 < x < s(t), \ t > 0 \qquad (3.9)$$

where $c_f = \phi[\alpha c_p + (1 - \alpha)c_l]$ and h is the heat released per unit time and volume by polymerization, while in the complementary region we have just conduction

$$(1 - \phi)c_s\frac{\partial u}{\partial t} - \frac{\partial}{\partial x}\left[(1 - \phi)\lambda_s\frac{\partial u}{\partial x}\right] = 0, \quad s(t) < x < 1, \ t > 0. \qquad (3.10)$$

The interface conditions are the continuity of temperature

$$[u] = 0 \quad , \quad x = s(t), \ t > 0, \qquad (3.11)$$

and of thermal flow

$$\left[\lambda\frac{\partial u}{\partial x}\right] = 0 \quad , \quad x = s(t), \ t > 0, \qquad (3.12)$$

$[\cdot]$ denoting the jump.

Curing is described by the first order p.d.e.

$$\frac{\partial \alpha}{\partial t} + \dot{s}\frac{\partial \alpha}{\partial x} = \mu(\alpha, u), \qquad (3.13)$$

when $\dot{s}$ is the molecular fluid velocity (remember that the fluid is incompressible and that porosity is constant), and $\mu(\alpha, u)$ is the polymerization rate, for which a possible form is

$$\mu = k_1(u)(1 - \alpha)^{n_1} + k_2(u)\alpha^m(1 - \alpha)^{n_2}, \qquad (3.14)$$

with k_1, k_2 positive functions n_1, m, n_2 positive exponents.

The initial conditions are

$$s(0) = b \geq 0 \qquad (3.15)$$
$$u(x, 0) = u_0(x), \qquad 0 < x < 1 \qquad (3.16)$$
$$\alpha(x, 0) = \alpha_0(x) \qquad \text{for } 0 < x < b, \quad \text{if } b > 0 \qquad (3.17)$$

(the typical situation is however $b = 0$, so that (3.17) is not needed). For simplicity temperature is prescribed at both sides of the layer

$$u(0, t) = u_1(t), \qquad t > 0 \qquad (3.18)$$
$$u(1, t) = u_2(t), \qquad t > 0 \qquad (3.19)$$

and finally

$$\alpha(0,t) = 0, \qquad t > 0. \tag{3.20}$$

Let us summarize the results of [8], [9], [7], illustrating first the approach not including the curing phenomenon

A)   The non-isothermal one-dimensional problem with no curing

In this simplified version α is set equal to zero in (3.4), (3.5), (3.7), the term h in (3.9) is also zero, equations (3.13), (3.17), (3.20) are suppressed. Thus we have just the flow problem, summarized by (3.4), and the thermal diffraction problem (3.9)-(3.12) with conditions (3.16), (3.18), (3.19).

In [8] [9] a classical technique is employed, noting that the diffraction problem can be solved if the boundary $x = s(t)$ is specified. This suggests the construction of a mapping from the set

$$\Sigma = \{s \in C([0,1]) \mid s(0) = b, \; |s|_\gamma < M, \; \gamma \in (\tfrac{1}{2}, 1)\}, \tag{3.21}$$

$|s|_\gamma$ denoting the Hölder norm of exponent γ, into itself. The idea is to solve the thermal diffraction problem corresponding to an interface in Σ and use the solution $u(x,t)$ to define the new boundary

$$\dot{\sigma}(t) = \frac{p_0}{\phi} \left\{ \int_0^{s(t)} R(u(x,t)) \, dx \right\}^{-1}, \quad \sigma(0) = b. \tag{3.22}$$

It must be remarked that since $s \in \Sigma$ is γ-Hölder continuous it seems not a suitable choice for equation (3.9), in which $\dot{s}$ appears explicitly. However this difficulty can be removed introducing the new coordinate

$$y = x - \phi \frac{c_l}{c_\phi}(s(t) - b) \tag{3.23}$$

in the wet region only, in which the equation reduces to

$$c \frac{\partial \tilde{u}}{\partial t} = \lambda \frac{\partial^2 \tilde{u}}{\partial y} \tag{3.24}$$

$(\tilde{u}(y,t) = u(x,t))$ while the boundary becomes $-\dfrac{\phi c_l}{c}(s(t) - b) < y < s(t) - \dfrac{\phi c_l}{c}(s(t) - b)$ which is a "good" boundary from the point of view of regularity (in the sense that it guarantees the continuity of $\dfrac{\partial \tilde{u}}{\partial y}$ up to the boundary).

At this point it is possible to use the classical representation of $u, \dfrac{\partial u}{\partial x}$ by means of heat potentials and eventually reduce the search of u to the solution of a system of four Volterra integral equations with weakly singular kernels, admitting one and only one solution. All this requires the assumption $b > 0$ and provides the existence of a unique solution of the diffraction problem in a time interval such that the interface remains at a positive distance from the lateral boundaries. Thus the definition (3.23) makes sense and it can be seen that Schauder's fixed point theorem can be applied to the mapping $s \to \sigma$, yielding existence. Uniqueness is provided by a finer analysis of the same mapping.

B)   The non-isothermal one-dimensional problem with curing

In [7] the full model described in the first part of this section is studied. Once more a fixed point argument is used, but the technique is totally different. Let us go through the main steps.

(1)   The linearized thermal problem ant its weak formulation.

Given an interface $s \in C^1$ and a Lipschitz continuous function α, again one has to solve the thermal problem, whose coefficients depend on s, α. This time such a linearized problem is given a weak formulation.

Thermal auxiliary problem: for s, α given as above look for $u \in L^2(0, T; H_0^1(\Omega))$ with $u_t \in L^2(Q)$ and $u(\cdot, s) = u_0$, such that

$$\int_Q \left(c \frac{\partial u}{\partial t} v + \lambda \frac{\partial u}{\partial x} \frac{\partial v}{\partial x} \right) dx\, dt + \int_{Q_1} \left(\dot{s} c_f \frac{\partial u}{\partial x} v + r u v \right) dx\, dt = \qquad (3.25)$$

$$\int_Q h v\, dx\, dt$$

for any test function $v \in L^2(0, T; H_0(\Omega))$.

Here $\Omega = (0, 1)$ $Q = (0, 1) \times (0, T)$ and Q_1 is the part of Q on the left of $x = s(t)$

$$r(x, t) = \frac{\partial c}{\partial t} + \dot{s} \frac{\partial c_f}{\partial x} = \phi(c_p - c_l) \left(\frac{\partial \alpha}{\partial t} + \dot{s} \frac{\partial \alpha}{\partial x} \right) \qquad (3.26)$$

(note that in our case $\dfrac{\partial \alpha}{\partial t}, \dfrac{\partial \alpha}{\partial x} \in L^\infty(Q_1)$).

Supposing u is a classical solution, equation (3.25) is obtained multiplying equations (3.9), (3.10) by a test function v and integrating over the respective domains, taking into account (3.11), (3.12) (in this formulation the boundary data for u are reduced to zero). In other words a classical solution is also a weak solution. Let us list some a priori properties of the function u. First of all we note that

(i) u is continuous w.r.t. x for a.a. $t \in (0, T)$

because of well known embedding theorems.
   Moreover

(ii) λu_x is weakly differentiable w.r.t. x and $(\lambda u_x)_x \in L^2(Q)$.

The latter property follows from (3.5), observing that it gives the equality

$$\int_Q \lambda u_x v_x \, dx \, dt = -\int_Q gv \, dx \, dt \qquad (3.27)$$

for some function $g \in L^2(Q)$, which has the role of $(\lambda u_x)_x$.

Thus we conclude that

(iii) $\lambda(\cdot, t)u_x(\cdot, t) \in H^1(\Omega)$ for a.a. $t \in (0, T)$

implying that for a solution u in the sense (3.25) the diffraction conditions are satisfied a.e. on the interface.

(2) Regularized problem and uniform estimates

The difficulty in dealing with the problem above arises mainly from the discontinuity of the coefficients across $x = s(t)$. If we introduce a smooth matching over the strip $s(t) - \epsilon < x < s(t)$ then we construct a regular approximation of the problem with a unique classical solution $u^\epsilon(x, t)$.

Uniform estimates for

$$\| u^\epsilon(\cdot, t) \|_{L^2(Q)}, \quad \text{a.a. } t \in (0, T), \quad \| u_x^\epsilon \|_{L^2(Q)} \qquad (3.28)$$

can be obtained by means of standard methods.

Much more difficult is to obtain uniform estimates for

$$\| u_x^\epsilon(\cdot, t) \|_{L^2(\Omega)}, \quad \text{a.a. } y \in (0, T), \quad \| u_t^\epsilon \|_{L^2(\Omega)} \qquad (3.29)$$

(see [7] for the details). Letting $\epsilon \to 0$ there is enough compactness to pass to the limit in the weak form of the regularized problem, showing that $u^\epsilon \to u$ (weakly in $H^1(Q)$), so that existence is proved and the uniform estimates for (3.29) are transferred to u.

Uniqueness follows from linearity and from the inequality $\| u(\cdot, t) \|_{L^2(\Omega)} \le C(\| h \|_{L^2(Q)} + \| u_0 \|_{L^2(\Omega)})$.

(3) Continuous dependence on the pair (s, α).

A fundamental step in the existence proof is to show the continuous dependence of the solution u to (3.25) on the pair (s, α) in the appropriate topology. The result shown in [7] is the following.

If $s \to \bar{s}$ in $C^1([0, T])$ and $\alpha \to \bar{\alpha}$ uniformly in Q with $\| \frac{\partial \alpha}{\partial x} \|_{L^\infty(Q)}, \| \frac{\partial \alpha}{\partial t} \|_{L^\infty(Q)}$ uniformly bounded, then

$$\| u - \bar{u} \|_{L^\infty(0,T;L^2(\Omega))} \to 0 \qquad (3.30)$$

$$\| \frac{\partial u}{\partial x} - \frac{\partial \bar{u}}{\partial x} \|_{L^2(Q)} \to 0 \qquad (3.31)$$

and eventually it can be shown that

$$\| u - \bar{u} \|_{C(\overline{Q})} \to 0, \qquad (3.32)$$

where of course $\bar{u}$ is the solution of (3.25) corresponding to the pair $(\bar{s}, \bar{\alpha})$.

(4) Existence of a solution to the free boundary problem.

Take a pair $(s, \alpha) \in \Sigma \times A$, where

$$\Sigma = \{ s \in C^1([0, T]) \mid s(0) = b > 0, \ s(t) \in (0, 1) \text{ for } t \in [0, T],$$
$$0 < \delta \leq \dot{s} \leq M_1 \},$$

$$A = \{ \alpha \in C(\overline{Q}) \mid \| \frac{\partial \alpha}{\partial x} \|_{L^\infty(Q)} \leq M_2, \ \| \frac{\partial \alpha}{\partial t} \|_{L^\infty(Q)} \leq M_3 \}.$$

To (s, α) there corresponds a unique solution u to (3.25), which is used to define the new pair (σ, β) as follows

$$\dot{\sigma} = \frac{p_0}{2} \left\{ \int\limits_0^{s(t)} R(\alpha, u) \, dx \right\}^{-1}, \quad \sigma(0) = b \tag{3.33}$$

$$\frac{\partial \beta}{\partial t} + \dot{\sigma} \frac{\partial \beta}{\partial x} = \mu(\beta, u), \qquad 0 < x < \sigma(t), \ t \in (0, T) \tag{3.34}$$
$$\beta(x, 0) = \alpha_0(x) \qquad\qquad 0 < x < b$$
$$\beta(0, t) = 0, \qquad\qquad t \in (0, T).$$

Making use of the previous results, it is possible to see that the mapping $(s, \alpha) \to (\sigma, \beta)$ in the topology specified in Σ, A satisfies the hypotheses of Schauder's fixed point theorem. Therefore the existence of a triple (s, α, u) solving the complete free boundary problem (with the thermal problem in the weak version (3.5)) is established.

Remark 3.1 Uniqueness remains an open question.

Remark 3.2 The restriction $b > 0$ can be removed showing the convergence of a family of solutions (s^b, α^b, u^b) as $b \to 0$, as in the previous case.

4 A one-dimensional injection problem in a largely deformable porous material

In the previous section we have considered the non-isothermal resin injection in a nondeformable preform. A more general model should include deformation (we refer once more to [45] and [19, 20] where the reader can find also the relevant literature). However, as far as I know there are no mathematical results about the comprehensive scheme. On the contrary, one can perform a nice analysis of the purely mechanical problem, leaving aside curing and non-uniformity of the thermal field. In some sense we could say that this approach is dual w.r.t. the one of Sect. 3 and provides interesting complementary information.

The basic experimental paper which we refer to is [47] but a more careful analysis of the model (including curing) can be found in [19], [20], [10]. The latter reference includes the theory we are going to describe very shortly, while in [19], [20] some computations are presented. For a more detailed literature see the review paper [21].

The physical situation is described in fig. 4.1.

Before compression the porous medium is dry and undeformed. At time $t = 0$ pressure is applied. Since the viscosity of the resin is quite large and the pressure is high, the porous medium shrinks before the liquid start filtrating. However while the resin flows into the medium, the wet region relaxes and therefore we have two boundaries: the wetting front $x = x_w(t)$, and the relaxation front $x = x_r(t)$, which bound the flow region.

The basic quantities appearing in the model are

- u_l, the velocity of the liquid component,

- u_s, the velocity of the solid component,

- ϕ, the porosity

- p, the pressure

and two experimental functions: $K(\phi)$, the permeability, and $\tau(\phi)$ the excess compression stress on the solid component relative to the flow direction x (i.e. the one added to the liquid pressure). Note that in our case the porosity is also a measure of the deformation.

Information about papers studying $K(\phi), \tau(\phi)$ are reported e.g. in [10]. For instance [47] proposes

$$K = K_0 \cdot 10^{-M(1-\phi)} \tag{4.1}$$

for polyurethane foam permeability, while $\tau(\phi)$ is a more complicated function with inflection points, satisfying $\tau'(\phi) < 0$ and which vanishes at the equilibrium porosity ϕ_e, prior to compression.

The one-dimensional model is based on the following equations

Darcy's law for the liquid molecular velocity relative to the solid velocity

$$u_l - u_s = -\frac{K(\phi)}{\mu\phi}\frac{\partial p}{\partial x}, \tag{4.2}$$

where μ is the liquid viscosity (here considered constant).

The solid mass balance

$$\frac{\partial}{\partial t}(1 - \phi) + \frac{\partial}{\partial x}[(1 - \phi)u_s] = 0. \tag{4.3}$$

The liquid mass balance

$$\frac{\partial\phi}{\partial t} + \frac{\partial}{\partial x}(\phi u_l) = 0. \tag{4.4}$$

The momentum balance for the solid

$$\frac{\partial p}{\partial x} + \frac{\partial\tau(\phi)}{\partial x} = 0, \tag{4.5}$$

where inertia is neglected (quasi-static approximation).

As a consequence of (4.3), (4.4) we note that the *compound velocity*

$$v = (1 - \phi)u_s + \phi u_l \tag{4.6}$$

is divergence free, i.e.

$$v = q_0(t). \tag{4.7}$$

The physical meaning of $q_0(t)$ is clear: since the flow is volume preserving $q_0(t)$ is just the volume injection rate per unit surface. Taking either the discharge or the pressure at the injection surface as a given quantity depends on the structure of the injection device. For the process under consideration the applied pressure p_0 is the prescribed quantity (one can easily check that the problem in which q_0 is given is much simpler). As we shall see, the applied pressure comes into play through the value of the porosity on the wetting front.

Now we eliminate $\dfrac{\partial p}{\partial x}$ between (4.2), (4.5) and multiply the resulting equation by ϕ, obtaining, thanks to (4.6), (4.7)

$$u_s = q_0(t) - \frac{K(\phi)}{\mu} \tau'(\phi)\frac{\partial \phi}{\partial x}. \tag{4.8}$$

Using (4.8) in the solid mass balance (4.3), we obtain the final equation for the porosity

$$\frac{\partial \phi}{\partial t} - \frac{\partial}{\partial x}\left[(1 - \phi)\frac{K(\phi)}{\mu}|\tau'(\phi)|\frac{\partial \phi}{\partial x}\right] + q_0(t)\frac{\partial \phi}{\partial x} = 0, \tag{4.9}$$

which contains the unknown velocity $q_0(t)$. We have used the assumption $\tau'(\phi) < 0$. Note that (4.9) is parabolic.

At this point we have to write down the boundary conditions and also say something more about the quantity $q_0(t)$.

As we said, we have two free boundaries in this problem, namely the wetting front $x = x_w(t)$, and the relaxation front $x = x_r(t)$. Initially they coincide

$$x_w(0) = x_r(0). \tag{4.10}$$

On the *wetting front* we have

$$\phi(x_w(t), t) = \phi_c, \tag{4.11}$$

$$\dot{x}_w(t) = \frac{K(\phi_c)}{\mu\phi_c}\tau'(\phi_c)\frac{\partial \phi}{\partial x}\bigg|_{x=x_w(t)}. \tag{4.12}$$

In (4.11) ϕ_c is the porosity reached after compression with the pressure p_0 and is an experimentally known constant. Equation (4.12) expresses, as usual, that the wetting front is a material surface obeying (4.2), where $u_s = 0$, and (4.5).

On the *relaxation front* the conditions are

$$\phi(x_r(t), t) = \phi_r, \tag{4.13}$$

$$\dot{x}_r(t) = q_0(t) - \frac{K(\phi_r)}{\mu}\tau'(\phi_r)\frac{\partial \phi}{\partial x}\bigg|_{x=x_r(t)}. \tag{4.14}$$

In the first condition ϕ_r is the porosity corresponding to the unstressed configuration of the medium $(\phi_r > \phi_c)$. The second equation is derived observing that $\dot{x}_r(t)$ is nothing but $u_s(x_r(t), t)$, so that (4.14) is simply (4.8) calculated for $x = x_r(t)$.

In order to complete the model we need one more equation for $q_0(t)$. On the wetting front the compound velocity (4.6) reduces to

$$q_0(t) = \phi_c u_l(w_w(t), t) = \phi_c \dot{x}_w(t). \tag{4.15}$$

Using (4.12), we get the desired expression:

$$q_0(t) = \frac{K(\phi)}{\mu} \tau'(\phi_e) \frac{\partial \phi}{\partial x}\bigg|_{x=x_w(t)}. \tag{4.16}$$

For the reader's convenience we summarize the final form of the model, which is a parabolic problem with two free boundaries:

Problem. Find the triple $(x_w, x_r, \phi(x, t))$ such that

$$\frac{\partial \phi}{\partial t} - \frac{\partial}{\partial x}\left[(1-\phi)\frac{K(\phi)}{\mu}|\tau'(\phi)|\frac{\partial \phi}{\partial x}\right] + \frac{K(\phi_c)}{\mu}\tau'(\phi_c)\frac{\partial \phi}{\partial x}\bigg|_{x=x_w(t)}\frac{\partial \phi}{\partial x} = 0 \tag{4,17}$$

$$x_w(0) = x_r(0) = 0, \tag{4.18}$$

$$\phi(x_r(t), t) = \phi_r, \tag{4.19}$$

$$\dot{x}_r(t) = \frac{K(\phi_c)}{\mu}\tau'(\phi_c)\frac{\partial \phi}{\partial x}\bigg|_{x=x_w(t)} - \frac{K(\phi_r)}{\mu}\tau'(\phi_r)\frac{\partial \phi}{\partial x}\bigg|_{x=x_r(t)} \tag{4.20}$$

$$\phi(w_w(t), t) = \phi_c(p_0) \tag{4.21}$$

$$\dot{x}_w(t) = \frac{K(\phi_c)}{\mu\phi_c}\tau'(\phi_c)\frac{\partial \phi}{\partial x}\bigg|_{x=x_w(t)} \tag{4.22}$$

The unusual features of this problem, compared to the classical Stefan problem, are

(i) the presence of the wetting front velocity in the p.d.e. (4.17),

(ii) the fact that the speed of the relaxation front is determined not only by the porosity gradient at the same front, but also by the porosity gradient at the opposite front.

However such a problem can be studied with the techniques of [30]

The interesting result obtained in [10] is the existence of a self-similar solution to (4.17)-(4.22). The authors adopt the Lagrangian point of view, making the transformation

$$x = x(\xi, t), \tag{4.23}$$

ξ being the Lagrangian coordinate of the solid particles, whose reference configuration is the compressed non-infiltrated state at $t = 0$: $x(\xi, 0) = \xi$. The obvious advantage is that the relaxation boundary $x_r(t)$ corresponds to $\xi = 0$. Introducing the linear elasticity assumption

$$\frac{\partial x}{\partial \xi} = \frac{1 - \phi_c}{1 - \phi} \tag{4.24}$$

and defining the *void ratio*

$$e(\phi) = \frac{\phi}{1 - \phi}, \quad \phi(e) = \frac{e}{1 + e} \tag{4.25}$$

we have

$$\frac{\partial e_E}{\partial x} = \frac{1 + e_c}{1 + e} \frac{\partial e}{\partial \xi} \tag{4.26}$$

where $e_c = e(\phi_c)$ and $e = e(\xi, t)$, $e_E(x, t) = e(\xi(x, t), t) = e(\phi(x, t))$. Thus we have

$$\frac{\partial e}{\partial t} = \frac{\partial e_E}{\partial t} + \frac{\partial e_E}{\partial x} u_s = \frac{\partial e_E}{\partial t} + \frac{\partial e_E}{\partial x} \left(q_0 - \frac{K}{\mu} \tau' \frac{\partial \phi}{\partial x} \right)$$

which gives

$$\frac{\partial \phi}{\partial t} = \phi'(e) \left\{ \frac{\partial e}{\partial t} - \frac{\partial e_E}{\partial x} \left(q_0 - \frac{K}{\mu} \tau' \frac{\partial \phi}{\partial x} \right) \right\}.$$

The last term in equation (4.17) reads

$$q_0 \phi'(e) \frac{\partial e_E}{\partial x}.$$

Therefore this convective term disappears in the Lagrangian formulation. The diffusion term in (4.17) is

$$\frac{\partial}{\partial x} \left[\frac{1}{1+e} \frac{K(\phi)}{\mu} \tau'(\phi) \frac{\partial \phi}{\partial x} \right] = -\frac{1}{(1+\phi)^2} \frac{\partial e_E}{\partial x} \frac{K}{\mu} \tau' \frac{\partial \phi}{\partial x}$$

$$+ \frac{1}{1+e} \frac{1+e_c}{1+e} \frac{\partial}{\partial \xi} \left[\frac{K(\phi(e))}{\mu} \tau'(\phi(e)) \phi'(e) \frac{1+e_c}{1+e} \frac{\partial e}{\partial \xi} \right]$$

$$= \phi'(e) \left\{ -\frac{\partial e_E}{\partial x} \frac{K}{\mu} \tau' \frac{\partial \phi}{\partial x} + (1+e_c)^2 \frac{\partial}{\partial \xi} \left[\frac{K(\phi(e))}{\mu(1+e)^3} \tau'(\phi(e)) \frac{\partial e}{\partial \xi} \right] \right\}$$

and the transformed equation takes the form

$$\frac{\partial e}{\partial t} - \frac{\partial}{\partial \xi} \left[F(e) \frac{\partial e}{\partial \xi} \right] = 0, \tag{4.27}$$

with $F(e) = \frac{K(\phi(e))}{\mu(1+e)^3} |\tau'(\phi(e))|(1+e_c)^2$. Thus the equation is greatly simplified, as well as (4.19), (4.20) which are replaced by the simple equation

$$e(0, t) = e_r = e(\phi_r), \tag{4.28}$$

since the relaxation boundary is $\xi = 0$.

The wetting front is the place where x and ξ coincide ($u_s = 0$) and $\dfrac{\partial \phi}{\partial x} = \phi'(e_c)\dfrac{\partial e}{\partial \xi} = \dfrac{1}{(1 + e_c)^2}\dfrac{\partial e}{\partial \xi}$ Thus (4.21), (4.22) become

$$e(\xi_w(t), t) = e_c = e(\phi_c),\tag{4.29}$$

$$\dot{\xi}_w(t) = \frac{K(\phi_c)}{\mu\phi_c}T'(\phi_c)(1 - \phi_c)^2\frac{\partial e}{\partial \xi}\bigg|_{\xi=\xi_w(t)},\tag{4.30}$$

$\xi = \xi_w(t)$ being the only remaining free boundary, satisfying the trivial condition

$$\xi_w(0) = 0.\tag{4.31}$$

In order to look for self-similar solutions the variable

$$\chi = \frac{\xi}{\xi_w(t)}, \qquad 0 < \chi < 1,\tag{4.32}$$

is introduced and we set

$$e(\xi, t) = u(\chi),\tag{4.33}$$

arriving at the following boundary value problem for a second order o.d.e.

$$\frac{F(e_c)}{e_c}u'(1)\chi u' - (F(u)u')' = 0, \qquad 0 < x < 1,\tag{4.34}$$

$$u(0) = e_r\tag{4.35}$$

$$u(1) = e_c.\tag{4.36}$$

The remarkable feature of (4.34) is that the unknown quantity $u'(1)$ appears in the equation.

The method used in [10] is a kind of shooting method, replacing (4.34)-(4.36) by the Cauchy problem

$$\frac{F(e_c)}{e_c}\alpha\chi u'_\alpha - (F(u_\alpha)u'_\alpha)' = 0, \qquad 0 < x < 1,\tag{4.37}$$

$$u_\alpha(1) = e\tag{4.38}$$

$$u'_\alpha(1) = \alpha,\tag{4.39}$$

and looking for α such that $u\alpha(0) = e_E$.

Analyzing the behaviour of $u_\alpha(\chi)$ for α in $(-\infty, +\infty)$ the existence of a unique α fulfilling the desired condition is established. For the details see [10].

The above analysis allows to derive the motion of the free boundaries in the original coordinates. Indeed from (4.33)

$$\frac{\partial e}{\partial \xi} = u'(\chi)\frac{1}{\xi_w(t)},\tag{4.40}$$

so that (4.30) yields

$$\xi_w\dot{\xi}_w = \frac{K(\phi_c)}{\mu\phi_c}T'(\phi_c)(1 - \phi_c)^2 u'(1)\tag{4.41}$$

and finally

$$x_w(t) = \xi_w(t) = \left[2\frac{K(\phi_c)}{\mu\phi_c}\tau'(\phi_c)(1-\phi_c)^2 u'(1)t\right]^{1/2}. \qquad (4.42)$$

Integrating the elasticity condition (4.24) over the wet region $0 < \xi < \xi_w(t)$ we get

$$
\begin{aligned}
x_w(t) - x_r(t) &= (1-\phi_c)\int_0^{\xi_w(t)} (1+e(\xi,t))\,d\xi = \\
&= (1-\phi_c)\xi_w(t)\left(1 + \int_0^1 u(\chi)\,d\chi\right),
\end{aligned}
$$

giving the relaxation boundary

$$x_r(t) = -x_w(t)\left(\phi_c + (1-\phi_c)\int_0^1 u(\chi)\,d\chi\right). \qquad (4.43)$$

Thus the two free boundaries are parabolas. It can be shown that the coefficient of x_w in (4.43) is positive and therefore the two boundaries move in opposite directions.

5 Flows in media with hydrophile granules

This problem has been studied in the early 90's by the Technomathematik group at the University of Kaiserslautern (see [50]).

A fluid flows through a porous medium, made of cellulose fibers, whose pores contain granules which can absorbe water reaching up to four times their original radius.

Also in this case we can say that we are dealing with not just one problem, but an entire class of problem, since, as we shall see, at the stage of modelling this process we have several different possibilities that, through all possible combinations, generate various free boundary problems of remarkable difficulty.

The model of [50] is based on the following reaction-diffusion system

$$\frac{\partial u}{\partial t} = \frac{1}{\theta(x)}\nabla\cdot(\theta(x)\,d(u)\nabla u) - A(u,v), \qquad x\in Q\subset\mathbf{R}^3,\ 0<t<T \quad (5.1)$$

$$\frac{\partial v}{\partial t} = A(u,v), \qquad (5.2)$$

with diffusivity $d(u)$ of the type

$$d(u) = a\,e^{bu} \quad (a>0,\ b>0), \qquad (5.3)$$

and the absorption rate $A(u, v)$ chosen as

$$A(u, v) = \gamma(u v_\infty - v u_\infty), \quad \gamma > 0. \tag{5.4}$$

Here u, v are the water concentrations outside and inside the granules, respectively, while u_∞, v_∞ denote their maxima. The function $\theta(x)$ describes the fine structure of the medium.

Part of the boundary is considered impermeable

$$\frac{\partial u}{\partial n} = 0, \tag{5.5}$$

and on the complementary part (a convex domain) the following flux condition is imposed

$$\theta(x) \, d(u) \, \frac{\partial u}{\partial n} = k(t)(u_\infty - u). \tag{5.6}$$

An existence and uniqueness theorem has been proved in [43].

Taking (non-degenerate) diffusion as the main transport mechanism with a linear absorption law is not adequate to describe the evolution of a wetting front, since u will be positive everywhere soon after injection.

Moreover it is apparent that convection has to play a major role. Therefore its looks more natural to use the classic Darcy's law (as in all the problems considered so far) as the pivotal transport mechanism, coupling the absorption process with the flow. Of course the most evident difficulty is the fact that porosity changes and is in fact history-dependent. However, as we shall see, there are other difficulties related to the boundary conditions.

A first sketch of Darcy-like flow has been presented in [23], while in [22] an existence theorem has been proved for the first stage of penetration in a largely simplified one-dimensional version. The paper [25] analyzes the model in greater detail, introducing the discussion about the inflow boundary conditions that will be summarized here, studying the first case of saturated flow and extending slightly the existence proof of [22].

Some improvement of the results of [22] are now available (see [36], [33]), but the great majority of the problems stated in [25] are open.

A)   The mathematical model

We denote by ϕ the porosity and by V the volume fraction occupied by the swelling granules. Their initial values are ϕ_0, V_0 and for simplicity here we suppose them constant.

As V increases, ϕ decreases according to

$$\phi(x, t) + V(x, t) = \phi_0 + V_0. \tag{5.7}$$

The process takes place in a bounded domain $\Omega \subset \mathbf{R}^3$, where Darcy's law is written as

$$\mathbf{q} = -k(S, \phi)\nabla p, \tag{5.8}$$

neglecting gravity for simplicity. The quantity S is the saturation and it is related to pressure by means of capillarity

$$S = S(p), \tag{5.9}$$

the function S being increasing and twice continuously differentiable for $p < p_S$, moreover S is continuous for all p and $S(p) = 1$, for $p \geq p_S$. That is to say that the medium is saturated when the pressure is larger than the saturation pressure p_S.

The rate at which granules absorb water is

$$\frac{\partial V}{\partial t} = f(V_{\max} - V)(S - S_0)_+, \tag{5.10}$$

where $V_{\max}$ is the maximum size of the granules, $f(\xi)$ is a C^1 function such that $f(0) = 0$ and $f'(\xi) > 0$ for $\xi > 0$. The saturation $S_0 > 0$ is a threshold for the absorption process and $(\cdot)_+$ denotes the positive part.

The water mass balance equation in the wet region is

$$\frac{\partial(S\phi)}{\partial t} + \operatorname{div} \mathbf{q} = -\frac{\partial V}{\partial t} = \frac{\partial \phi}{\partial t}, \tag{5.11}$$

which combined with (5.8), gives

$$\phi S'(p)\frac{\partial p}{\partial t} + (S - 1)\frac{\partial \phi}{\partial t} - \operatorname{div}(k(S, \phi)\nabla p) = 0. \tag{5.12}$$

In the saturated region it reduces to

$$\operatorname{div}(k(1, \phi)\nabla p) = 0. \tag{5.13}$$

Note that (5.13) becomes $\Delta p = 0$ only after the granules have reached their maximum volume (which may not happen in a finite time)

In conclusion we may expect that during a first stage of injection we have an unsaturated region bounded by an advancing front Γ_w. After some time part of the medium may become saturated and a saturation front will propagate.

Let us discuss the conditions on such free boundaries.

(Γ_w) We suppose that the front advances only if saturation has reached some minimum value (the minimum moisture content) which is reasonably identifiable with the threshold saturation S_0. Accordingly, if v_n denotes the normal velocity of Γ_w, the propagation of the boundary is associated to the volumetric velocity $\phi_0 S_0 v_n$, which must be provided by an appropriate pressure gradient:

$$\phi_0 S_0 v_n = -k(S_0, \phi_0)\frac{\partial p}{\partial n}, \tag{5.14}$$

and the pressure assumes the reference value

$$p|_{\Gamma_w} = p_0. \tag{5.15}$$

Conditions (5.14), (5.15) are parallel to the Stefan conditions in the classical phase change problem in one phase.

Remark 5.1 If penetration is fast enough the above conditions may be modified in the sense that the saturation on the front becomes a function of the front speed. The modified conditions take the form

$$\phi_0 S_w(v_n)v_n = -k(S_w(v_n), \phi_0)\frac{\partial p}{\partial n}, \tag{5.16}$$

$$p = p_w(v_n) \geq p_0, \tag{5.17}$$

with $S_w(v_n)$ a nondecreasing function with values in $[S_0, 1]$ and $p_w(v_n)$ such that $S(p_w(v_n)) = S_w$. In this case the phase change analog is the so-called kinetic undercooling. In the limit case $S_w = 1$ we have the collapse of the unsaturated region, also a problem of practical importance.

(Γ_S) Denoting the jump across Γ_S by $[\ ]_{\Gamma_S}$, the continuity of volumetric velocity across Γ_S (which is not a material surface) is simply expressed by

$$\left[\frac{\partial p}{\partial n}\right]_{\Gamma_S} = 0, \tag{5.18}$$

while the pressure must be continuous and equal to the saturation pressure

$$p = p_S \quad \text{on both sides of } \Gamma_S. \tag{5.19}$$

The discussion about the conditions on the external boundary $\partial\Omega$ is divided into two parts, one permeable (Γ_1) and one impervious (Γ_2).

In the simplest situation (fig. 5.1) injection takes place through a simply connected, convex subset Γ_0 of Γ_1 at a known rate, so that the boundary conditions are expressed by the single equation

$$k(S, \phi)\frac{\partial p}{\partial n} = \chi(\Gamma_0)q_0(t) \tag{5.20}$$

(n outer normal), where $q_0(t)$ is a positive given function and $\chi(\Gamma_0)$ is the characteristic of Γ_0.

We call (5.20) the *unconstrained boundary condition*. However it may occur that the medium is not able to accept the imposed injection rate through Γ_0. In such a case we must introduce some constraint (a possibility envisaged in [23] and discussed to some extent in [25]).

A simple (but less physically significant) kind of constraint is on the volumetric velocity. We refer to this case as the *discharge constrained inflow*. This is defined by the inequality

$$-\mathbf{q} \cdot \mathbf{n} \leq q_{max} \quad \text{on } \Gamma_1, \tag{5.21}$$

$q_{max} > 0$ being a prescribed upper bound.

Adopting this point of view, the constraint does not come in to play as long as the total discharge imposed $Q_c(t)$ is such that $Q_c \leq q_{max}\text{meas}\,\Gamma_0$. In that case in (5.20) we mean that $q_0(t) = Q_c(t)/\text{meas}\,\Gamma_0$.

If however for some t the above restriction on $Q_c(t)$ is violated, then we modify (5.20) as follows

$$k(S, \phi)\frac{\partial p}{\partial n} = \chi(\Gamma(t))q_{max},\qquad (5.22)$$

where $\Gamma_1 \supset \Gamma(t) \supset \Gamma_0$ and

$$\text{meas}\,\Gamma(t) = \frac{Q_c(t)}{q_{max}}.\qquad (5.23)$$

Of course (5.23) does not define $\Gamma(t)$ completely. We may impose for instance that all points of $\partial\Gamma(t)$ have the same distance from Γ_0.

Another kind of constraint that can limit the inflow conditions is to impose an upper bound for pressure (then the same bound will then be respected inside the medium because of the maximum principle):

$$p \leq p_{max} \quad \text{on } \Gamma_1,\qquad (5.24)$$

(of course $p_{max} > p_S$). Such a constraint leads to a far more complicated situation, characterized by the presence of two unknowns moving curves on Γ_1. Indeed if the coincidence set $\Gamma_p(t) = \{x \in \Gamma_1 \mid p(x, t) = p_{max}\}$ is not empty, the set Γ_0 cannot bear the whole discharge $Q_c(t)$.

Correcting the inflow condition requires now the introduction of the unknown volumetric flux $Q_p(t)$ through the coincidence set $\Gamma_p(t)$. Then we can define the enlarged inflow domain $\Gamma(t)$ through its measure

$$\text{meas}\,\{\Gamma(t) \setminus \Gamma_p(t)\}q_0(t) + Q_p(t) = Q_c(t),\qquad (5.25)$$

and the geometric condition that the points of $\partial\Gamma(t)$ have the same distance from Γ_0 (fig. 5.3).

In (5.25) $q_0(t) = \dfrac{Q_c(t)}{\text{meas}\,\Gamma_0}$ and of course because of the constraint (5.24) we will have $\dfrac{Q_p(t)}{\text{meas}\,\Gamma_p(t)} < q_0(t)$, which implies that $\text{meas}\,\Gamma(t) > \text{meas}\,\Gamma_0$.

Thus the pair of *pressure constrained inflow conditions* are

$$-\mathbf{q} \cdot \mathbf{n} = \chi(\Gamma(t) \setminus \Gamma_p(t))q_0(t), \quad q_0(t) = \frac{Q_c(t)}{\text{meas}\,\Gamma_0},\qquad (5.26)$$

$$p = p_{max} \quad \text{on } Q_p(t).\qquad (5.27)$$

Remark 5.2 The above choice of constrained conditions certainly does not represent the general solution to the problem of the impingement of a water stream on the surface of a porous medium. On the contrary, we suggested a radically simplified solution which however looks reasonable for our case (certainly it would not be so if the highly permeable napkin is replaced by a low permeability medium).

Remark 5.3 In the complement of $\Gamma(t)$ we imposed that no flow occurs. Moreover, if such condition forces the pressure above p_S then it looks reasonable to allow outflow. However these are obviously not the normal operating conditions of the device and we do not consider such a possibility here.

In conclusion, there are many different combinations of possible conditions on the external boundary and on the wetting front. More generalizations are presented in [25].

B) Expressing porosity as a functional of pressure and of the wetting front

We can integrate (5.10) formally, assuming, as we did, that the threshold saturation for absorption coincides with that for penetration.
Introducing the function

$$\Phi(V) = \int_{V_0}^{V} \frac{dy}{f(V_{\max} - y)}, \quad V_0 < V < V_{\max} \tag{5.28}$$

and the time $\theta(x)$ taken by the water to reach the point x

$$\theta(x) = \sup\{t \mid S(x,t) < S_0\}, \tag{5.29}$$

from (5.10) we deduce

$$\Phi(V) = \Theta(S, \theta, t), \tag{5.30}$$

with

$$\Theta(S, \theta, t) = \int_{\theta(x)}^{t} [S(x,\tau) - S_0]\, d\tau, \quad \forall\, t > \theta(x). \tag{5.31}$$

By the assumptions made on f we have $\Phi' > 0$ and we can write

$$V(x,t) = \psi(\Theta), \tag{5.32}$$

with $\psi = \Phi^{-1}$. Finally from (5.7) we obtain

$$\phi(x,t) = \phi_0 + V_0 - \psi(\Theta). \tag{5.33}$$

We can also write the derivatives of ϕ to be used in equation (5.12)

$$\frac{\partial \phi}{\partial t} = -\psi'(\Theta)(S - S_0) \leq 0, \quad t > \theta(x), \tag{5.34}$$

$$\nabla \phi = -\psi'(\Theta) \int_{\theta(x)}^{t} \nabla S\, d\tau, \quad t > \theta(x) \tag{5.35}$$

(if e.g. $\theta(x)$ is differentiable).

It is useful to remark that $-\dfrac{\psi''}{\psi'^2} = \dfrac{\Phi''}{\Phi'}$. In the case

$$f(\xi) = \lambda\xi^n, \quad n > 0, \ \lambda > 0 \tag{5.36}$$

we deduce easily

$$-\frac{\psi''}{\psi'^2} = \frac{n}{V_{\max} - V}. \tag{5.37}$$

The ratio $-\dfrac{\psi''}{\psi'^2}$ is going to play an important role in the study of the unsaturated regime.

C) The one-dimensional problem

As we have seen, the model presented above leads to a class of free boundary problems, since there are numerous possibility of selecting the type of inflow conditions, free boundary conditions, etc.

None of such problems have been studied in the multidimensional case. In [22], [25] the one-dimensional problem is considered in two cases: (i) the limit case of no capillarity, (ii) the Stefan-like case for the unsaturated regime only. The first case is relatively simple, even if a constraint is imposed on the inflow condition. On the contrary (ii) is a very difficult problem for which only an existence result is known.

Let us review (i) and (ii) very briefly.

Case (i): *saturated flow (no capillarity)*
In absence of capillarity the problem is just a generalization of the well known Green-Ampt model [38] (penetration of an incompressible fluid in a dry medium in a saturation regime).

Still neglecting gravity, the problem is formulated as follows

$$\frac{\partial}{\partial x}\left[k(1,\phi)\frac{\partial p}{\partial x}\right] = 0, \qquad 0 < x < s(t), \ t > 0 \tag{5.38}$$

$$-k(1,\phi)\frac{\partial p}{\partial x} = q_0(t) > 0 \qquad x = 0, \ t > 0, \tag{5.39}$$

$$s(0) = 0, \tag{5.40}$$

$$p(s(t),t) = 0, \qquad t > 0 \tag{5.41}$$

$$\phi_0(s)\dot{s}(t) = q_0(t), \qquad t > 0 \tag{5.42}$$

$$\phi(x,t) = \phi_0 + V_0 - \psi[(t - \theta(x))(1 - S_0)], \quad 0 < x < s(t), \ t > 0. \tag{5.43}$$

In [25] the more general case $\phi_0 = \phi_0(x)$ is considered. In (5.41) the value of pressure is the saturation pressure. The velocity $\dot{s}(t)$ of the front is the molecular velocity, i.e. $-\dfrac{K(1,\phi_0)}{\phi_0}\dfrac{\partial p}{\partial x}$. However a consequence of (5.38) is that the volumetric velocity is $q_0(t)$ throughout the saturated layer and this explains (5.42). As to (5.43), it follows directly from (5.33) setting $S = 1$ in (5.31). It is quite obvious that in the one-dimensional case the boundary conditions at the inflow surface is simpler. The only possibility of imposing a constraint on discharge is directly limiting $q_0(t)$ below the desired bound. More delicate is the introduction of a constraint on pressure. In the unconstrained case, we just note that (5.42) defines the penetration front

$$s(t) = \frac{1}{\phi_0} \int_0^t q_0(\tau)\, d\tau, \tag{5.44}$$

so that $\theta = s^{-1}$ becomes known and consequently $\phi(x,t)$ is immediately known via (5.43). At this point the function $K(x,t) = k(1,\phi(x,t))$ is also known and we get the expression for pressure

$$p(x,t) = \int_x^{s(t)} \frac{q_0(t)}{K(\xi,t)}\, d\xi, \tag{5.45}$$

under quite obvious assumptions on k

Having obtained the explicit expression for $p(0,t)$, if we have imposed a constraint $p(0,t) \le p_{max}$ we can check directly if such inequality is satisfied. If it becomes violated beyond some $t^* > 0$, we shift to the boundary condition $p(0,t) = p_{max}$. The new boundary value problem is not as simple as the previous one. The search of the penetration front $x = s(t)$ is reduced to the solution of the equation

$$p_{max} = \phi_0 \dot{s}(t) \int_0^s \frac{1}{K(\xi,t)}\, d\xi, \quad s(t^*+) = s(t^*-) \tag{5.46}$$

where K is defined as before and contains the unknown $\theta = s^{-1}$.

An existence and uniqueness theorem for (5.46) is given in [25] ($k(1,s)$ must be bounded, bounded away from zero, and continuously differentiable).

Once the function $s(t)$ is found, one can check whether the quantity $\int_0^s \frac{q_0(t)}{K(\xi,t)}\, d\xi$ is not less than p_{max} and when this happen the original boundary condition (5.29) has to be imposed

We summarize the above discussion in the following statement

Theorem 5.1 *The one-dimensional problem in purely saturation regime (5.38)-(5.43) has one and only one classical solution global in time. The result extends to the case of constrained inflow conditions.*

Remark 5.4 The problem has been considered in a semi-infinite medium. The presence of an impermeable wall will terminate the solution at the time at which the layer becomes fully saturated. Then the absorption process (if not already extinct) will continue until completion, but the layer cannot accept more liquid.

Case (ii): *first stage of penetration* (with capillarity).
When we have capillarity the flow is unsaturated during some time interval. Even assuming the simplest possible situation (no inflow constraint, plain Stefan-like-condition on the wetting front, identifications of the thresholds for penetration and absorption) the corresponding flow problem is remarkably difficult.

Instead of repeating the whole proof described in [22] in the special case $S - S_0 = cp$ and in [25] in a more general situation, we just outline here the main difficulties.

(α) Statement of the problem

Find (s, p, ϕ) such that the following system is satisfied in the classical sense:

$$\phi S'(p)\frac{\partial p}{\partial t} + (1 - S)\psi'(\Theta)(S(p) - S_0) - \frac{\partial}{\partial x}\left[k(\phi)\frac{\partial p}{\partial x}\right] = (5.47)$$

$$-k(\phi)\frac{\partial p}{\partial x} = q_0(t), \qquad x = 0, \ 0 < t < t_S \tag{5.48}$$

$$s(0) = 0, \tag{5.49}$$

$$p(s(t), t) = 0, \tag{5.50}$$

$$\dot{s}(t) = -\frac{k(\phi_0)}{\phi_0 s_0}\frac{\partial p}{\partial x}\bigg|_{x=s(t)}, \qquad 0 < t < t_S, \tag{5.51}$$

and (5.33). Here t_S is the (unknown) saturation time, i.e. the first time at which the pressure p_S is reached. In (5.50) $p = 0$ is taken as the reference pressure defined by $S(0) = S_0$.

(β) Assumptions

Let us list the assumptions required in [25] for the existence theorem, starting with conductivity

$$\frac{\partial k}{\partial S} = 0, \tag{5.52}$$

$$0 < k_m \le k(\phi) \le k_M \quad \text{for } 0 < \phi_m \le \phi \le \phi_0, \tag{5.53}$$

where ϕ_0 is the initial (maximum) porosity and $\phi_m = \phi_0 + V_0 - V_{\max}$.

$$\phi\frac{k'}{k} \ge 2, \quad \phi_m \le \phi \le \phi_0, \quad k \in C^3. \tag{5.54}$$

The independence of k on S is taken for the sake of simplicity and is an unphysical limitation, which quite recently has been remover [36]. Should k depend on S but not on ϕ the problem would propably be less difficult (in particular uniqueness would not present any truble). However the dependence to be emphasized is the one on ϕ, owing to the particular structure of the physical problem. The bounds (5.53) are quite natural. Condition (5.54) may look artificial, but is in agreement with specific examples found in the literature.

The function $S(p)$ is also assumed to be in C^3 and such that

$$S'(p) > 0 \quad \text{for } p \in [0, p_S).$$ (5.55)

Furthermore it is required that

$$\frac{S''}{(S')^2}(S - S_0) \leq 1 \quad \text{for } p \in [0, p_S).$$ (5.56)

and

$$\frac{\phi_M}{\inf \psi'} \sup \frac{S''}{S'} \frac{q_0^2(0)}{k(\phi_0)\phi_0} > -(1 - S_0),$$ (5.57)

Note that the latter condition is automatically satisfied if $S'' > 0$. If (5.57) is required then $\inf \psi' = \inf f$ must stay away from zero, i.e. the time interval considered must be reduced so that the absorption process is not completed. Also, if $S'(p_S^-) = 0$ p is required not to reach p_S. In addition, if $S'' \neq 0$ condition (5.54) must be strenghtened replacing 2 by 3 on the r.h.s.

Remark 5.5 In our approach it is important that S' never vanishes, since this preserves the uniform parabolicity of the equation for p. If we want to allow p approach p_S than the conditions above must be modified accordingly.

Concerning the absorption kinetics described by (5.33), besides the assumptions $f \in C^2$, $f(0) = 0$, $f'(\xi) > 0$ for $\xi > 0$, we need that the corresponding function ψ defined in (5.32) satisfies

$$\frac{1}{\phi_m}\psi'^2 + \psi'' \leq 0.$$ (5.58)

which is equivalent to $\dfrac{\Phi''}{\Phi'} \geq \dfrac{1}{\phi_m}$.

The physical meaning of (5.58) can be discovered if one considers the particular case

$$f(\xi) = \lambda \xi^n, \quad n > 0, \ \lambda > 0,$$ (5.59)

which is consistent with (5.58) if

$$\phi_m \geq \frac{\phi_0}{n + 1}.$$ (5.60)

Thus it can be interpreted as a condition on the minimum porosity (needed to prevent the so-called gel–blocking effect). It is interesting to note that the condition is more stringent for fast reactions (n small) than for slow reactions (n large).

Finally $q_0(t)$ is supposed to be C^1 and strictly positive.

(γ) Structure of the existence theorem

Theorem 5.2 *Problem (5.47)-(5.51), (5.33) has at least one classical solution under the assumptions listed in (β). The solution is local in time if $S'' \neq 0$.*

The proof is based on a fixed point argument.

Taken a boundary $x = s(t)$ in a suitable Lipschitz class of increasing functions it is possible to show that the nonlinear system (5.47), (5.48), (5.50) and (5.33) has a unique classical solution. The proof can be found in [22], [25]. Next a mapping $\mathcal{M} : s \to \sigma$ is defined using

$$\dot{\sigma}(t) = -\frac{k_0}{s_0 \phi_0} \frac{\partial p}{\partial x}\bigg|_{x=s(t)}, \quad \sigma(0) = 0, \tag{5.61}$$

so that the solutions of our problem are associated to the fixed points of $\mathcal{M}$. Existence of at least one fixed point is proved by means of the Schauder theorem.

The necessary estimates are the following, which are fulfilled in $0 < x < s(t)$, $0 < t < t_S$

$$0 < p(x,t) < p_S, \tag{5.62}$$

$$-k(\phi)\frac{\partial p}{\partial x} > 0, \tag{5.63}$$

$$\frac{\partial p}{\partial x}\bigg|_{x=s(t)} \geq -\frac{1}{k_m} \sup_{0<\tau<t} q_0(\tau), \tag{5.64}$$

$$-k(\phi)\frac{\partial p}{\partial x} \leq M_1(T), \tag{5.65}$$

$$\left|\frac{\partial p}{\partial t}\right|, \ \left|\frac{\partial^2 p}{\partial x^2}\right| \leq M_2(T), \tag{5.66}$$

$M_1(T), M_2(T)$ being a-priori computable increasing functions.

The proof of such estimates is quite long. They are based on the use of maximum principle. The difficulty comes from the necessity of controlling the sign of source terms or of some basic coefficient in the parabolic equations satisfied by p, $\frac{\partial p}{\partial x}$, $\frac{\partial p}{\partial t}$.

For instance, (5.63) looks physically trivial since it expresses that the flow proceeds constantly in the same direction. However it relies on the study of the equation

$$\frac{\phi S'}{k} \frac{\partial w}{\partial t} + \left(\frac{S''}{S'}\frac{1}{k}w - \frac{1}{\phi}\psi'(\Theta)\Xi\right)\frac{\partial w}{\partial x} - \frac{\partial^2 w}{\partial x^2} \tag{5.67}$$

$$+ \frac{1}{k}\psi'(\Theta)S'\left\{\left(\frac{\phi k'}{k} - 1\right)(S - S_0) + (1 - S)\left(1 - (S - S_0)\frac{S''}{S'^2}\right)\right\}w$$

$$= -(1 - S)(S - S_0)\Xi\left(\frac{1}{\phi}\psi'^2 + \psi''\right),$$

where $w = k(\phi)\dfrac{\partial p}{\partial x}$ and

$$\Xi = \int_{\theta(x)}^{t} \frac{w(x,\tau)}{k(\phi(x,\tau))} S'(p(x,\tau))\, d\tau. \tag{5.68}$$

The necessity of controlling the sign of the r.h.s. explains the adoption of (5.58), while (5.51) guarantees that the sign of the coefficient of w is the correct one in order to apply the maximum principle. The equation for $\dfrac{\partial p}{\partial t}$ is much more complicated

$$\phi S' \frac{\partial^2 p}{\partial t^2} - \frac{\partial}{\partial x}\left(k \frac{\partial}{\partial x}\frac{\partial p}{\partial t}\right) + \tag{5.69}$$

$$\psi'S'\left\{(1-S) + \left(\phi\frac{k'}{k} - 2\right)(S - S_0)\right\}\frac{\partial p}{\partial t} + \phi S''\left(\frac{\partial p}{\partial t}\right)^2 =$$

$$= -(1-S)(S - S_0)^2\psi'' + \frac{\partial p}{\partial x}(S - S_0)\Xi\left\{k''\psi'^2 + k'\psi'' + \frac{k'}{k}\psi'^2\right\}$$

$$-k'\psi'S'\left(\frac{\partial p}{\partial x}\right)^2.$$

because it contains a quadratic term, if $S'' \neq 0$. This explains the necessity of additional requirements and implies the use of great care in the application of maximum principle as well as a possible restriction of the time interval. Note the full use of (5.55) (in the stronger version if $S'' \neq 0$).

It is unnecessary to report the complete calculations and details of the proofs for which we refer to [25].

The most delicate step of the existence proof is showing the continuity of the mapping $\mathcal{M}$ in the sup-norm.

This starts from the integration of (5.47) written as

$$\frac{\partial}{\partial t}[(1 - S)\phi] + \frac{\partial}{\partial x}\left(k(\phi)\frac{\partial p}{\partial x}\right) = 0 \tag{5.70}$$

over the domain $0 < x < s(t)$, $0 < \tau < t$.

The resulting equation

$$\phi_0 S_0(\sigma(t) - s(t)) = \int_0^t q_0(\tau)\, d\tau + \int_0^{s(t)} (\phi(x,t) - \phi_0)\, dx \tag{5.71}$$

$$- \int_0^{s(t)} \phi(x,t) S(x,t)\, dx$$

for a fixed point reduces to

$$\int_0^{s(t)} \phi(x,t)S(x,t)\,dx = \int_0^t q_0(\tau)\,d\tau + \int_0^{s(t)} (\phi(x,t) - \phi_0)\,dx \qquad (5.72)$$

describing the partition of injected liquid volume $\int_0^t q_0(\tau)\,d\tau$ into the pore volume occupied by the liquid $\int_0^{s(t)} \phi(x,t)S(x,t)\,dx$ and the volume increase of the granules $\int_0^{s(t)} (\phi(x,t) - \phi_0)\,dx$. Equation (5.71) allows to compare the two images σ_1, σ_2 of two elements s_1, s_2 in the selected subset of $C([0,T])$:

$$\sigma_1(t) - \sigma_2(t) = s_1(t) - s_2(t) + \frac{1}{\phi_0 S_0}\int_0^{\alpha(t)} (\phi_1(x,t) - \phi_2(x,t))\,dx \qquad (5.73)$$

$$-\frac{1}{\phi_0 S_0}\int_0^{\alpha(t)} (\phi_1(x,t)S_1(x,t) - \phi_2(x,t)S_2(x,t))\,dx - \frac{1}{S_0}(s_1(t) - s_2(t))$$

$$-\frac{(-1)^j}{S_0\phi_0}\int_{\alpha(t)}^{\beta(t)} \phi_j(x,t)\,dx + \frac{(-1)^j}{S_0\phi_0}\int_{\alpha(t)}^{\beta(t)} \phi_j(x,t)S_j(x,t)\,dx,$$

where α and β represents the min and max between s_1, s_2 respectively, and $j = i$ if $\beta = s_i$.

The equation above shows the necessity of estimating $\| S_1 - S_2 \|$ in terms of $\| s_1 - s_2 \|$ ($\| \cdot \|$ denotes the sup-norm).

Observing the equation satisfied by the difference $u = p_1 - p_2$

$$\phi_2 S'(p_2)\frac{\partial u}{\partial t} - \frac{\partial}{\partial x}\left(k_2\frac{\partial u}{\partial x}\right) + \qquad (5.74)$$

$$+ \{(\phi_1 - \phi_2)S'(p_1) + \phi_2[S'(p_1) - S'(p_2)]\}\frac{\partial p_1}{\partial t} +$$

$$+ (S_1 - S_2)\{(1 - S_2)\psi'(\Theta_2) - (S_1 - S_0)\psi'(\Theta_1)\} +$$

$$+ (1 - S_2)(S_1 - S_0)[\psi'(\Theta_1) - \psi'(\Theta_2)] - \frac{\partial}{\partial x}\left[(k_1 - k_2)\frac{\partial p_1}{\partial x}\right] = 0$$

one understands that the nontrivial term is the one containing $\frac{\partial}{\partial x}(\phi_1 - \phi_2)$ through $\frac{\partial}{\partial x}(k_1 - k_2)$:

$$\frac{\partial}{\partial x}(\phi_1 - \phi_2) = -[\psi'(\Theta_1) - \psi'(\Theta_2)]\int_{\theta_1(x)}^t S'(p_1)\frac{\partial p_1}{\partial x}\,d\tau \qquad (5.75)$$

$$-\psi'(\Theta_2)\left[\int_{\eta(x)}^{t}\frac{\partial}{\partial x}(S_1-S_2)\,d\tau-(-1)^j\int_{\xi(x)}^{\eta(x)}S'(p_j)\frac{\partial p_j}{\partial x}\,d\tau\right],$$

where ξ,η are min and max between θ_1,θ_2, respectively. Clearly (5.75) requires an estimate of $\int_{\theta(x)}^{t}S'(\bar{p})\frac{\partial u}{\partial x}\,d\tau$, with $\bar{p}\in(p_1,p_2)$. Instead of looking for it we aim at the less difficult target of an energy estimate for u.

Multiplying (5.74) by u and integrating over $0<x<\alpha(\tau)$, $0<\tau<t$ one gets

$$\frac{1}{2}\int_{0}^{\alpha(t)}\phi_2 S'(p_2)u^2(x,t)\,dx+\int_{0}^{t}d\tau\int_{0}^{\alpha(\tau)}dx\,k_2\left(\frac{\partial u}{\partial x}\right)^2= \tag{5.76}$$

$$\frac{1}{2}\int_{0}^{t}\phi_2 S'(p_2)u^2(\alpha(\tau),\tau)\dot{\alpha}(\tau)\,d\tau+\int_{0}^{t}u(\alpha(\tau),\tau)\left[k_1\frac{\partial p_1}{\partial x}-k_2\frac{\partial p_2}{\partial x}\right]_{x=\alpha(\tau)}d\tau$$

$$+\int_{0}^{t}d\tau\int_{0}^{\alpha(\tau)}dx\,u^2(x,\tau)\left\{\frac{1}{2}\left[\frac{\partial\phi_2}{\partial t}S'(p_2)+\phi_2 S''(p_2)\frac{\partial p_2}{\partial t}\right]\right.$$

$$\left.-S'(\bar{p})[(1-S_2)\psi'(\Theta_2)-(S_1-S_0)\psi'(\Theta_1)-\phi_2 S''(\bar{p})\frac{\partial p_2}{\partial t}\right\}$$

$$+\int_{0}^{t}d\tau\int_{0}^{\alpha(\tau)}dx\,u(x,\tau)\left\{[\psi(\Theta_1)-\psi(\Theta_2)]\left[S'(p_1)\frac{\partial p_1}{\partial t}-(1-S_2)(S_1-S_0)\right]\right.$$

$$-\int_{0}^{t}d\tau\int_{0}^{\alpha(\tau)}dx\,(k_1-k_2)\frac{\partial u}{\partial x}\frac{\partial p_1}{\partial x}.$$

Noting that

$$|\Theta_1(x,\tau)-\Theta_2(x,\tau)|\leq\{M\sup_{0<\tau'<\tau}|u(x,\tau')|+1\}(\eta(x)-\xi(x)). \tag{5.77}$$

we can obtain from (5.76) a Gronwall inequality from the quantity

$$\int_{0}^{\alpha(t)}\sup_{0<\tau<t}u^2(x,\tau)\,dx,$$

leading to the estimate

$$\int_{0}^{\alpha(t)}u^2(x,t)\,dx\leq C\int_{0}^{t}\|s_1-s_2\|_\tau\,d\tau. \tag{5.78}$$

At this point (5.73) gives the estimate

$$\| \sigma_1 - \sigma_2 \| \le C \| s_1 - s_2 \|^{1/2}, \tag{5.79}$$

with C depending on the data and on T.

The final inequality (5.79) concludes the existence proof. Quite recently, uniqueness has been obtained following a different approach [33].

D) Open questions

The material presented here provides a very partial response to the problem, the only case analyzed from the mathematical point of view is the one dimensional case either in the extreme relatively simple case of no capillarity, or in the first stage of unsaturated penetration which, even in the simplest possible setting, presents remarkable difficulties.

Stiil in the 1-D framework, open problems are the continuation of the solution beyond the appearance of the saturation front at the inflow surface (with possible pressure constraint) and the possible formation of a saturated (stagnant) region in the proximity of the impermeable wall.

Also variant of the wetting front conditions, such as the kinetic undercoling analog, have not yet been considered.

Totally open are the multidimensional problems, which can be formulated in several different ways, depending on the choice of the initial conditions, the conditions on the wetting front, and on the inflow conditions.

A large amount of work has to be done from the numerical point of view.

References

[1] R.J. Arzenault (eds.) R.K. Everett. *Treatize on Material Science and Technology*. Academic Press, 1991.

[2] G. Baldini and M. Petracco. Models for water percolation during the preparation of espresso-coffee. *Proc. 7th ECMI Conf., A. Fasano, M. Primicerio eds., Teubner Stuttgart*, (1992), 137–140.

[3] I. Barenblatt, V.M. Entov, and V.M. Ryzhik. *Theory of fluid flows through natural rocks*. Kluwer, 1990.

[4] J. Bear. *Dynamics of Fluids in Porous Media*. America Elsevier, New York, 1972.

[5] J. Bear and A. Verruijt. *Modelling Ground Water Flow and Pollution*. Reidel, Dordrecht, N. Y., 1987.

[6] D. Bertaccini. Simulation of a filtration process in a deformable porous medium. *Nonlinear Analysis* **30** (1997), 663–668.

[7] L. Billi. Non-isothermal flows in porous media with curing. *EJAM* **8** (1997), 623–637.

[8] L. Billi. Non-isothermal incompressible flows in porous media. Ph. D. Thesis (SISSA, Trieste 1996).

[9] L. Billi. Incompressible flows through porous media with temperature-dependent parameters. *Nonlinear Analysis, Theory, Meth. Appl.* **31**, (1998), 363–383.

[10] L. Billi and A. Farina. Unidirectional infiltration in deformable porous media: mathematical modelling and self-similar solutions. To appear in Quart. Appl. Math.

[11] D.C. Blest, B.R. Duffy, S. Mc Kee, and A.K. Zulkifle. Curing simulation of thermoset composites. *Strathclyde Math. res. Rep.* **37** (1998).

[12] J. Chadam, Xinfu Chen, E. Comparini and R. Ricci. Traveling wave solutions of a reaction-infiltration problem and a related free boundary problem. *European J. Appl. Math.* To appear.

[13] J. Chadam, Xinfu Chen, R. Gianni and R. Ricci. A reaction-infiltration problem: Existence, uniqueness and regularity of solutions in two space dimensions. *Math. Models Methods Appl. Sci.* To appear.

[14] J. Chadam, Xinfu Chen, R. Gianni and R. Ricci. A reaction-infiltration problem: Calssical solutions. *Proc. Edinburgh Math. Soc.* **40** (1997), 275–291.

[15] E. Comparini and P. Mannucci. Penetration of a wetting front in a porous medium interacting with the flow. To appear.

[16] E. Comparini and M. Ughi. Shock propagation in a one dimensional flow through deformable porous media (1998), submitted to *Interfaces and Free Boundaries (Modelling, Analysis ana Computation)*.

[17] E. Comparini and M. Ughi. Existence of shock propagation in a flow through deformable porous media (1998), in preparation.

[18] E.L. Cussler. *Diffusion and Mass Transfer in Fluid Systems*. Cambridge Univ. Press, 1984.

[19] A. Farina and L. Preziosi. Free boundary problems in the production of composites. Proc. Int. Conference on Free Boundary Problems (Creta, 1997), I. Athanasoupulos ed., to appear.

[20] A. Farina and L. Preziosi. Non-isothermal injection moulding with resin cure and preform deformability. To appear.

[21] A. Farina and L. Preziosi. Infiltration process in composite materials manufacturing: modelling and qualitative results. In *Complex Flows in Industrial Processes*, A. Fasano ed., Birkhäuser, to appear.

[22] A. Fasano. A one-dimensional flow problem in porous media with hydrophile grains. To appear on Math. Meth. Appl. Sc.

[23] A. Fasano. Some two-scale processes involving parabolic equations. Proc. Int. Conference on Free Boundary Problems (Creta, 1997), I. Athanasoupulos ed., to appear.

[24] A. Fasano. Some non-standard one-dimensional filtration problems. *Bull. Fac. Ed. Chiba Univ.* **44**, (1996), 5–29.

[25] A. Fasano. Porous media with hydrophile granules. In *Complex Flows in Industrial Processes*. A. Fasano ed., Birkhäuser, In press.

[26] A. Fasano, A. Mikelič, and M. Primicerio. Homogenization of flows through porous media with permeable grains. to appear in Adv. Math. Sci. Appl.

[27] A. Fasano, M. Petracco, and F. Talamucci. The espresso-coffee problem. In *Complex Flows in Industrial Processes*, A. Fasano ed., Birkhäuser, to appear.

[28] A. Fasano and M. Primicerio. Mathematical models for filtration through porous media interacting with the flow. Proc. "Nonlinear Mathematical Problems in Industry", H. Kawerada, N. Kenmochi, N. Yanagihara eds., *Math. Sci. Appl.* **1** (1993), 61–85.

[29] A. Fasano and M. Primicerio. Flows through saturated mass exchanging porous media with high pressure gradients. In "Calculus of Variations and Computations", C. Bandle et al. eds. (1994), Pitman Res. Notes Math. **326**, Lougman (1995), 109–129.

[30] A. Fasano and M. Primicerio. Free-boundary problems for nonlinear parabolic equations with nonlinear free boundary conditios. *J. Math. Anal. Appl.* **72**, (1979), 247-273.

[31] A. Fasano, M. Primicerio and G. Baldini. Mathematical models for espresso-coffee preparation *Proc. 6th ECMI Conf.*, F. Hodnett ed., Teubner, Stuttgart (1992), 137-140.

[32] A. Fasano, M. Primicerio, and A. Watts. On a filtration problem with flow-induced displacement of fine particles. In *Boundary Control and Boundary Variations*. (J.P. Zolesio ed.), M. Dekker, New York, (1994), 205–232.

[33] A. Fasano and V. Solonnikov. An existence and uniqueness theorem for a flow problem through absorving porous media. To appear.

[34] A. Fasano and F. Talamucci. A comprehensive mathematical model for a multi-species flow through ground coffee. To appear.

[35] A. Fasano and P. Tani. Penetration of a wetting front in a porous medium with flux dependent hydraulic parameters. In *Nonlinear Problems in Applied Mathematics, (K. Cooke et al eds.), SIAM*, (1995).

[36] R. Gianni and P. Mannucci. The propagation of a wetting front through an absorving porous material with saturation dependent permeability. To appear.

[37] R. Gianni and R. Ricci. Existence and uniqueness for a reaction-diffusion problem in infiltration. *Ann. Mat. Pura Appl.* To appear.

[38] W.H. Green and G.A. Ampt. Studies on soil physics. The flow of air and water through soils. *J. Agric. Sci.* **4**, (1911), 1–24.

[39] E. Illy and T. Bullo. Considérations sur le procedé d'extraction. *Caffé, Cacao, Thé* **7**, (1963), 395–399.

[40] A.I. Isayev. *Injection and compression of moleting fundamental.* Marcel Dekker, 1987.

[41] M.S. Johnson, C.D. Rudd, and D.J. Hill. Microwave assisted resin transfer moulding. IV Int. Conf. on Flow Processes in Composite Materials, *Composite* (special issue) 1996.

[42] G.M. Maistros and I.K. Partridge. Autoclave cure of composites: validation of models using dynamic dyelectric analysis. IV Int. Conf. on Flow Processes in Composite Materials (1996), to appear on *Composite*.

[43] P. Mannucci. Study of the mathematical model for absorption and diffusion in ultra-napkins. *Le Mathematiche* **50** (1995), 3–14.

[44] M. Petracco. Physico-chemical and structural characterization of espresso coffee brew. *13th Colloquiun ASIC*, Paipa 1989, Pavia.

[45] L. Preziosi. The theory of deformable porous media and its applications to composite material manufacturing. *Surveys Math. Ind.* **6**, (1996), 167–214.

[46] K.R. Rajagopal and L. Tao. *Mechanics of Mixtures.* World Scientific (1995).

[47] J.L. Sommer and A. Mortensen. Forced unidirectional infiltration of deformable porous media. *J. Fluid Mechanics* **311**, (1996), 193–217.

[48] F. Talamucci. Analysis of the coupled heat-mass transport in freezing porous media. *Survey in Mathematics for Industry* **7**, (1997), 93–139.

[49] F. Talamucci. Flow throgh porous medium with mass removal and diffusion. To appear on *NODEA*.

[50] J. Weickert. A mathematical model for diffusion and exchange phenomena in ultra napkins. *Math. Meth. Appl. Sci.* **16**, (1973), 759–777.

HOMOGENIZATION THEORY AND APPLICATIONS TO FILTRATION THROUGH POROUS MEDIA

by

Andro Mikelić *

UFR Mathématiques, Analyse Numérique
Université Lyon 1, Bât. 101,
43, bd. du 11 novembre 1918
69622 Villeurbanne Cedex, FRANCE

0. Introduction

Homogenization applied to porous media is a mathematical method that allows to " upscale " the equations of fluid mechanics, being valid at the pore level. This way we are not obliged to solve nonlinear PDEs of the fluid mechanics in the complicated geometry of a porous medium, which is in addition usually unknown. The **homogenization theory** of porous media studies the effects of the micro-structure (i.e. of the pore structure) upon solutions of PDEs of the fluid mechanics.

Even in the simplest case of an viscous single phase flow through a porous medium, we are given a PDE with two natural **length scales** : a macroscopic scale (the scale of the reservoir) of order 1 and a microscopic scale (the pore scale) of order ε, the later measuring the scale of oscillations. This disparity in length scales is what provides us with our expansion parameter ε. For fixed, but small, characteristic pore length $\varepsilon > 0$ the solutions u^ε of the flow equations will in general be complicated, having different behaviors on the two length scales. A closed-form solution is unachievable and a numerical solution would be nearly impossible to calculate.

In the practical simulations of the flows through porous media, we use PDEs at the macroscale. Informations about the pore structure are only kept through some averaged quantities as porosity and permeability. Therefore, one of the fundamental questions in the modeling of flows through porous media is how to get the "averaged" or "upscaled" equations. Next we wish to calculate the effective coefficients describing the influence of the microstructure. Also, as we shall see in the modeling of the inertia effects for flows in porous media, it is of interest to know whether our derived model is correct, in the sense that it should approximate the original problem involving the micro-structure.

Homogenization theory studies the limiting behavior $u^\varepsilon \to u$ as $\varepsilon \to 0$. The idea is that in this limit the micro-structure (generating the high-frequency

---

* This review article was partly written during author's sabbatical stay at IWR, SFB 359, Universität Heidelberg, from October 1, 1998 to March 31, 1999

oscillations) will " average out ", and there will be a simple " averaged " or " homogenized " PDE, which will represent a filtration law.

As even the very simple example of Darcy's law cofirms, the homogenized PDE can differ very much from the original one. In overcoming this fundamental difficulty it is useful to use **formal multiscale expansions** in ε, containing behavior on different length scales. The idea is to suppose u^ε has the following expansion:

$$u^\varepsilon = \varepsilon^\beta \left\{ u_0(x, \frac{x}{\varepsilon}) + \varepsilon u_1(x, \frac{x}{\varepsilon}) + \varepsilon^2 u_2(x, \frac{x}{\varepsilon}) + \dots \right\} \qquad (0.1)$$

Now we plug (1.1) into the PDE and try to determine the functions $u_i, i = 1, \dots$. Nevertheless, before plugging the expansion (1.1) into the PDE, we should somehow determine β.

In order to answer all those questions, we establish the following strategy, which we are going to apply in the sections which follow:

A) We give a *description of the porous medium* , which can be periodic, statistically homogeneous etc.

B) We establish *a priori estimates* for solutions of the PDE, uniform with respect to ε. For the flow problems we usually need:

B1) A priori estimates for the velocity.

B2) A priori estimates for the pressure.

C) Having obtained a priori estimates, we establish *a formal multiscale expansion* . We shall see that for the linear and monotone problems it corresponds to passing to the homogenization limit in the sense of the *2-scale convergence*.

D) It is necessary to *study the upscaled problem*. We prove uniqueness, regularity etc.

This introductory text will try to initiate the reader to the homogenization applied to porous media problems. In the first section, it contains a reasonably detailed rigorous derivation of Darcy's law. The second section contains an unpublished result on homogenization of a nonstationary quasi-Newtonian flow through a porous medium. It can be considered as an application of the methods developped in Section 1. The third and the fourth section contain a review of the author's recent results on the inertia effects for flows through porous media and on the law by Beavers and Joseph, respectively.

In connection with the homogenization in porous media, we recommend to the reader the recent book edited by U. Hornung [Horn 1997]. It contains number of articles, and we mention the section on the 2-scale convergence by G. Allaire, which we are going to quote frequently. Also there is a somewhat different derivation of Darcy's law, written by G. Allaire, and being much closer to the original proof of Luc Tartar (see [Ta1980]). This text is much shorter, does not discuss the same multitude of subjects, but, nevertheless, contains the material on the inertia effects and on the law by Beavers and Joseph which is posterior to the book edited by Hornung.

As general references on homogenization we recommend the classic text by E. Sanchez-Palencia [SP80], containing as an appendix the fundamental result by L. Tartar [Ta1980]). More recent reference is the book by Jikov, Kozlov and Oleinik [JKO]. It is a very complete text on homogenization, but does not discuss flows through porous media.

Formal multiscale expansions for the incompressible Navier-Stokes system in a porous medium can be found in § 2 of [LiPe]. This reference has strongly motivated Section 3.

1. An introduction to the derivation of Darcy's law by a multi-scale expansion in a porous medium

1.1 Introduction

In most applications, a slow fluid flow through a rigid porous medium is modeled by Darcy's law, linking by a linear relation the filtration velocity with the pressure drop. Its derivation is nontrivial since we don't deal with a problem in some fixed domain, but with a sequence of problems in varying geometries. Ene and Sanchez-Palencia seem to be first to give a derivation of it, from the Stokes system, using a formal multiscale expansion (see [ESP]). This derivation was made rigorous in the case of a 2D periodic porous medium by L. Tartar in [Ta1980]. This result was generalized in number of other papers. We mention only the generalization to 3D by Allaire [ALL89] and to a random statistically homogeneous porous medium by Beliaev and Kozlov [Beko95]. Qualitatively, the case of a random statistically homogeneous porous medium leads to the same type of law, as the periodic model. That's reason why we give here a rigorous derivation only for the periodic case. Also this way we avoid some difficult technicalities, which are not completely settled down mathematically.

As already mentioned, we start with the description of a porous medium and we consider a periodic porous medium as an example of a random statistically homogeneous medium. In order to avoid the complications with the outer boundaries, we suppose that $\Omega = (0, L)^n$ and set the periodic boundary conditions at $\partial\Omega$.

1.2 Description of a periodic porous medium

We consider a periodic porous medium with a periodic arrangement of the pores. The formal description goes along the following lines:

Firstly we define the geometrical structure inside the unit cell $Y =]0, 1[^n$, $n = 2, 3$. Let Y_s (the solid part) be a closed subset of $\bar{Y}$ and $Y_F = Y \backslash Y_s$ (the fluid part). Now we make the periodic repetition of Y_s all over $\mathbb{R}^n$ and set $Y_s^k = Y_s + k$, $k \in \mathbb{Z}^n$. Obviously the obtained set $E_s = \bigcup_{k \in \mathbb{Z}^n} Y_s^k$ is a closed subset of $\mathbb{R}^n$ and $E_F = \mathbb{R}^n \backslash E_s$ in an open set in $\mathbb{R}^n$. Following Allaire [ALL89] we make the following assumptions on Y_F and E_F:

(i) Y_F is an open connected set of strictly positive measure, with a Lipschitz boundary and Y_s has strictly positive measure in $\bar{Y}$, as well .

(ii) E_F and the interior of E_s are open sets with the boundary of class $C^{0,1}$, which are locally located on one side of their boundary. Moreover E_F is connected.

Now we see that Ω is covered with a regular mesh of size ε, each cell being a cube Y_i^ε, with $1 \le i \le N(\varepsilon) = |\Omega|\varepsilon^{-n}[1 + O(1)]$. Each cube Y_i^ε is homeomorphic to Y, by linear homeomorphism Π_i^ε, being composed of translation and an homothety of ratio $1/\varepsilon$.

We define

$$Y_{S_i}^\varepsilon = (\Pi_i^\varepsilon)^{-1}(Y_s) \qquad \text{and} \qquad Y_{F_i}^\varepsilon = (\Pi_i^\varepsilon)^{-1}(Y_F)$$

For sufficiently small $\varepsilon > 0$ we consider the set

$$T_\varepsilon = \{k \in \mathbb{Z}^n | Y_{S_k}^\varepsilon \subset \Omega\}$$

and define

$$O_\varepsilon = \bigcup_{k \in T_\varepsilon} Y_{S_k}^\varepsilon, \quad S^\varepsilon = \partial O_\varepsilon, \quad \Omega^\varepsilon = \Omega \backslash O_\varepsilon = \Omega \cap \varepsilon E_F$$

Obviously, $\partial \Omega^\varepsilon = \partial \Omega \cup S^\varepsilon$. The domains O_ε and Ω^ε represent, respectively, the solid and fluid parts of a porous medium Ω. For simplicity we suppose $L/\varepsilon \in \mathbb{N}$.

A very important property of the porous media is a variant of Poincaré's inequality. In the random case it lead Beliaev and Kozlov to introduce the following definition of a random porous medium:

Definition 1.1. We call a random domain $E_F(\omega)$ porous if there exists a random variable $h(\omega) > 0$ such that $\mathbb{E}\{h^{-1}\} < +\infty$ and

$$\int_{\mathbb{R}^n} h(T(y)\omega)|\varphi(y)|^2 \, dy \le \int_{\mathbb{R}^n} |\nabla\varphi(y)|^2 \, dy, \quad \forall \varphi \in C_0^\infty(E_F(\omega)), \ (a.s) \quad (1.1)$$

where T is an n-dimensional dynamical system describing the random structure.

In the periodic case the measure space is $\Omega = Y$, the probability measure is $d\mu = dy$ and the n-dimensional dynamic system T is given by $T(y)(\omega) = (y + \omega) \bmod 1$. The following result confirms that in the periodic case E_F is porous in the sense of Definition 1.1:

Lemma 1.2. Let $W^\varepsilon = \{z \in H^1(\Omega^\varepsilon)^n, \ z = 0$ on $\partial \Omega^\varepsilon \backslash \partial \Omega$ and z is $L-$ periodic $\}$. Then in the periodic case the function h takes the form $\dfrac{\lambda_1(Y_F)}{\varepsilon^2}$, i.e. we have

$$\int_\Omega |w|^2 dx \le \frac{\varepsilon^2}{\lambda_1(Y_F)} \int_\Omega |\nabla_x w|^2 dx \qquad \forall w \in W^\varepsilon, \quad (1.2)$$

where $\lambda_1(Y_F)$ is the smallest eigenvalue of $-\Delta$ on W^ε.

Proof: In Y_F we have $\int_{Y_F} |w|^2 dy \leq \frac{1}{\lambda_1(Y_F)} \int_{Y_F} |\nabla_y w|^2 dy$, for all $w \in H^1(Y_F)$, $w = 0$ on $\partial Y_F \backslash \partial Y$ (Friedrichs inequality). Change of variables $x = \varepsilon y$; $dx = \varepsilon^n dy$ dans $\mathbb{R}^n$; $\frac{\partial}{\partial y} = \varepsilon \frac{\partial}{\partial x}$ gives:

$$\int_{\varepsilon Y_F} |w|^2 dx \leq \frac{\varepsilon^2}{\lambda_1(Y_F)} \int_{\varepsilon Y_F} |\nabla_x w|^2 dx$$

After adding all the integral over different cells, we get (1.2) (the constant $\lambda_1(Y_F)$ is the same for all cells). ∎

1.3 Steady Stokes equations in a porous medium. The a priori estimates for the velocity

Having defined the geometrical structure of the porous medium, we precise the flow problem. Here we deal with slow viscous incompressible flow of a single fluid through a porous medium. We suppose the no-slip condition at the boundaries of the pores. Then we describe it by the following steady Stokes system in Ω^ε (the fluid part of the porous medium Ω):

$$-\mu \Delta v^\varepsilon + \nabla p^\varepsilon = f \quad \text{in} \quad \Omega^\varepsilon \qquad (1.3)$$

$$\text{div } v^\varepsilon = 0 \quad \text{in} \quad \Omega^\varepsilon \qquad (1.4)$$

$$v^\varepsilon = 0 \quad \text{on} \quad \partial \Omega^\varepsilon \backslash \partial \Omega, \quad \{v^\varepsilon, p^\varepsilon\} \quad \text{is } L - \text{periodic} \qquad (1.5)$$

Here μ is the viscosity, f stands for the effects of external forces or an injection at the boundary or a given pressure drop. v^ε denotes the velocity and p^ε is the pressure.

The variational form of the problem (1.3)-(1.5) is:
Find $v^\varepsilon \in W^\varepsilon$, div $v^\varepsilon = 0$ in Ω^ε and $p^\varepsilon \in L^2(\Omega^\varepsilon)$ such that

$$\mu \int_{\Omega^\varepsilon} \nabla v^\varepsilon \nabla \varphi \, dx - \int_{\Omega^\varepsilon} p^\varepsilon \text{ div } \varphi \, dx = \int_{\Omega^\varepsilon} f\varphi \, dx \quad \forall \varphi \in W^\varepsilon. \qquad (1.6)$$

Then for $f \in L^2(\Omega^\varepsilon)^n$, the elementary elliptic variational theory gives the existence of the unique velocity field $v^\varepsilon \in W^\varepsilon$, div $v^\varepsilon = 0$ in Ω^ε, which solves (1.6) for every $\varphi \in W^\varepsilon$, div $\varphi = 0$ in Ω^ε. The construction of the pressure field goes through De Rham's theorem. We note a very simple construction of the pressure field $p^\varepsilon \in L^2(\Omega^\varepsilon)$ for Lipschitz boundaries by L. Tartar (see the classic textbook by Temam [Te1]). Obviously, the pressure is unique up to a constant. Usually, it is fixed by supposing $\int_{\Omega^\varepsilon} p^\varepsilon \, dx = 0$.

We would like to upscale the system (1.3)-(1.5), i.e. to pass to the limit as $\varepsilon \to 0$. Thus we switch to the step **B** of our strategy.

The first part is finding the *a priori* estimates for the velocity v^ε. This is fairly simple. We test (1.6) with $\varphi = v^\varepsilon$ and obtain

$$\mu \int_{\Omega^\varepsilon} |\nabla v^\varepsilon|^2 \, dx = \int_{\Omega^\varepsilon} f v^\varepsilon \, dx \leq \|f\|_{L^2(\Omega^\varepsilon)^n} \|v^\varepsilon\|_{L^2(\Omega^\varepsilon)^n} \qquad (1.7)$$

Now, we apply Poincaré's inequality (1.2) to (1.7) and obtain

$$\|v^\varepsilon\|_{L^2(\Omega^\varepsilon)^n} \leq C\varepsilon^2 \tag{1.8}$$

$$\|\nabla v^\varepsilon\|_{L^2(\Omega^\varepsilon)^{n^2}} \leq C\varepsilon \tag{1.9}$$

The optimality of the *a priori* estimates (1.8)-(1.9) will be confirmed by the steps **C** and **D**. We extend v^ε by zero in $\Omega \setminus \Omega^\varepsilon$. It is well known that extension by zero preserves L^q and $W_0^{1,q}$ norms for $1 < q < \infty$. Therefore, we can replace Ω^ε by Ω in (1.8)-(1.9).

Contrary to the velocity, *a priori* estimates for the pressure are not easy to obtain. If we use the variational equation (1.6), then

$$\langle \nabla p^\varepsilon, \varphi \rangle_{\Omega^\varepsilon} = \int_{\Omega^\varepsilon} f\varphi \, dx - \mu \int_{\Omega^\varepsilon} \nabla v^\varepsilon \nabla \varphi \, dx \qquad \forall \varphi \in W^\varepsilon \tag{1.10}$$

and we obtain the estimate

$$\|\nabla p^\varepsilon\|_{(W^\varepsilon)'} = \sup_{\varphi \in W^\varepsilon} \frac{|\langle \nabla p^\varepsilon, \varphi \rangle_{\Omega^\varepsilon}|}{\|\varphi\|_{W^\varepsilon}} \leq C\varepsilon \tag{1.11}$$

It is not clear how to use this estimate, since only the estimates in Ω are useful. We note that in the functional space H^{-1} extension of the gradient by zero is not necessarily gradient of some L^2-function. The estimate (1.11) implies, through the Nečas inequality (see e.g. [Te1]), an estimate for the L^2-norm of the pressure, but with the constant depending on ε on an unknown way. This difficulty with the *a priori* estimate for the pressure makes the upscaling of the flow problems much more difficult than upscaling of diffusion processes or linear elasticity problems.

Classical way of overcoming this difficulty is due to L. Tartar (see [Ta1980]). He used a restriction operator and then a duality argument gave the required estimate. We are going to proceed differently and in Section 1.4 a direct extension is constructed and the corresponding *a priori* estimate obtained.

1.4 A priori estimates for the pressure

In this Section we follow the approach from [Zh94] and extend the pressure field. The notation $\Omega \cap \varepsilon E_F$ is used rather than Ω^ε.

We start with an auxiliary result

Lemma 1.3. *We have*

$$\int_{\Omega \cap \varepsilon E_F} \left| \varphi - \frac{1}{|\Omega \cap \varepsilon E_F|} \int_{\Omega \cap \varepsilon E_F} \varphi \right|^2 dx \leq k \int_{\Omega \cap \varepsilon E_F} |\nabla \varphi|^2 \, dx, \qquad \forall \varphi \in C_{per}^\infty(\Omega) \tag{1.12}$$

where $\varepsilon \leq \varepsilon_0(\Omega)$.

Proof: It is based on the following result from [Ace92], which we give for the convinience of the reader:

Extension theorem for periodic Lipschitz sets ([Ace92]). *Let E_F be the periodic Lipschitz domain defined above and let $1 < r < \infty$. Then for every open bounded set D in $\mathbb{R}^n$ and $\forall \delta > 0$, there is a linear and continuous extension operator,*

$$T_\varepsilon : W^{1,r}(D \cap \varepsilon E_F) \to W^{1,r}_{loc}(D)$$

and positive constants k_0, k_1 and k_2, such that for every $u \in W^{1,r}(D \cap \varepsilon E_F)$

$$T_\varepsilon u = u \qquad \text{a.e. in} \quad D \cap \varepsilon E_F$$

$$\int_{D(\delta k_0)} |T_\varepsilon u|^r \, dx \leq k_1 \int_{D \cap \varepsilon E_F} |u|^r \, dx$$

$$\int_{D(\delta k_0)} |\nabla(T_\varepsilon u)|^r \, dx \leq k_2 \int_{D \cap \varepsilon E_F} |\nabla u|^r \, dx$$

where $D(\delta k_0) = \{x \in D : \text{dist}\,(x, \partial D) > \delta k_0\}$. Constants k_0, k_1 and k_2 depend on E, n and r, but they are independent of δ, D and ε.

Hence, according to the above *Extension theorem*, it is possible to extend an H^1-function defined on $\Omega \cap \varepsilon E_F$, and L-periodic, to a function $\tilde{u} \in H^1_{per}(\Omega)$ in the way that

$$\int_\Omega |\nabla \tilde{u}|^2 \, dx \leq k_0 \int_{\Omega \cap \varepsilon E_F} |\nabla u|^2 \, dx \tag{1.13}$$

Now we use Poincaré's inequality for H^1 and get

$$\int_{\Omega \cap \varepsilon E_F} |u - \frac{1}{|\Omega \cap \varepsilon E_F|} \int_{\Omega \cap \varepsilon E_F} u|^2 \, dx \leq$$

$$\int_{\Omega \cap \varepsilon E_F} |u - \frac{1}{|\Omega|} \int_\Omega \tilde{u}|^2 \, dx + |\frac{1}{|\Omega \cap \varepsilon E_F|} \int_{\Omega \cap \varepsilon E_F} \left(u - \frac{1}{|\Omega|} \int_\Omega \tilde{u} \right) dx|^2$$

$$\leq k_1 \int_\Omega |\tilde{u} - \frac{1}{|\Omega|} \int_\Omega \tilde{u}|^2 \, dx \leq c_0 \int_\Omega |\nabla \tilde{u}|^2 \, dx \leq c_0 k_0 \int_{\Omega \cap \varepsilon E_F} |\nabla u|^2 \, dx$$

The estimate (1.12) follows. ∎

Normally, we think of the elements of H^{-1} as derivatives of L^2-functions and we get informations about them through duality arguments. With this motivation, let us fix a function g, defined on $\mathbb{R}^n$, and obtained by the periodic extension from a given function from $L^2(\Omega)$, such that supp $g \subset \Omega \cap \varepsilon E_F$ and $\int_\Omega g \, dx = 0$. We have the following result

Proposition 1.4.

a) *The function g admits the representation $g = \text{div}\, F^\varepsilon$ in Ω, with*

$$F^\varepsilon \in L^2(\Omega)^n, \qquad \int_\Omega |F^\varepsilon|^2 \, dx \leq k \int_\Omega g^2 \, dx. \tag{1.14}$$

b) *Furthermore, $F^\varepsilon \in H^1_{per}(\Omega)^n$ and*

$$\int_\Omega |\nabla F^\varepsilon|^2 \, dx \leq \frac{C_1}{\varepsilon^2} \int_\Omega g^2 \, dx. \tag{1.15}$$

Proof:

a) We consider the Neumann problem

$$\Delta u^\varepsilon = g \quad \text{in } \Omega \cap \varepsilon E_F; \qquad \frac{\partial u^\varepsilon}{\partial \nu} = 0 \quad \text{on } \partial\Big(\Omega \cap \varepsilon E_F\Big)\setminus\partial\Omega, \quad u^\varepsilon \text{ is } L-\text{periodic.}$$
(1.16)

Since $g \in L^2(\Omega)$, supp $g \subset \Omega \cap \varepsilon E_F$ and $\int_\Omega g = 0$, the problem (1.16) has a solution $u^\varepsilon \in H^1(\Omega \cap \varepsilon E_F)$, unique up to a constant. Furthermore, by Lemma 1.3

$$\|\nabla u^\varepsilon\|^2_{L^2(\Omega \cap \varepsilon E_F)^n} = -\int_{\Omega \cap \varepsilon E_F} g u^\varepsilon \, dx =$$

$$-\int_{\Omega \cap \varepsilon E_F} g\Big(u^\varepsilon - \frac{1}{|\Omega \cap \varepsilon E_F|} \int_{\Omega \cap \varepsilon E_F} u^\varepsilon \, dx\Big)\, dx \le \sqrt{k}\|g\|_{L^2(\Omega)}\|\nabla u^\varepsilon\|_{L^2(\Omega \cap \varepsilon E_F)^n}$$

implying the estimate

$$\|\nabla u^\varepsilon\|_{L^2(\Omega \cap \varepsilon E_F)^n} \le \sqrt{k}\|g\|_{L^2(\Omega)} \tag{1.17}$$

We set now $F^\varepsilon = \nabla u^\varepsilon$ in $\Omega \cap \varepsilon E_F$ and 0 elsewhere. The function F^ε satisfies the estimate (1.14) and has the properties required in **a)**.

b) Contrary to **a)**, this part requires some effort. We need the following well-known result (see e.g. [Tel])

Representation of a function as divergence of a vector field. *Let D be a bounded Lipschitz domain. Let $g \in L^2(D)$ and $\int_D g \, dx = 0$. Then there is a $w \in H^1_0(D)^n$, such that $g = \operatorname{div} w$ and*

$$\|\nabla w\|_{L^2(D)^{n^2}} \le c_0\|g\|_{L^2(D)} \tag{1.18}$$

Now we consider a smooth partition of unity related to E_F, i.e. a family of functions $\{\psi_j\} \subset C^\infty(\bar{E}_F)$ such that
(i) $1 = \sum_j \psi_j(x), \qquad \forall x \in \bar{E}_F$;
(ii) $0 \le \psi_j \le 1; \qquad |\nabla \psi_j| \le M$;
(iii) for all $x \in \bar{E}_F$, the sum contains not more than N nonzero terms;
(iv) $E_F = \cup_j E^j$, and supp $\psi_j \subset \bar{E}^j$, where $E^j \subset E_F$ is a bounded Lipschitz domain.
Since E_F is obtained by the periodic repetition of Y_F, we can suppose the functions $\{\psi_j\}$ 1-periodic.

By introducing the functions $\psi_j^\varepsilon(x) = \psi_j(\frac{x}{\varepsilon})$ and and domains $E_\varepsilon^j = \varepsilon E^j$, we obtain a decomposition of unity related to εE_F.

Since E_F is obtained by the periodic repetition of Y_F, we can suppose the functions $\{\psi_j\}$ 1-periodic.

Let $g \in L^2(\Omega)$ be such that $\int_\Omega g \, dx = 0$. Then (1.14) implies

$$g = \operatorname{div} F^\varepsilon = \sum_j \operatorname{div}(\psi_j^\varepsilon F^\varepsilon) = \sum_j g_j^\varepsilon$$

Also, we note that $F^\varepsilon \cdot \nu = 0$ on $\partial(\varepsilon E_F)$ and $\psi_j^\varepsilon = 0$ on $\varepsilon E_F \cap \partial E_\varepsilon^j$. Consequently, for every j such that $E_\varepsilon^j \subset\subset \Omega$ we have

$$\int_{E_\varepsilon^j} g_j^\varepsilon \, dx = \int_{\partial E_\varepsilon^j} \psi_j^\varepsilon F^\varepsilon \cdot \nu \, dS = 0.$$

Let $\bar{g}_j^\varepsilon(y) = g_j^\varepsilon(\varepsilon y)$, $y \in E^j$. Then

$$\int_{E^j} |\bar{g}_j^\varepsilon(y)|^2 \, dy = \varepsilon^{-n} \int_{E_\varepsilon^j} |g_j^\varepsilon(x)|^2 \, dx \le$$

$$C\varepsilon^{-n} \int_{E_\varepsilon^j} \left\{ |F^\varepsilon|^2 |\nabla \psi_j^\varepsilon|^2 + |g|^2 |\psi_j^\varepsilon|^2 \right\} \le C\varepsilon^{-n-2} \int_{E_\varepsilon^j} \left\{ |F^\varepsilon|^2 + \varepsilon^2 |g|^2 \right\} \quad (1.19)$$

Now we use the surjectivity of the divergence operator (the above representation theorem from [Te1]) and there is a vector field $z^j \in H_0^1(E^j)^n$ such that

$$\operatorname{div}_y z^j = \bar{g}_j^\varepsilon \quad \text{in } E^j \quad \text{and} \quad \int_{E^j} |\nabla_y z^j|^2 \le C_0 \int_{E^j} |\bar{g}_j^\varepsilon|^2 \qquad (1.20)$$

After the local construction, we define

$$z_\varepsilon^j(x) = \varepsilon z^j\left(\frac{x}{\varepsilon}\right) \qquad \text{in } E_\varepsilon^j \qquad (1.21)$$

Then

$$\operatorname{div}_x z_\varepsilon^j = g_j^\varepsilon \qquad \text{in } E_\varepsilon^j \quad \text{and}$$

$$\int_{E_\varepsilon^j} |\nabla_x z_\varepsilon^j|^2 \, dx = \varepsilon^n \int_{E^j} |\nabla_y z^j|^2 \, dy \le C_0 \varepsilon^{-2} \int_{E_\varepsilon^j} \left\{ |F^\varepsilon|^2 + \varepsilon^2 |g|^2 \right\} \quad (1.22)$$

It remains to consider the indices j for which $\bar{E}_\varepsilon^j \cap \partial\Omega \ne \emptyset$, i.e. having a non-empty intersection with the boundary of Ω. For simplicity, let us suppose that $\bar{E}_\varepsilon^j$ neither intersect vertices nor more than one side of Ω. Then there is a minimal $k(j) \in \mathbb{N}^n$, $|k(j)| = 1$, such that $g_j^\varepsilon(x + k(j)L) = g_j^\varepsilon(x)$, $\forall x$. Then there is $l \in \mathbb{N}$ such that $(E_\varepsilon^j + k(j)L) \cap \Omega \subset E_\varepsilon^l$. We consider now a new domain $E_\varepsilon^{j,l} = (E_\varepsilon^j \cap \bar{\Omega}) \cup (E_\varepsilon^l \cap \Omega)$ and a new piece of g given by

$$\tilde{g}_j^\varepsilon = \begin{cases} g_j^\varepsilon & \text{in } E_\varepsilon^j \cap \bar{\Omega} \\ g_l^\varepsilon & \text{in } E_\varepsilon^l \cap \Omega \end{cases} \qquad (1.23)$$

Then $\int_\Omega \tilde{g}_j^\varepsilon \, dx = 0$, supp $\tilde{g}_j^\varepsilon \subset E_\varepsilon^{j,l}$ and after repeating the above argument we obtain (1.22) once more. For $\bar{E}_\varepsilon^j$ intersecting more than one side (or containing a vertex) we take, on an analogous way, the union of corresponding pieces. Since their number is always inferior or equal to 2^n, we obtain again the estimate (1.22).

In the final step we set $z^\varepsilon = \sum_j z_\varepsilon^j$. Then z^ε satisfies (1.14)-(1.15). ∎

Theorem 1.5 (a priori estimate for the pressure field). *Let v^ε be the solution for (1.6) and p^ε defined by (1.10). Then we have*

$$\|p^\varepsilon - \frac{1}{|\Omega \cap \varepsilon E_F|} \int_{\Omega \cap \varepsilon E_F} p^\varepsilon \, dx\|_{L^2(\Omega \cap \varepsilon E_F)} \le \frac{C}{\varepsilon} \|\nabla p^\varepsilon\|_{(W^\varepsilon)'} \qquad (1.24)$$

and, consequently,

$$\|p^\varepsilon - \frac{1}{|\Omega \cap \varepsilon E_F|} \int_{\Omega \cap \varepsilon E_F} p^\varepsilon \, dx\|_{L^2(\Omega \cap \varepsilon E_F)} \le C. \qquad (1.25)$$

Proof: By Proposition 1.4, we have

$$\int_{\Omega \cap \varepsilon E_F} |p^\varepsilon - \frac{1}{|\Omega \cap \varepsilon E_F|} \int_{\Omega \cap \varepsilon E_F} p^\varepsilon \, dx|^2 =$$
$$\int_{\Omega \cap \varepsilon E_F} (p^\varepsilon - \frac{1}{|\Omega \cap \varepsilon E_F|} \int_{\Omega \cap \varepsilon E_F} p^\varepsilon \, dx)p^\varepsilon = \int_{\Omega \cap \varepsilon E_F} \text{div } F^\varepsilon p^\varepsilon,$$

where F^ε and z^ε are defined by (1.20)-(1.22) with

$$g = p^\varepsilon - \frac{1}{|\Omega \cap \varepsilon E_F|} \int_{\Omega \cap \varepsilon E_F} p^\varepsilon \, dx.$$

Hence

$$\|p^\varepsilon - \frac{1}{|\Omega \cap \varepsilon E_F|} \int_{\Omega \cap \varepsilon E_F} p^\varepsilon \, dx\|^2_{L^2(\Omega \cap \varepsilon E_F)} = -\langle \nabla p^\varepsilon, F^\varepsilon \rangle_{\Omega \cap \varepsilon E_F}$$
$$\le \frac{C}{\varepsilon} \|\nabla p^\varepsilon\|_{(W^\varepsilon)'} \|p^\varepsilon - \frac{1}{|\Omega \cap \varepsilon E_F|} \int_{\Omega \cap \varepsilon E_F} p^\varepsilon \, dx\|_{L^2(\Omega \cap \varepsilon E_F)} \qquad (1.26)$$

and (1.24) follows.

(1.25) is a consequence of (1.11) and (1.26). ∎

We have already seen that, after extension by zero to the rigid part, the velocity v^ε satisfies the *a priori* estimates (1.8)-(1.9), with Ω^ε replaced by Ω. Furthermore, it would be more comfortable to work with the pressure field p^ε defined on Ω. Following the approach from [LA], we define the pressure extension $\tilde{p}^\varepsilon$ by

$$\tilde{p}^\varepsilon = \begin{cases} p^\varepsilon & \text{in} & \Omega_\varepsilon \\ \frac{1}{|Y_{F_i}^\varepsilon|} \int_{Y_{F_i}^\varepsilon} p^\varepsilon & \text{in the} & Y_{S_i}^\varepsilon & \text{for each } i \end{cases} \qquad (1.27)$$

where $Y_{F_i}^\varepsilon$ is the fluid part of the cell Y_i^ε. Note that solid part of the porous medium is a union of all $Y_{S_i}^\varepsilon$.

Then we have

Corollary 1.6 (a priori estimate for the pressure field in Ω). *Let $\tilde{p}^\varepsilon$ be defined by (1.27). Then it satisfies the estimates*

$$\|\tilde{p}^\varepsilon - \frac{1}{|\Omega|}\int_\Omega \tilde{p}^\varepsilon dx\|_{L^2(\Omega)} \le$$

$$\frac{1}{|Y_F|}\|p^\varepsilon - \frac{1}{|\Omega \cap \varepsilon E_F|}\int_{\Omega \cap \varepsilon E_F} p^\varepsilon dx\|_{L^2(\Omega \cap \varepsilon E_F)} \le C \qquad (1.28)$$

$$\|\nabla \tilde{p}^\varepsilon\|_{W'} \le C, \qquad (1.29)$$

where $W = \{z \in H^1_{per}(\Omega)^n : \int_\Omega z\, dx = 0\}$.

We conclude this subsection with a result implying the strong L^2-compactness of the family $\{\tilde{p}^\varepsilon\}$. Usually, one makes use of fine properties of Tartar's restriction operator (see [Ta1980], [ALL89] or [All97]). Here, the following weaker result is sufficient

Proposition 1.7. *There exists a continuous linear operator $R_\varepsilon \in \mathcal{L}(W, W^\varepsilon)$, such that*

$$\text{div}\,(R_\varepsilon\varphi) = \text{div}\,\varphi + \sum_i \frac{1}{|Y^\varepsilon_{F_i}|}\chi_{Y^\varepsilon_{F_i}}\int_{Y^\varepsilon_{F_i}} \text{div}\,\varphi\, dx, \qquad \forall \varphi \in W \quad (1.30)$$

$$\text{div}\,\varphi = 0 \Rightarrow \text{div}\,(R_\varepsilon\varphi) = 0, \quad \forall \varphi \in W \qquad (1.31)$$

$$\|R_\varepsilon\varphi\|_{L^2(\Omega^\varepsilon)^n} \le C\{\|\nabla u^\varepsilon\|_{L^2(\Omega^\varepsilon)^n} + \varepsilon\|\,\text{div}\,(R_\varepsilon\varphi)\|_{L^2(\Omega)}\} \quad \forall \varphi \in W \quad (1.32)$$

$$\|\nabla(R_\varepsilon\varphi)\|_{L^2(\Omega^\varepsilon)^{n^2}} \le$$

$$\frac{C}{\varepsilon}\{\|\nabla u^\varepsilon\|_{L^2(\Omega^\varepsilon)^n} + \varepsilon\|\,\text{div}\,(R_\varepsilon\varphi)\|_{L^2(\Omega)}\} \,\forall \varphi \in W, \qquad (1.33)$$

where u^ε is a solution for (1.16) with $g = \text{div}\,(R_\varepsilon\varphi)$.

Proof : We use the proof of Proposition 1.4. First, we set

$$g = \text{div}\,\varphi + \sum_i \frac{1}{|Y^\varepsilon_{F_i}|}\chi_{Y^\varepsilon_{F_i}}\int_{Y^\varepsilon_{F_i}} \text{div}\,\varphi\, dx. \qquad (1.34)$$

Then $\int_{\Omega^\varepsilon} g = 0$ and $g \in L^2(\Omega)$. Then for $\varphi \in W$, we define $R_\varepsilon\varphi$ by $R_\varepsilon\varphi = z^\varepsilon$. By construction it satisfies (1.30)-(1.31). A priori estimates (1.32)-(1.33) are direct consequences of the proof of Proposition 1.4 and (1.22). ∎

Lemma 1.8. *Let $\{\psi^\varepsilon\} \subset L^2(\Omega)$, $\int_\Omega \psi^\varepsilon = 0$, $\forall \varepsilon$, be a sequence converging weakly to 0. Then there is a subsequence of $\{\tilde{p}^\varepsilon\}$ such that*

$$\int_\Omega (\tilde{p}^\varepsilon - \frac{1}{|\Omega|}\int_\Omega \tilde{p}^\varepsilon\, dx)\psi^\varepsilon\, dx \to 0. \qquad (1.35)$$

Proof : Since the mean of ψ^ε is zero, there is a sequence $\{\varphi^\varepsilon\} \subset W$, such that div $\varphi^\varepsilon = 0$ and $\varphi^\varepsilon \to 0$ weakly in W. Then we have

$$|\int_\Omega (\tilde{p}^\varepsilon - \frac{1}{|\Omega|}\int_\Omega \tilde{p}^\varepsilon\, dx)\psi^\varepsilon\, dx| = |\int_{\Omega^\varepsilon} (p^\varepsilon - \frac{1}{|\Omega^\varepsilon|}\int_{\Omega^\varepsilon} p^\varepsilon)\psi^\varepsilon\, dx +$$

$$\sum_i (\int_{Y^\varepsilon_{S_i}} \text{div}\,\varphi^\varepsilon\, dx)(\frac{1}{|Y^\varepsilon_{F_i}|}\int_{Y^\varepsilon_{F_i}} p^\varepsilon\, dx - \frac{1}{|\Omega^\varepsilon|}\int_{\Omega^\varepsilon} p^\varepsilon\, dx)| = |\langle \nabla p^\varepsilon, R_\varepsilon\varphi^\varepsilon\rangle_{\Omega^\varepsilon}|$$

$$\le \|\nabla p^\varepsilon\|_{(W^\varepsilon)'}\|R_\varepsilon\varphi\|_{W^\varepsilon} \le C\{\|\nabla u^\varepsilon\|_{L^2(\Omega^\varepsilon)^n} + \varepsilon\|\,\text{div}\,(R_\varepsilon\varphi^\varepsilon)\|_{L^2(\Omega)}\}, (1.36)$$

where u^ε is a solution, unique up to a constant, of the problem (1.16) with $g^\varepsilon = \mathrm{div}\,(R_\varepsilon \varphi^\varepsilon)$. Since $\chi_{\Omega^\varepsilon} g^\varepsilon \to 0$ weakly in $L^2(\Omega)$, we easily conclude that $\chi_{\Omega^\varepsilon} \nabla u^\varepsilon \to 0$ in $L^2(\Omega)^n$ strongly and (1.35) follows. ∎

For the convinience of the reader we now repeat the following elementary result from the theory of Banach spaces

Characterisation of the strong convergence in the reflexive Banach spaces. *Let X be a reflexive Banach space and X' its dual. Let $\{x_m\} \subset X'$ be a sequence weakly converging towards $x \in X'$ and let*

$$\langle x_m, \zeta_m \rangle_{X',X} \to \langle x, \zeta \rangle_{X',X}, \quad \forall \{\zeta_m\} \subset X,$$

converging weakly in X towards $\zeta \in X$. Then $x_m \to x$ in X' strongly.

The above result and Lemma 1.8 imply the following result

Corollary 1.9. *Sequences $\{\nabla \tilde{p}^\varepsilon\}$ and $\{\tilde{p}^\varepsilon - \frac{1}{|\Omega|} \int_\Omega \tilde{p}^\varepsilon\}$ are strongly relatively compact in W' and $L^2(\Omega)$, respectively.*

1.5 Formal multiscale expansion

We have established a priori estimates for the velocity and the pressure. As a consequence of (1.8), (1.9) and (1.28), we postulate the following asymptotic expansion

$$v^\varepsilon(x) = \varepsilon^2 v^0(x,y) + \varepsilon^3 v^1(x,y) + \dots, \qquad y = \frac{x}{\varepsilon} \tag{1.37}$$

$$p^\varepsilon(x) = p^0(x,y) + \varepsilon p^1(x,y) + \dots, \qquad y = \frac{x}{\varepsilon}. \tag{1.38}$$

This expansion takes care of the disparity of the two length scales in the problem. Furthermore, since the geometry is periodic, it is natural to suppose a **periodic** dependence on the fast scale y.

Having two scales we should transform the derivatives. We have

$$\nabla = \nabla_x + \frac{1}{\varepsilon}\nabla_y; \quad \mathrm{div} = \mathrm{div}_x + \frac{1}{\varepsilon}\,\mathrm{div}_y \quad \text{and} \quad \Delta = \Delta_x + \frac{2}{\varepsilon}\,\mathrm{div}_x \nabla_y + \frac{1}{\varepsilon^2}\Delta_y,$$

where the subscript indicates the variable involved in the differentiation.

Substituting the expansions (1.37)-(1.38) into (1.3)-(1.5), the following equations are obtained:

$O(\varepsilon^{-1})$:

$$\nabla_y p^0(x,y) = 0 \quad \text{in} \quad \Omega \times Y_F \tag{1.39}$$

$O(1)$:

$$-\mu\Delta_y v^0(x,y) + \nabla_y p^1(x,y) + \nabla_x p^0(x,y) = f(x) \quad \text{in} \quad \Omega \times Y_F \tag{1.40}$$

$O(\varepsilon)$:

$$\text{div}_y v^0(x,y) = 0 \quad \text{in} \quad \Omega \times Y_F \tag{1.41}$$
$$-\mu\Delta_y v^1(x,y) + \nabla_y p^2(x,y) + \nabla_x p^1(x,y) = 0 \quad \text{in} \quad \Omega \times Y_F \tag{1.42}$$

$O(\varepsilon^2)$:

$$\text{div}_x v^0(x,y) + \text{div}_y v^1(x,y) = 0 \quad \text{in} \quad \Omega \times Y_F \tag{1.43}$$
$$v^0(x,y) = 0 \text{ on } \Omega \times (\partial Y_F \setminus \partial Y); \tag{1.44}$$
$$\{v^0(x,y), p^1(x,y)\} \text{ is } 1 - \text{periodic in } y, \tag{1.45}$$
$$-\mu\Delta_y v^2(x,y) + \nabla_y p^3(x,y) + \nabla_x p^2(x,y) = 0 \quad \text{in} \quad \Omega \times Y_F \tag{1.46}$$

$O(\varepsilon^3)$: ...

We are interested only in the lowest order approximation, corresponding to the homogenized problem. First, we note that (1.39) is equivalent to

$$p^0 = p^0(x),$$

i.e. the zeroth order approximation of the pressure doesn't depend on y . In the next step we repeat the classical lemma on surjectivity of the operator div:

Surjectivity of *div* **in** Y_F. Let $\Phi \in L^2(Y_F)$. Then there is $\varphi \in H^1(Y_F)^n$, such that

$$\begin{cases} \text{div}_y \varphi = \Phi & \text{in} & Y_F \\ \varphi = 0 & \text{on } \partial Y_F \setminus \partial Y, \quad \varphi \text{ is } 1 - \text{periodic in } y \end{cases} \tag{1.47}$$

if and only if

$$\int_{Y_F} \Phi \, dy = 0$$

Hence, there is v^1 satisfying (1.43) if and only if

$$\text{div}_x \int_{Y_F} v^0(x,y) \, dy = 0 \quad \text{in} \quad \Omega. \tag{1.48}$$

We can now summarize the conditions that $\{v^0, p^0, p^1\}$ should satisfy in $\Omega \times Y_F$. We take (1.39), (1.40), (1.41), (1.44) and (1.48) and obtain the following problem

$$\begin{cases} -\mu\Delta_y v^0(x,y) + \nabla_y p^1(x,y) + \nabla_x p^0(x) = f(x) & \text{in} \quad \Omega \times Y_F \\ \text{div}_y v^0(x,y) = 0 & \text{in} \quad \Omega \times Y_F \\ v^0(x,y) = 0 & \text{on } \Omega \times (\partial Y_F \setminus \partial Y) \quad (1.49) \\ \{v^0(x,y), p^1(x,y)\} & \text{is } 1 - \text{periodic in } y \\ \text{div}_x \int_{Y_F} v^0(x,y) \, dy = 0 & \text{in} \quad \Omega \end{cases}$$

The quantity $q^0(x) = \int_{Y_F} v^0(x,y)\, dy$ is the **seepage velocity** (or the specific discharge) for the filtration through the porous medium Ω. $|Y_F|$ is the **porosity** of Ω and the average velocity is the seepage velocity divided by the porosity.

We still miss another boundary condition in x-variable. It is reasonable to impose

$$\{p^0, \int_{Y_F} v^0\, dy\} \quad \text{is} \quad L - \text{periodic}. \tag{1.50}$$

In general fixing the boundary conditions at the outer boundary $\partial\Omega$ is a difficult problem. The early reference on this problem is [LiPe], where boundary layers for Laplace's operator in a perforated domain were constructed. The exhaustive references for the boundary conditions satisfied by the velocity and the pressure, on the flat interface between a porous medium and a non-perforated domain, are the papers [JaMi1], [JaMi2], [JaMi3] and [JM5]. The detailed study of the boundary layers for the homogenization of the Stokes flow in a porous medium, with a general boundary, is in [MPM]. In this introductory paper we don't discuss those questions and this is the reason why we have choosen the periodic conditions at the outer boundary, for the ε-problem.

System (1.49)-(1.50) is called the **Stokes system with two pressures**. We are going to show that it has a unique solution $\{v^0, p^0, p^1\}$ in an appropriate functional space. Then it is natural to consider it as the homogenized problem corresponding to (1.3)-(1.5) and we shall justify the approximation in the next subsection. Furthermore, one should find a relationship between (1.49)-(1.50) and Darcy's law in theories of groundwater flows, stating that the seepage velocity q^0 is proportional to the pressure gradient $\nabla_x p^0$. We start with the study of the problem (1.49)-(1.50). It was already studied in [LiPe] and we give details for the convinience of the reader.

We start with **elimination of two pressures**. In order to write a variational formulation for (1.49)-(1.50), we introduce the following Hilbert spaces

$$\mathcal{W}(Y_F) = \{\varphi \in H^1(Y_F)^n : \ \text{div}_y\varphi = 0 \ \text{in} \ Y_F,$$
$$\varphi = 0 \ \text{on} \ \partial Y_F \setminus \partial Y, \ \varphi \ \text{is} \ 1 - \text{periodic}\} \tag{1.51}$$

$$\mathcal{V} = \{v \in L^2(\Omega; \mathcal{W}(Y_F)); \quad \text{div}_x \int_{Y_F} v(x,y)\, dy = 0 \ \text{in} \ \Omega,$$

$$\nu(x) \int_{Y_F} v(x,y)\, dy \ \text{is} \ H^{-1/2} - \text{antiperiodic}\, \}, \tag{1.52}$$

where ν is the unit outward normal on Ω. Also we need the corresponding bilinear and linear forms

$$a_{Y_F}(u,v) = \int_{Y_F} \nabla_y u \nabla_y v\, dy, \quad u,v \in \mathcal{W}(Y_F) \tag{1.53}$$

$$a_\Omega(u,v) = \int_\Omega a_{Y_F}(u,v)\, dx, \quad u,v \in \mathcal{V} \tag{1.54}$$

$$(f,v)_{\Omega \times Y_F} = \int_\Omega \int_{Y_F} fv\, dx\, dy \tag{1.55}$$

We have now

Theorem 1.10 (Solvability of the homogenized problem). a) *Problem*

$$
\begin{cases}
\text{Find } v^0 \in \mathcal{V} \quad \text{such that} \\
a_\Omega(v^0, \varphi) = (f, \varphi)_{\Omega \times Y_F}, \quad \forall \varphi \in \mathcal{V},
\end{cases}
\tag{1.56}
$$

admits a unique solution.

b) *Let* $v^0 \in \mathcal{V}$ *be the solution for (1.56). Then there exist unique* $p^0 \in H^1_{per}(\Omega)$, $\int_\Omega p^0(x)\, dx = 0$ *and* $p^1 \in L^2(\Omega \times Y_F)$, $\int_{Y_F} \int_\Omega p^1(x,y)\, dx dy = 0$ *such that (1.49)-(1.50) holds in the sense of distributions.*

Proof : a) Obviously, $a_\Omega(\psi, \varphi)$ defines a scalar product and a norm on $\mathcal{V}$. $\mathcal{V}$ is a Hilbert space with the respect to the scalar product induced by a_Ω. Consequently, Lax-Milgram's lemma implies the unique solvability of (1.56).

b) Here we are going to show that an interpretation of the problem (1.56) gives p^0 and p^1, as the interpretation of the weak solution of the Stokes system gives the pressure field. Construction of p^0 and p^1 by the penalty method is given in [LiPe]. However, since $f = f(x)$ we can do it directly.

Let $\Upsilon = \{\varphi \in C^\infty_{per}(\Omega)^n : \text{div } \varphi = 0 \text{ in } \Omega\}$ and $H = $ closure of Υ in $L^2(\Omega)^n$. Then $H = \{z \in L^2(\Omega)^n : \text{div } z = 0,\ z\nu \text{ is } L - \text{antiperiodic}\}$ and $H^\perp = \{z \in L^2(\Omega)^n : z = \nabla w,\ w \in H^1_{per}(\Omega)\}$.

We consider the following auxiliary problems in Y_F:

$$
\begin{cases}
\text{For } 1 \leq i \leq n, \quad \text{find } \{w^i, \pi^i\} \in H^1_{per}(Y_F)^n \times L^2(Y_F) \quad \text{such that} \\
\qquad\qquad -\Delta_y w^i(y) + \nabla_y \pi^i(y) = e^i \quad \text{in } Y_F \\
\qquad\qquad\qquad\qquad \text{div}_y w^i(y) = 0 \quad \text{in } Y_F \\
\qquad\qquad\qquad\qquad\qquad w^i(y) = 0 \quad \text{on } (\partial Y_F \setminus \partial Y) \\
\qquad\qquad\qquad\qquad \int_{Y_F} \pi^i(y)\, dy = 0
\end{cases}
\tag{1.57}
$$

Obviously, these problems always admit a unique solution. Let us introduce the **permeability matrix** K by

$$
K_{ij} = \int_{Y_F} \nabla_y w^i \nabla_y w^j \, dy = \int_{Y_F} w^i_j \, dy, \ 1 \leq i, j \leq n.
\tag{1.58}
$$

Permeability tensor K is symmetric and positive definite. Consequently, the **drag tensor** K^{-1} is also positive definite. Now, let $\theta \in L^2(\Omega)^n$. Folowing [All92] we set $\Phi(x,y) = \sum_{i=1}^n (K^{-1}\theta, e^i) w^i$. Then $\Phi \in L^2(\Omega; W(Y_F))$ and satisfies

$$
\int_{Y_F} \Phi(x,y)\, dy = \theta(x) \text{ in } \Omega, \quad \text{and } \|\Phi\|_{L^2(\Omega; W(Y_F))} \leq C\|\theta\|_{L^2(\Omega)^n}.
\tag{1.59}
$$

Hence, the mean value over Y_F is a bounded and surjective operator between $L^2(\Omega; W(Y_F))$ and $L^2(\Omega)^n$.

Now, let $\theta \in H$. Then there is a $\varphi_\theta \in \mathcal{V}$ such that $\int_{Y_F} \varphi_\theta \, dy = \theta$ and

$$
a_\Omega(v^0, \varphi_\theta) = (f, \varphi_\theta)_{\Omega \times Y_F} = \int_\Omega f(x)\left(\int_{Y_F} \varphi_\theta \, dy\right) dx = \int_\Omega f(x)\theta(x)\, dx.
$$

Therefore, there is $p^0 \in H^1_{per}(\Omega)$, $\int_\Omega p^0 \, dx = 0$ such that $\forall \theta \in L^2(\Omega)^n$ we have

$$a_\Omega(v^0, \varphi_\theta) = \int_\Omega (f(x) - \nabla_x p^0(x)) \Big(\int_{Y_F} \varphi_\theta \, dy \Big) \, dx = \int_\Omega (f(x) - \nabla_x p^0(x)) \theta(x) \, dx. \tag{1.60}$$

Condition $\int_\Omega p^0 \, dx = 0$ implies uniqueness of p^0.

(1.60) is equivalent to

$$a_{Y_F}(v^0(x,.), \varphi) = (f(x) - \nabla_x p^0(x)) \int_{Y_F} \varphi \, dy, \qquad \forall \varphi \in W(Y_F), \text{ (a.e.) on } \Omega. \tag{1.61}$$

Existence of a unique $p^1 \in L^2(\Omega \times Y_F)$, $\int_{Y_F} \int_\Omega p^1(x,y) \, dxdy = 0$, such that (1.49)-(1.50) holds in the sense of distributions, is now classic and we refer to the classic textbook [Te1]. ∎

Corollary 1.11 (Darcy's law). *We have*

$$v^0(x,y) = \frac{1}{\mu} \sum_{j=1}^n w^j(y)(f_j(x) - \frac{\partial p^0(x)}{\partial x_j}) \qquad x \in \Omega, \, y \in Y_F \tag{1.62}$$

$$p^1(x,y) = \sum_{j=1}^n \pi^j(y)(f_j(x) - \frac{\partial p^0(x)}{\partial x_j}) \qquad x \in \Omega, \, y \in Y_F \tag{1.63}$$

$$q^0(x) = \int_{Y_F} v^0(x,y) \, dy = \frac{K}{\mu}(f(x) - \nabla_x p^0(x)), \text{ (\textbf{Darcy's law})}, \tag{1.64}$$

where $\{w^j, \pi^j\}$ are given by (1.57) and the permeability matrix K by (1.58).
Furthermore, if $f \in H^k_{per}(\Omega)^n$, $k \in \mathbb{N}$, then $p^0 \in H^{k+1}_{per}(\Omega)^n$.

Proof : Pressure field p^0 is uniquely defined by

$$\begin{cases} \text{Find } p^0 \in H^1_{per}(\Omega), \, \int_\Omega p^0 \, dx = 0, \quad \text{such that} \\ \text{div}_x \{ \dfrac{K}{\mu}(f(x) - \nabla_x p^0(x)) \} = 0 \quad \text{in } \Omega. \end{cases} \tag{1.65}$$

Then v^0 and p^1, given by (1.62) and (1.63), respectively, satisfy the system (1.49)-(1.50). By uniqueness, they are the solutions. ∎

1.6 Review of results about the two-scale convergence

We recall properties of the two-scale convergence in L^q-spaces, $1 < q < \infty$, which will be crucial for our proof of convergence. For the proofs we refer mainly to the article [All92] and to the Appendix A of the book [Horn 1997]. We start with the definition of the two-scale convergence, introduced in [Ng89] :

Definition 1.11. *The sequence $\{w^\varepsilon\} \subset L^q(\Omega)$, $1 < q < +\infty$ is said to two-scale converge in L^q to a limit $w \in L^q(\Omega \times Y)$ if for any $\sigma \in C_0^\infty(\Omega; C_{per}^\infty(Y))$ ("per" denotes Y-periodicity) one has*

$$\lim_{\varepsilon \to 0} \int_\Omega w^\varepsilon(x) \sigma\left(x, \frac{x}{\varepsilon}\right) \, dx = \int_\Omega \int_Y w(x,y)\sigma(x,y) \, dy \, dx \tag{1.66}$$

Next, we give various useful properties of two-scale convergence in L^q-setting.

Proposition 1.12.

(i). *From each bounded sequence in $L^q(\Omega)$ one can extract a subsequence which two-scale converges in L^q to a limit $w \in L^q(\Omega \times Y)$.*

(ii). *Let w^ε and $\varepsilon\nabla w^\varepsilon$ be bounded sequences in $L^q(\Omega)$. Then there exists a function $w \in L^q(\Omega; W^{1,q}_{per}(Y))$ and a subsequence such that both w^ε and $\varepsilon\nabla w^\varepsilon$ two-scale converge in L^q to w and $\nabla_y w$, respectively.*

(iii). *Let w^ε two-scale converge in L^q to $w \in L^q(\Omega \times Y)$. Then w^ε converges weakly in L^q to $\int_Y w(x,y)\,dy$.*

After recalling these basic properties we give a sequential lower semicontinuity result for two-scale convergence in L^q. Let Φ be a continuous functional on $I\!\!R^n$ satisfying

$$0 \le \Phi(\xi) \le C[1 + |\xi|^q], \qquad \forall \xi \in I\!\!R^n. \tag{1.67}$$

Proposition 1.13. *Let Φ be a convex function on $I\!\!R^n$ satisfying (1.67). Let w^ε be a bounded sequence from $L^q(\Omega)^n$ which two-scale converges in L^q to $w \in L^q(\Omega \times Y)^n$. Then*

$$\lim_{\varepsilon \to 0} \inf \int_\Omega \Phi(v^\varepsilon)\,dx \ge \int_\Omega \int_Y \Phi(v)\,dy\,dx. \tag{1.68}$$

Proof. The proof of the proposition is given in [Bing] . In somewhat different form it is also given in [All92]. ∎

Proposition 1.14.

(i). *Let $\sigma \in L^\infty_{per}(Y), \sigma^\varepsilon = \sigma(x/\varepsilon)$ and let a sequence $\{w^\varepsilon\} \subset L^q(\Omega)$ two-scale converge in L^q to a limit $w \in L^q(\Omega \times Y)$. Then $\sigma^\varepsilon w^\varepsilon$ two-scale converges in L^q to the limit σw.*

(ii). *Let Φ and Φ_1 be continuous functionals defined on $I\!\!R^n$ and $I\!\!R^{n^2}$, respectively, and let $w \in C_0^\infty\big(\Omega; C^\infty_{per}(Y)\big)^n$. Then*

$$\int_\Omega \Phi(w^\varepsilon)\,dx \to \int_\Omega \int_Y \Phi(w)\,dy\,dx \text{ and } \int_\Omega \Phi(\varepsilon\nabla w^\varepsilon)\,dx \to \int_\Omega \int_Y \Phi(\nabla_y w)\,dy\,dx$$

as $\varepsilon \to 0$, where $w^\varepsilon(x) = w(x, \frac{x}{\varepsilon})$.

We conclude this subsection by noting that a generalization of the above results to the random case was developped in [BMW].

1.7 Convergence of the homogenization process

In this subsection we prove the two-scale convergence in L^2 of the sequence $\{v^\varepsilon\}$ towards v^0 and the strong L^2-convergence of $\tilde{p}^\varepsilon$ towards p^0. We shall immediately see that it follows the formal two-scale expansion from §1.5 and in fact the main difficulties were overcomed in §1.4. For simplicity, we choose $\int_{\Omega^\varepsilon} p^\varepsilon = 0$. Then we have

Theorem 1.15 (Convergence theorem). *Let $\{v^\varepsilon, p^\varepsilon\}$ be defined by (1.3)-(1.5) and $\int_{\Omega^\varepsilon} p^\varepsilon = 0$. We suppose v^ε extended by zero to $\Omega \setminus \Omega^\varepsilon$ and the pressure extension $\tilde{p}^\varepsilon$ given by (1.27). Let v^0 and p^0 be given by (1.49)-(1.50). Then*

$$\frac{v^\varepsilon}{\varepsilon^2} \to v^0(x, y) \qquad \text{in the two-scale sense in } L^2 \qquad (1.69)$$

$$\nabla \frac{v^\varepsilon}{\varepsilon} \to \nabla_y v^0(x, y) \qquad \text{in the two-scale sense in } L^2 \qquad (1.70)$$

$$\tilde{p}^\varepsilon \to p^0(x) \qquad \text{in } L^2(\Omega) \qquad (1.71)$$

Proof : A priori estimates (1.8)-(1.9) for the velocity field and the two-scale compactness result from Proposition 1.12 imply that there is $u^0 \in L^2(\Omega; H^1(Y)^n)$ and a subsequence of $\{v^\varepsilon\}$, denoted by the same superscript, such that

$$\frac{v^\varepsilon}{\varepsilon^2} \to u^0(x, y) \qquad \text{in the two-scale sense in } L^2 \qquad (1.72)$$

$$\nabla \frac{v^\varepsilon}{\varepsilon} \to \nabla_y u^0(x, y) \qquad \text{in the two-scale sense in } L^2. \qquad (1.73)$$

Furthermore, the *a priori* estimates (1.28)-(1.29) and Corollary 1.9, imply that there is a $\mathcal{P} \in L^2(\Omega)$, $\int_\Omega \mathcal{P}\, dx = 0$, and a subsequence of $\{\tilde{p}^\varepsilon\}$ denoted by the same superscript, such that

$$\tilde{p}^\varepsilon \to \mathcal{P}(x) \qquad \text{in } L^2(\Omega). \qquad (1.74)$$

We are looking for equations satisfied by $\{u^0, \mathcal{P}\}$. First we should establish that $u^0 \in V$. In order to prove it, we take $\varphi \in C_0^\infty(\Omega; C_{per}^\infty(Y)^n)$, $\varphi(x, y) = 0$ $\forall y \in Y_F$. Let $\varphi^\varepsilon(x) = \varphi(x, x/\varepsilon)$. Then

$$0 = \int_\Omega \frac{v^\varepsilon}{\varepsilon^2} \varphi^\varepsilon \, dx \to \int_\Omega \int_{Y_s} u^0(x, y)\varphi(x, y) \, dxdy = 0, \qquad \text{as } \varepsilon \to 0,$$

and $u^0 = 0$ (a.e.) on $\Omega \times Y_s$. Next, for $\zeta \in C_0^\infty(\Omega; C_{per}^\infty(Y))$, we have

$$0 = \int_\Omega \operatorname{div} v^\varepsilon \frac{\zeta^\varepsilon}{\varepsilon} \, dx = - \int_\Omega \frac{v^\varepsilon}{\varepsilon^2} \left(\varepsilon \nabla_x \zeta^\varepsilon + \nabla_y \zeta^\varepsilon \right) dx \to$$

$$- \int_\Omega \int_Y u^0(x, y)\nabla_y \zeta(x, y) \, dxdy = 0$$

and $u^0 \in L^2(\Omega; \mathcal{W}(Y_F))$. It remains to establish the properties of $\int_{Y_F} u^0(x, y) \, dy$. Let $\varphi \in C_{per}^\infty(\Omega)$. Then

$$0 = \int_\Omega \operatorname{div} \frac{v^\varepsilon}{\varepsilon^2} \varphi \, dx = - \int_\Omega \frac{v^\varepsilon}{\varepsilon^2} \nabla\varphi \, dx \to - \int_\Omega \int_Y u^0(x, y)\nabla\varphi(x) \, dxdy = 0$$

implying $\operatorname{div}_x \int_{Y_F} u^0(x, y) \, dy = 0$ in Ω and $\nu(x) \int_{Y_F} u^0(x, y) \, dy$ is $H^{-1/2}$-periodic. Therefore, $u^0 \in V$.

Now, let $\psi \in C_0^\infty(\Omega; C_{per}^\infty(\bar{Y}_F)^n)$, $\text{div}_y \psi = 0$, $\psi = 0$ for $y \in \partial Y_F \setminus \partial Y$. Let $\psi^\varepsilon(x) = \psi(x, x/\varepsilon)$. We test the variational formulation (1.6) by ψ^ε and get

$$\int_\Omega \nabla \frac{v^\varepsilon}{\varepsilon} \left(\varepsilon \nabla_x \psi^\varepsilon + \nabla_y \psi^\varepsilon\right) dx - \int_\Omega \chi_{\Omega^\varepsilon} \tilde{p}^\varepsilon \, \text{div}_x \psi^\varepsilon \, dx = \int_{\Omega^\varepsilon} f \psi^\varepsilon \, dx. \quad (1.75)$$

Using definition of the two-scale convergence and (1.72)-(1.75), we obtain

$$\int_\Omega \int_{Y_F} \nabla_y u^0(x, y) \nabla_y \psi(x, y) \, dx dy - \int_\Omega \int_{Y_F} P(x) \, \text{div}_x \psi(x, y) \, dx dy =$$

$$\int_\Omega \int_{Y_F} f(x) \psi(x, y) \, dx dy, \quad \forall \psi \in C_0^\infty(\Omega; \mathcal{W}(Y_F)). \quad (1.76)$$

Now surjectivity of the mean over Y_F and (1.76) imply $P \in H_{per}^1(\Omega)$. (1.76) implies that u^0 satisfies (1.56). Hence $u^0 = v^0$. Finally, as in the proof of Theorem 1.10, we get $p^0 = P$. The uniqueness implies convergence of the whole sequence. ∎

1.8 Corrector for the velocity field and concluding remarks

We have established in § 1.7 the two-scale convergence of $\{v^\varepsilon/\varepsilon^2\}$ towards v^0. This is stronger than the weak convergence of $\{v^\varepsilon/\varepsilon^2\}$ towards $q^0 = \int_{Y_F} v^0 \, dy$, but weaker than the strong convergence. Since we know the form of oscillations of the highest amplitude in v^ε (see (1.62)), it is natural to get more. The first proof of the strong convergence of $r^\varepsilon = (v^\varepsilon/\varepsilon^2 - v^0(x, x/\varepsilon)) \to 0$ in L^2 was given by G. Allaire in [All91]. Here we give a stronger result : we prove that the L^2-norm of r^ε is bounded as $C\varepsilon$. The idea is to estimate it using the Stokes equations with appropriate forces. However, the presence of *bad* incompressibility effects in the velocity $v^0(x, x/\varepsilon)$ forces us to add a corrector for $\text{div}\, v^0$ as in [JaMi2], [JaMi1] and in [MPM]. Consequently, we establish an error estimate in

$$H_{per}(\Omega, \text{div}) = \{z \in L^2(\Omega)^n \mid \text{div}\, z = 0 \text{ in } \Omega, \ z \cdot \nu \text{ is } \Omega \text{ anti- periodic}\}$$

for velocities.

Now we suppose $f \in H_{per}^1(\Omega)^n$ and set

$$\begin{cases} v^{0,\varepsilon}(x) &= v^0(x, x/\varepsilon) & x \in \Omega^\varepsilon \\ v^{0,\varepsilon}(x) &= 0 & x \in \Omega \setminus \Omega^\varepsilon \\ p^{1,\varepsilon}(x) &= p^1(x, x/\varepsilon) & x \in \Omega^\varepsilon. \end{cases} \quad (1.77)$$

As mentioned before $\text{div}\, v^{0,\varepsilon}$ introduces important compressibility effects. We follow [JaMi1] and [JaMi2] and introduce the following auxiliary problem :

$$\begin{cases} \text{Find } Q \in H_{per}^1(\Omega; H_{per}^2(Y_F)^n) & \text{such that} \\ \text{div}_y Q = \text{div}_x v^0(x, y) & \text{in } \Omega \times Y_F \\ Q = 0 & \text{on } \Omega \times \partial Y_F \setminus \partial Y. \end{cases} \quad (1.78)$$

Owing to the regularity of f, (1.62), the equation

$$\int_{Y_F} \mathrm{div}_x v^0(x,y)dy = 0$$

and to the assumptions on the geometry, we are in situation to use the results of Bogovski (see e.g. [BorSo]) and get existence of at least one function $Q \in H^1_{per}(\Omega; H^1_{per}(Y_F)^n)$, satisfying (1.78).

Now we set $L^2_0(\Omega) = \{z \in L^2(\Omega) : \int_\Omega z\, dx = 0\}$ and

$$Q^\varepsilon(x) = \begin{cases} \varepsilon Q(x, x/\varepsilon), & x \in \Omega^\varepsilon \\ 0, & x \in \Omega \backslash \Omega^\varepsilon \end{cases} \tag{1.79}$$

and state the result :

Theorem 1.16. *Let $f \in H^1_{per}(\Omega)^n$, $n = 2, 3$ and $\mathrm{div}\, f = 0$. Then we have*

$$\|\mu \frac{v^\varepsilon}{\varepsilon^2} - v^{0,\varepsilon} + Q^\varepsilon\|_{H_{per}(\Omega, div)} \le C\varepsilon \tag{1.80}$$

where v^ε is a solution for (1.3)-(1.5) extended by zero to Ω. Furthermore, there exists an extension $\widetilde{\Pi}^\varepsilon$ of $\Pi^\varepsilon = p^\varepsilon - p^0 - \varepsilon p^{1,\varepsilon}$ such that

$$\|\widetilde{\Pi}^\varepsilon\|_{L^2_0(\Omega)} \le C\varepsilon, \tag{1.81}$$

In order to prove the theorem we mimic the ideas of [JaMi2] and find the equations for $v^{0,\varepsilon}$, $v^{0,\varepsilon} - Q^\varepsilon$ and $v^\varepsilon - v^{0,\varepsilon} + Q^\varepsilon$, respectively. Clearly $\{v^{0,\varepsilon}, p^{1,\varepsilon}\}$ satisfies the Stokes system

$$-\varepsilon^2 \Delta v^{0,\varepsilon} + \nabla(p^0 + \varepsilon p^{1,\varepsilon}) = f - \Psi^\varepsilon_1 \quad \text{in } \Omega^\varepsilon \tag{1.82}$$

$$\mathrm{div}\, v^{0,\varepsilon} = \mathrm{div}_x v^{0,\varepsilon} \quad \text{in } \Omega^\varepsilon \tag{1.83}$$

$$v^{0,\varepsilon} = 0 \text{ on } \partial\Omega^\varepsilon \backslash \partial\Omega \ , \ \{v^{0,\varepsilon}, \varepsilon p^{1,\varepsilon} + p^0\} \text{ is } L-\text{periodic} \tag{1.84}$$

where the subscript x denotes the differentiation only with respect to the slow variable and

$$\Psi^\varepsilon_{1,j} = \varepsilon^2 \Delta_x v^{0,\varepsilon}_j - 2\varepsilon^2 \mathrm{div}(\nabla_x v^{0,\varepsilon}_j) + \varepsilon \frac{\partial p^{1,\varepsilon}}{\partial x_j} \ . \tag{1.85}$$

It is easy to verify that

$$\|\nabla_x v^{0,\varepsilon}\|_{L^2(\Omega^\varepsilon)^{n^2}} \le C, \ \|\nabla_x p^{1,\varepsilon}\|_{L^2(\Omega^\varepsilon)^n} \le C; \ \|\nabla v^{0,\varepsilon}\|_{L^2(\Omega^\varepsilon)^{n^2}} \le \frac{C}{\varepsilon}. \tag{1.86}$$

and consequently

$$| < \Psi^\varepsilon_1, \varphi >_{\Omega^\varepsilon} | \le C\varepsilon^2 \|\nabla\varphi\|_{L^2(\Omega^\varepsilon)^{n^2}} \ , \ \forall \varphi \in W^\varepsilon. \tag{1.87}$$

At this stage we see that the tempting idea of subtracting the equations for v^ε and $v^{0,\varepsilon}$ cannot work since $\operatorname{div} v^{0,\varepsilon}$ is in the order of 1. Consequently, we derive an appropriate Stokes system for $v^{0,\varepsilon} - Q^\varepsilon$. It reads

$$-\varepsilon^2 \Delta(v^{0,\varepsilon} - Q^\varepsilon) + \nabla(p^0 + \varepsilon p^{1,\varepsilon}) = f + \Psi^\varepsilon \text{ in } \Omega^\varepsilon \tag{1.88}$$

$$\operatorname{div}(v^{0,\varepsilon} - Q^\varepsilon) = -\operatorname{div}_x Q^\varepsilon \text{ in } \Omega^\varepsilon \tag{1.89}$$

$$v^{0,\varepsilon} - Q^\varepsilon = 0 \text{ on } \partial\Omega^\varepsilon \backslash \partial\Omega, \; \{v^{0,\varepsilon} - Q^\varepsilon, \varepsilon p^{1,\varepsilon} + p^0\} \text{ is } L - \text{periodic} \tag{1.90}$$

with

$$\Psi^\varepsilon = -\Psi_1^\varepsilon + \varepsilon^2 \Delta Q^\varepsilon \text{ in } \Omega^\varepsilon. \tag{1.91}$$

In view of (1.78)-(1.79) we obtain

$$\| \operatorname{div}_x Q^\varepsilon \|_{L^2(\Omega)^n} \leq C\varepsilon, \; \|\nabla Q^\varepsilon\|_{L^2(\Omega)^{n^2}} \leq C, \; \|\nabla_x Q^\varepsilon\|_{L^2(\Omega)^{n^2}} \leq C\varepsilon \tag{1.92}$$

and consequently

$$| < \Psi^\varepsilon, \varphi >_{\Omega^\varepsilon} | \leq C\varepsilon^2 \|\nabla\varphi\|_{L^2(\Omega^\varepsilon)^{n^2}}, \; \forall\varphi \in W^\varepsilon. \tag{1.93}$$

Now we get a Stokes system in Ω^ε for $w^\varepsilon = \mu\dfrac{v^\varepsilon}{\varepsilon^2} - v^{0,\varepsilon} + Q^\varepsilon$. In view of (1.82)-(1.85) we have the following equations

$$-\varepsilon^2 \Delta w^\varepsilon + \nabla\Pi^\varepsilon = -\Psi^\varepsilon \text{ in } \Omega^\varepsilon \tag{1.94}$$

$$\operatorname{div} w^\varepsilon = \operatorname{div}_x Q^\varepsilon \text{ in } \Omega^\varepsilon \tag{1.95}$$

$$w^\varepsilon = 0 \text{ on } \partial\Omega\backslash\partial\Omega, \; \{w^\varepsilon, \Pi^\varepsilon\} \text{ is } L - \text{periodic} \tag{1.96}$$

where $\Pi^\varepsilon = p^\varepsilon - p^0 - \varepsilon p^{1,\varepsilon}$.

We observe immediately that the forcing term Ψ^ε is estimated on a satisfactory way in (1.93). It remains to estimate the pressure term using ∇w^ε.

Then, in complete analogy with (1.10)-(1.11), Theorem 1.5 and Corollary 1.6, we have

Lemma 1.17. *Assume that f satisfies the conditions from Theorem 1.16. Then there exists an extension $\widetilde{\Pi^\varepsilon}$ of Π^ε such that*

$$\|\widetilde{\Pi^\varepsilon}\|_{L_0^2(\Omega)} \leq C\varepsilon\{1 + \|\nabla w^\varepsilon\|_{L^2(\Omega^\varepsilon)^{n^2}}\}. \tag{1.97}$$

Now, the proof of Theorem 1.6 is straightforward :

Proof of Theorem 1.16. We insert w^ε as a test function in (1.94)-(1.96). Owing to (1.93) and (1.97)

$$\varepsilon^2 \int_{\Omega^\varepsilon} \|\nabla w^\varepsilon\|^2 = \int_{\Omega^\varepsilon} \widetilde{\Pi^\varepsilon} \operatorname{div} w^\varepsilon - < \Psi^\varepsilon, w^\varepsilon >_{\Omega^\varepsilon} \leq C\varepsilon^2(1 + \|\nabla w^\varepsilon\|_{L^2(\Omega^\varepsilon)^{n^2}}). \tag{1.98}$$

(1.98) implies

$$\|\nabla w^\varepsilon\|_{L^2(\Omega)^{n^2}} \leq C \tag{1.99}$$

which implies the theorem. ∎

Remark 1.18. *Correctors for the no-slip condition at the outer boundary were constructed in [MPM]. In addition to the divergence corrector (1.79), in this case one also has to construct outer boundary layers. They don't contribute to the effective filtration law, but seriously deteriorate the error estimate. In general it reads $C\varepsilon^{1/(3n)}$ and it not clear if one could get $C\sqrt{\varepsilon}$ as in the case of Laplace's operator.*

Remark 1.19. *If the rigid part of the porous medium Ω has a critical size, much smaller than $O(1)$, then it is possible to obtain* **Brinkman's law** *by homogenization. For details we refer to [All97] and references therein. Also, it is possible to obtain the effective filtration law for an* **inviscid** *flow through a porous medium. For a derivation of a filtration law for the acceleration, having the form of Darcy's law, but without viscosity and with a totally different permeability tensor, we refer to [MiPa].*

Remark 1.20. *In many applications (underwater acoustics, consolidation theory, ...), the solid part is supposed to be elastic. Then the effective filtration law is known as* **Biot's law.** *It has a much more complicated structure and it depends on the contrast of property number. There are 2 models which are frequently used in applications : the overall monophasic viscoelastic behavior and the overall diphasic behavior. In the monophasic case, the effective deformation satisfies the equations of the linear viscoelasticity, with the kernels defined by the geometry of the porous medium. The averaged pressure depends non-locally on the effective deformation and its derivatives. For a rigorous study by using the two-scale convergence and decomposition we refer to [GiMi99]. The diphasic case corresponds to the original theory of M. Biot and the filtration is described by an elliptic-parabolic system. For more details we refer to [CFGM].*

2. Non-Newtonian single phase flow through a porous medium

2.1. Introduction

Many mathematical and engineering papers study the filtration of Newtonian fluids through porous media. However, in applied problems involving production of oil and gas from underground reservoirs non-Newtonian fluids are extensively involved. Important examples are the flow of heavy oils and Enhanced Oil Recovery (EOR). In EOR applications, non-Newtonian fluids such as low concentration polymer solutions are injected to increase the viscosity of driving agents that displace oil. Similarly many heavy oils behave as Bingham viscoplastic fluids.

Flow of a non-Newtonian fluid through a porous medium can be viewed as a dynamics of a complex 2-phase mixture involving the fluid and the solid phase. Therefore establishing filtration laws for non-Newtonian flows through porous media is equivalent to finding the effective (or averaged) stress tensor for the mixture. As in the Newtonian case homogenization method enables us to simplify solving the non-Newtonian flow equations in extremly complicated geometric structures (porous media) by deriving constitutive filtration laws. As in the classic case, we keep only some overall properties of the porous medium, entering the averaged coefficients on a nonlinear way.

For the single phase Newtonian fluid flow the method of homogenization has confirmed the well-known Darcy law, giving us in addition formulas for permeability and porosity. With even the simplest non-Newtonian flow situation is different. The filtration laws are not known a priori. In the engineering literature porous media are frequently modelled as a bundle of capillary tubes. Then for an individual capillary tube a sort of "dual" functional relationship between shear stress and shear rate is usually found (see e.g. [BSL], [CMi], [SY] and references therein for more details). After assembling such "local" laws, one obtains non-linear Darcy-type laws at macro level.

These results, at engineering level of rigour, were only partially justified in recent articles. Namely, for the flow through an individual capillary tube, the laws from the engineering references are always confirmed (see e.g. [BMT], [MT] and [ClM]). To the contrary, results obtained for the periodic porous media don't agree with the laws from the engineering references. The averaged system contains a velocity depending on the fast variable and on the slow variables. We determine the velocity and the effective pressure by solving the full two-scale system of PDE's. Then the filtration velocity is the spatial average over a caracteristic pore volume of the two-scales velocity. This structural complexity comes from nonlinear viscosity and should not be seen as a drawback. We can not expect to have always simple laws, as Darcy's, describing complex natural phenomena! When flow is directional, we get nonlinear power laws for a filtration.

2.2. Equations governing the flow of a quasi-Newtonian fluid

The rigorous results concern the creeping flow of a quasi-Newtonian fluid. For quasi-Newtonian fluids, the viscosity is shear rate dependent. Three typical examples are : the "power law" or Ostwald-de Waele model, Carreau - Yasuda law and the Bingham law.

The case of the creeping stationary Bingham flow is described by the viscosity law

$$\eta_r(\xi) = \begin{cases} \infty, & |\xi| \leq \tau_0 \\ \mu_0 + \tau_0|\xi|^{-1}, & |\xi| \geq \tau_0 \end{cases} \qquad \xi \in \mathbb{R}^{n^2}, \qquad (2.1)$$

where τ_0 is the "yield stress" and $\mu_0 > 0$. The matrix norm $|\cdot|$ is defined by $|\xi|^2 = Tr(\xi\xi^*)$. It was considered in the article [LiSP] by J.L. Lions and E. Sanchez-Palencia. They gave a detailed study of the corresponding multi-scale expansion and of the homogenized equations. Convergence towards the solutions to the homogenized system is proved in [Bing].

The stationary creeping flows obeying the power law and Carreau-Yasuda's law (2) were considered in the article [BM1]. The effective equations are established in the function of Reynolds number and convergence proofs of the homogenization procedures are given. The same problem was reconsidered in the review article [MH], where a fluid injection at the boundary is supposed.

Studying only quasi-Newtonian fluids is based on the assumption that Deborah number is small. From the a priori estimates, we see that that the rate of strain tensor is much smaller in the mixture fluid/solid than in the non-Newtonian fluid itself. This observation justifies the approach but it should be remarked that in porous media with very non-uniform cross sectional areas many pseudoplastic fluids exhibit extensional viscosity effects, especially when the pressure drop is high enough. Dealing with equations of viscoelastic flows in porous media is an open mathematical problem.

In this review article our goal is to consider a nonstationary flow of the quasi-Newtonian fluid obeying the Bird-Carreau law. Namely, Carreau-Yasuda's law reads

$$\eta_r(\xi) = (\eta_0 - \eta_\infty)\left(1 + \lambda|\xi|^2\right)^{\frac{r}{2}-1} + \eta_\infty, \qquad \xi \in \mathbb{R}^{n^2}, \qquad (2.2)$$

with $\eta_0 > \eta_\infty > 0, \lambda > 0, r > 1$. η_0 is the zero-shear-rate viscosity, λ is a time constant and $(r-1)$ is a dimensionless constant describing the slope in the "power law region" of $\log \eta_r$ versus $\log (D(v))$. In most applications η_∞ is very small. Consequently, in the applications it is often set to zero. Now, the viscosity law (2.2) reads

$$\eta_r(\xi) = \eta_0(1 + \lambda|\xi|^2)^{\frac{r}{2}-1}, \qquad \xi \in \mathbb{R}^{n^2}, \qquad (2.3)$$

and it is called the law of Bird and Carreau.

For more details about the viscosity laws we refer to [BAH].

For simplicity we restrict ourselves to the case when the porous medium Ω is a square in $\mathbb{R}^2$, i.e. $\Omega = (0, L)^n, n = 2, 3$. It consists of a subdomain Ω^ε (the part filled with the fluid) and a closed set O_ε, given by $O_\varepsilon = \overline{\Omega \setminus \Omega^\varepsilon}$ (the solid part of Ω). ε denotes the typical pore size.

Our starting model is given by

$$\partial_t v^\varepsilon + (v^\varepsilon \nabla) v^\varepsilon - \text{Div} \{\eta_r(D(v^\varepsilon))D(v^\varepsilon)\} + \nabla p^\varepsilon = \varepsilon^\gamma f \text{ in } \Omega^\varepsilon \times (0,T) \quad (2.4)$$

$$\text{div } v^\varepsilon = 0 \quad \text{in} \quad \Omega^\varepsilon \times (0,T) \quad (2.5)$$

$$v^\varepsilon = 0 \quad \text{on} \quad \partial\Omega^\varepsilon \times (0,T) \quad (2.6)$$

$$v^\varepsilon(0) = 0 \quad \text{in} \quad \Omega^\varepsilon \quad (2.7)$$

$$\{v^\varepsilon, p^\varepsilon\} \quad \text{is} \quad L - \text{periodic} \quad (2.8)$$

with $2 > r > 1$ and η_r obeying (2.3). $\varepsilon^\gamma f$ stands for the effects of external forces or an injection at the boundary or a given pressure drop. $D(v^\varepsilon) = (\nabla v^\varepsilon +^\tau \nabla v^\varepsilon)/2$ is the rate-of-strain tensor and p^ε is the pressure.

Before studying the limit behavior of the above system when $\varepsilon \to 0$, we should discuss solvability of the system (2.4) -(2.8).

The mathematical difficulties with this model are well known. Usually the authors suppose the power law, but the difficulties are almost identical. The corresponding stationary problem is known to have at least one variational solution in $W^{1,r}(\Omega)$ for $2 \geq r > 3n/(n+2)$ (see e.g. [Lno]) and recently the result was extended to $2 \geq r > 2n/(n+2)$ (see [FMS]).

At the other hand, the complete mathematical theory for the non-stationary problem is developed only in the case of periodic geometries, i.e. in domains with no boundaries (see [BBN]).

In [Ci] D.Cioranescu has given a complete theory for variants of (2.4)-(2.8) with regularized viscosity laws. She has used the classical monotonicity-compactness approach from [Lno].

At present, there are no mathematical results for the non–stationary Navier-Stokes equations, with the viscosity obeying either power law or the Bird-Carreau law (2.3) with $1 < r < 2$, in the presence of boundaries. Furthermore, as it was pointed out in the classic reference Lions [Lno], estimates in time are insufficient for passing to the limit in the inertial term and the monotonicity-compactness approach doesn't work.

From the other side, even if the uniqueness and regularity questions for the 3D non-stationary incompressible Newtonian Navier-Stokes system are in general open, for "thin" domains the regularity was proved by G.Raugel and G.Sell in [RS]. Motivated by this result, in [CIM] the global existence, regularity and uniqueness results for the incompressible quasi-Newtonian flows in the thin slab, with the viscosity obeying the Bird-Carreau law (2.3), was established.

This seems to be the first result of this type and its generalization to the system (2.4)-(2.8) is straightforward. We give a sketch of the proof and uniform a priori estimates.

Let us introduce some notations which we use throughout the paper. We suppose Ω^ε defined as in §1.2 . Let

$$V_r(\Omega^\varepsilon) = \{v | v \in W^{1,r}(\Omega^\varepsilon)^n, \text{ div } v = 0 \text{ in } \Omega^\varepsilon, v = 0 \text{ on } \partial\Omega^\varepsilon, v \text{ is } L - \text{periodic}\}.$$

In order to construct Galerkin's approximation we take a basis $\{w_j\}_{j=1}^{\infty} \subset V_r(\Omega^\varepsilon)$ with $\bigcup_{n=1}^{\infty} V_n = V_r(\Omega^\varepsilon)$ where $V_n = \text{spann } \{w_1, \ldots, w_n\}$.

We denote $r' = r/(r-1)$ and $Q_T^\varepsilon = \Omega^\varepsilon \times (0,T)$.

Let us list some well known and useful inequalities.

Lemma 2.1 (Poincaré's inequality). *For $u \in V_r(\Omega^\varepsilon)$, $1 \le r < \infty$*

$$\|u\|_{L^r(\Omega^\varepsilon)^n} \le C\varepsilon \, \|\nabla v\|_{L^r(\Omega^\varepsilon)^n} \, . \tag{2.9}$$

Lemma 2.2 (Korn's inequality). *For $u \in V_r(\Omega^\varepsilon)$, $1 < r < \infty$*

$$\|\nabla u\|_{L^r(\Omega^\varepsilon)^{n \times n}} \le C \, \|D(u)\|_{L^r(\Omega^\varepsilon)^{n \times n}} \tag{2.10}$$

where C is independent of u and ε.

Let us now consider the following initial value problem.

Definition 2.3. $v^{\varepsilon,m} = \sum_{j=1}^{m} g_{mi}(t)w_j$ *is called the approximative Galerkin solution of (2.4)-(2.8) if*

$$\int_{\Omega^\varepsilon} \partial_t v^{\varepsilon,m}(t)w \, dx + \int_{\Omega^\varepsilon} \left(1 + \lambda |D(v^{\varepsilon,m})|^2\right)^{\frac{r}{2}-1} D(v^{\varepsilon,m}) : D(w) \, dx$$

$$+ \int_{\Omega^\varepsilon} (v^{\varepsilon,m}(t)\nabla) \, v^{\varepsilon,m}(t)w \, dx = \varepsilon^\gamma \int_{\Omega^\varepsilon} f(t)w \, dx \qquad \forall w \in V_m \tag{2.11}$$

$$v^{\varepsilon,m}(0) = 0. \tag{2.12}$$

From the theory of Cauchy problem for the ordinary differential equations it is well-known that there exist $t_m \in (0,T]$ and continuously differentiable functions g_{mj} on $[0, t_m)$ which satisfy (2.11)-(2.12) a.e. on $[0, t_m)$.

The global existence of the Galerkin solution is a consequence of the following proposition.

Proposition 2.4. *Let $1 < r \le 2$ and $f \in L^{r'}\left(Q_T^\varepsilon\right)^n$. Then (2.11)–(2.12) has a unique Galerkin solution on $[0,T]$ satisfying*

$$\|v^{\varepsilon,m}\|_{L^\infty(0,T;L^2(\Omega^\varepsilon)^n)} \le C\left\{\varepsilon^{\gamma+1} + \varepsilon^{\frac{r}{2}(\gamma+1)}\right\} \qquad \text{and} \tag{2.13}$$

$$\|D(v^{\varepsilon,m})\|_{L^r(0,T,L^r(\Omega^\varepsilon)^{n2})} \le C\left\{\varepsilon^{\gamma+1} + \varepsilon^{\frac{r+1}{r-1}}\right\} \tag{2.14}$$

where C is independent of m and ε.

Proof:

After having multiplied (2.11) by $v^{\varepsilon,m}$, and integrating in time, we get for all $t \in [0, t_m)$

$$\frac{1}{2}\|v^{\varepsilon,m}(t)\|_{L^2(\Omega^\varepsilon)^n}^2 + \int_0^t \int_{\Omega^\varepsilon} \left(1 + \lambda|D(v^{\varepsilon,m})|^2\right)^{\frac{r}{2}-1}|D(v^{\varepsilon,m})|^2 \, dx d\tau \le$$

$$\varepsilon^\gamma \left| \int_0^t \int_{\Omega^\varepsilon} f v^{\varepsilon,m} \, dx d\tau \right| \le C\varepsilon^{\gamma+1}\|D(v^{\varepsilon,m})\|_{L^r(0,t,L^r(\Omega^\varepsilon)^{n2})}\|f\|_{L^{r'}(Q_T^\varepsilon)^n} \tag{2.15}$$

After noting that $0 < 1 - \frac{r}{2} \leq 1$ and after the application of Jensen's inequality to the convex function $\Phi(x) = \frac{x^{\frac{2}{r}}}{1+(\lambda x)^{\frac{2-r}{r}}}$ we obtain

$$\int_0^T \int_{\Omega^\varepsilon} \left(1 + \lambda|D(v^{\varepsilon,m})|^2\right)^{\frac{r}{2}-1} |D(v^{\varepsilon,m})|^2 \, dx d\tau \geq \int_0^T \int_{\Omega^\varepsilon} \frac{|D(v^{\varepsilon,m})|^2 \, dx d\tau}{1 + \lambda^{\frac{2-r}{2}}|D(v^{\varepsilon,m})|^{2-r}}$$

$$\geq C \frac{\|D(v^{\varepsilon,m})\|^2_{L^r(Q_T^\varepsilon)^{n^2}}}{1 + \lambda^{\frac{2-r}{2}}|Q_T^\varepsilon|^{\frac{r-2}{r}} \|D(v^{\varepsilon,m})\|^{2-r}_{L^r(Q_T^\varepsilon)^{n^2}}} \tag{2.16}$$

where $|Q_T^\varepsilon|$ is the Lebesgue measure of Q_T^ε.

Using the estimates (2.15) and (2.16), we get

$$\|D(v^{\varepsilon,m})\|^2_{L^r(Q_T^\varepsilon)^{n^2}} \leq C\varepsilon^{\gamma+1}\|D(v^{\varepsilon,m})\|_{L^r(Q_T^\varepsilon)^{n^2}} \left(1 + \|D(v^{\varepsilon,m})\|^{2-r}_{L^r(Q_T^\varepsilon)^{n^2}}\right). \tag{2.17}$$

Now Young's inequality implies (2.14). Then (2.13) is an obvious consequence of (2.11), (2.14) and (2.15). ∎

The later estimate implies that the functions g_{mj} are uniformly bounded on $[0, t_m]$ and may therefore be continued to the whole interval $[0, T]$.

For passing to the limit in the convection term we need more uniform estimates in time for the Galerkin solution.

Proposition 2.5. Let $\frac{3n}{n+2} < r \leq 2$ and $\gamma+1 > \frac{r-1}{r(3-r)}(3n-(n+2)r)$, we suppose that $f \in L^\infty(Q_T^\varepsilon)^n$, $\partial_t f \in L^{r'}(Q_T^\varepsilon)^n$. Moreover we suppose that $f(0) = 0$. Then there exist ε_0 such that for $\varepsilon < \varepsilon_0$ the unique approximative Galerkin solution on $[0, T]$ from Proposition 1 satisfies

$$\|\partial_t v^{\varepsilon,m}\|_{L^\infty(0,T,L^2(\Omega^\varepsilon)^n)} \leq C\left\{\varepsilon^{\gamma+1} + \varepsilon^{(\gamma+1)r'/2}\right\} \tag{2.18}$$

$$\|D(v^{\varepsilon,m})\|_{L^\infty(0,T,L^r(\Omega^\varepsilon)^{n^2})} \leq C\left\{\varepsilon^{\gamma+1} + \varepsilon^{\frac{\gamma+1}{r-1}}\right\} \tag{2.19}$$

and $\quad \|D(\partial_t v^{\varepsilon,m})\|_{L^2(0,T,L^r(\Omega^\varepsilon)^{n^2})} \leq C\left\{\varepsilon^{\gamma+1} + \varepsilon^{\frac{\gamma+1}{r-1}}\right\} \tag{2.20}$

where C is independent of m and ε.

Proof:

The assumption $f(0) = 0$ imply $\partial_t v^{\varepsilon,m}(0) = 0$.

Now we take the derivative with respect to t of (2.11). After multiplying by $\partial_t v^{\varepsilon,m}$ we get

$$\frac{1}{2}\partial_t\|\partial_t v^{\varepsilon,m}(t)\|^2_{L^2(\Omega^\varepsilon)^n} + \int_{\Omega^\varepsilon} \left(\partial_t v^{\varepsilon,m}(t)\nabla\right)v^{\varepsilon,m}(t)\partial_t v^{\varepsilon,m}(t) \, dx +$$

$$\lambda(r-2)\int_{\Omega^\varepsilon} \left(1 + \lambda|D(v^{\varepsilon,m})|^2\right)^{\frac{r}{2}-2}\left(D(\partial_t v^{\varepsilon,m}) : D(v^{\varepsilon,m})\right)^2 \, dx +$$

$$\int_{\Omega^\varepsilon} \left(1 + \lambda|D(v^{\varepsilon,m})|^2\right)^{\frac{r}{2}-1}|D(\partial_t v^{\varepsilon,m})|^2 dx = \varepsilon^\gamma \int_{\Omega^\varepsilon} \partial_t f(t)\partial_t v^{\varepsilon,m}(t)dx \tag{2.21}$$

Firstly we note that $(r-2)$ is negative and $(r-1)$ positive. Consequently

$$(r-1)\int_{\Omega^\varepsilon}\left(1+\lambda|D(v^{\varepsilon,m})|^2\right)^{\frac{r}{2}-1}|D(\partial_t v^{\varepsilon,m})|^2\,dx\leq$$
$$\int_{\Omega^\varepsilon}\left(1+\lambda|D(v^{\varepsilon,m})|^2\right)^{\frac{r}{2}-1}|D(\partial_t v^{\varepsilon,m})|^2$$
$$+\lambda(r-2)\int_{\Omega^\varepsilon}\left(1+\lambda|D(v^{\varepsilon,m})|^2\right)^{\frac{r}{2}-2}\left(D(\partial_t v^{\varepsilon,m}):D(v^{\varepsilon,m})\right)^2\,dx$$

and the Hölder inequality gives

$$\int_{\Omega^\varepsilon}|D(\partial_t v^{\varepsilon,m})|^r dx\leq\left(\int_{\Omega^\varepsilon}\left(1+\lambda|D(v^{\varepsilon,m})|^2\right)^{\frac{r}{2}-1}|D(\partial_t v^{\varepsilon,m})|^2\,dx\right)^{r/2}$$
$$\left(\int_{\Omega^\varepsilon}\left(1+\lambda|D(v^{\varepsilon,m})|^2\right)^{\frac{r}{2}}\,dx\right)^{(2-r)/2}. \tag{2.22}$$

Now, (2.21) becomes

$$\frac{1}{2}\partial_t\|\partial_t v^{\varepsilon,m}(t)\|^2_{L^2(\Omega^\varepsilon)^n}+\frac{(r-1)\|D(\partial_t v^{\varepsilon,m})\|^2_{L^r(\Omega^\varepsilon)^{n2}}}{1+\lambda^{\frac{2-r}{2}}\|D(v^{\varepsilon,m})\|^{2-r}_{L^r(\Omega^\varepsilon)^{n2}}}\leq$$
$$\varepsilon^\gamma\left|\int_{\Omega^\varepsilon}\partial_t f(t)\partial_t v^{\varepsilon,m}(t)\,dx\right|+\left|\int_{\Omega^\varepsilon}(\partial_t v^{\varepsilon,m}(t)\nabla)v^{\varepsilon,m}(t)\partial_t v^{\varepsilon,m}(t)\,dx\right|. \tag{2.23}$$

The second step concerns the inertial term. Using the assumption $\frac{3n}{n+2}<r$, Riesz-Thorin's interpolation theorem and Poincaré's inequality, we obtain

$$\left|\int_{\Omega^\varepsilon}(\partial_t v^{\varepsilon,m}(t)\nabla)v^{\varepsilon,m}(t)\partial_t v^{\varepsilon,m}(t)\,dx\right|$$
$$\leq C_1\,\varepsilon^\theta\|D(v^{\varepsilon,m})\|_{L^r(\Omega^\varepsilon)^{n2}}\|D(\partial_t v^{\varepsilon,m})\|^2_{L^r(\Omega^\varepsilon)^{n2}}, \tag{2.24}$$

with $\theta=\frac{(n+2)r-3n}{r}$. For details we refer to [ClM].

In the third step we perform a classical estimate for the source term

$$\varepsilon^\gamma\left|\int_{\Omega^\varepsilon}\partial_t f(t)\partial_t v^{\varepsilon,m}(t)\,dx\right|\leq C\varepsilon^{\gamma+1}\|D(\partial_t v^{\varepsilon,m})\|_{L^r(\Omega^\varepsilon)^{n2}}\|\partial_t f\|_{L^{r'}(\Omega)^n}$$
$$\leq C_0\varepsilon^\delta\|D(\partial_t v^{\varepsilon,m})\|^2_{L^r(\Omega^\varepsilon)^{n2}}+C(r)\varepsilon^{2(\gamma+1)-\delta}\|\partial_t f\|^2_{L^{r'}(\Omega)^n} \tag{2.25}$$

Now (2.23), (2.24) and (2.25) give

$$\frac{1}{2}\partial_t\|\partial_t v^{\varepsilon,m}(t)\|^2_{L^2(\Omega^\varepsilon)^n}+\Lambda(t)\,\|D(\partial_t v^{\varepsilon,m}(t))\|^2_{L^r(\Omega^\varepsilon)^{n2}}$$
$$\leq C\varepsilon^{2(\gamma+1)-\delta}\|\partial_t f(t)\|^2_{L^{r'}(\Omega)^n} \tag{2.26}$$

with

$$\Lambda(t) = \frac{(r-1)}{1 + \lambda^{\frac{2-r}{2}} \|D(v^{\varepsilon,m}(t))\|_{L^r(\Omega^\varepsilon)^{n^2}}^{2-r}} - C_1\,\varepsilon^\theta \|D(v^{\varepsilon,m}(t))\|_{L^r(\Omega^\varepsilon)^{n^2}} - C_0\varepsilon^\delta.$$

$$(2.27)$$

$\delta \geq 0$ is to be determined. Let us note that

$$\Lambda(0) = r - 1 - C_0\varepsilon^\delta > 0, \quad \text{for} \quad \varepsilon \leq \varepsilon_0.$$

We deduce by continuity argument that there exists a time $t'_m \in [0,T]$ for which Λ is positive on $[0,t'_m]$. Then (2.26) gives

$$\|\partial_t v^{\varepsilon,m}(t)\|_{L^2(\Omega^\varepsilon)^n} \leq C\varepsilon^{\gamma+1-\delta/2}\left(\int_0^t \|\partial_t f(\tau)\|_{L^{r'}(\Omega)^n}^2\, d\tau\right)^{1/2} \quad \forall t \in [0,t'_m].$$

$$(2.28)$$

Then (2.11) and (2.16) give for all $t \in [0,t'_m]$

$$\frac{\|D(v^{\varepsilon,m}(t))\|_{L^r(\Omega^\varepsilon)^{n^2}}^2}{|\Omega^\varepsilon|^{\frac{2-r}{r}} + \lambda^{\frac{2-r}{2}}\|D(v^{\varepsilon,m}(t))\|_{L^r(\Omega^\varepsilon)^{n^2}}^{2-r}} \leq \|\partial_t v^{\varepsilon,m}(t)\|_{L^2(\Omega^\varepsilon)^n}\|v^{\varepsilon,m}(t)\|_{L^2(\Omega^\varepsilon)^n}$$

$$+\varepsilon^\gamma\|f(t)\|_{L^{r'}(\Omega)^n}\|v^{\varepsilon,m}(t)\|_{L^r(\Omega^\varepsilon)^n} \qquad (2.29)$$

Now after multiplying the estimate (2.29) by the denominator of the left hand side, and after estimating $\|v^{\varepsilon,m}(t)\|_{L^2(\Omega^\varepsilon)}$ by $C\varepsilon^{\frac{(n+2)r-2n}{2r}}\|D(v^{\varepsilon,m}(t))\|_{L^r(\Omega^\varepsilon)}$ we obtain

$$\|D(v^{\varepsilon,m}(t))\|_{L^r(\Omega^\varepsilon)^{n^2}} \leq C\Big\{\varepsilon^{\frac{(n+2)r-2n}{2r}}\|\partial_t v^{\varepsilon,m}(t)\|_{L^2(\Omega^\varepsilon)^n} + \varepsilon^{\gamma+1}\|f(t)\|_{L^{r'}(\Omega)^n}\Big\}$$

$$\Big\{1 + \|D(v^{\varepsilon,m}(t))\|_{L^r(\Omega^\varepsilon)^{n^2}}^{2-r}\Big\}. \qquad (2.30)$$

Due to the assumption $\frac{3n}{n+2} < r \leq 2$ and to the estimate (2.28), (2.30) implies

$$\|D(v^{\varepsilon,m}(t))\|_{L^r(\Omega^\varepsilon)^{n^2}} \leq C\varepsilon^{\gamma+1}\Big\{1 + \|D(v^{\varepsilon,m}(t))\|_{L^r(\Omega^\varepsilon)^{n^2}}^{2-r}\Big\}.$$

Now, after applying Young inequality, we obtain

$$\|D(v^{\varepsilon,m}(t))\|_{L^r(\Omega^\varepsilon)^{n^2}} \leq C\Big\{\varepsilon^{\gamma+1} + \varepsilon^{\frac{\gamma+1}{r-1}}\Big\} \quad \forall t \in [0,t'_m]. \qquad (2.31)$$

We have two different possibilities:

a) If $\gamma + 1 \geq 0$, we choose $\delta = 0$. Then, with an appropriate choice of C_0, Λ is always positive (for $\varepsilon < \varepsilon_0$). Hence (2.30) is valid on $[0,T]$ and (2.18)-(2.19) is proved.

b) If $0 > \gamma + 1 > \frac{r-1}{r(3-r)}(-3n + (n+2)r)$, then our choice is $\delta = \frac{\gamma+1}{r-1}(r-2)$. We get (2.18)-(2.19) as in a).

It remains to obtain (2.20). We start once again from (2.26) and obtain for all $t \in [0, T]$ the inequality

$$\frac{1}{2}\|\partial_t v^{\varepsilon,m}(t)\|^2_{L^2(\Omega^\varepsilon)^n} + C' \frac{1}{1+\Theta^{2-r}} \int_0^t \|D(\partial_t v^{\varepsilon,m})\|^2_{L^r(\Omega^\varepsilon)^{n^2}} d\tau$$

$$\leq C\varepsilon^{\gamma+1} \int_0^t \|\partial_t f\|_{L^{r'}(\Omega)^n} \|D(\partial_t v^{\varepsilon,m})\|_{L^r(\Omega^\varepsilon)^{n^2}} d\tau \qquad (2.32)$$

where $\Theta = \|D(v^{\varepsilon,m})\|_{L^\infty(0,T;L^r(\Omega^\varepsilon))^{n^2}}$. Then, (2.20) follows from the estimate (2.19). ∎

As a direct consequence of Proposition 1 and Proposition 2, we have the following global existence result, which gives simultaneously uniform a priori estimates with respect to ε.

Theorem 2.6. *Let $\frac{3n}{n+2} < r \leq 2$ and $\gamma + 1 > \frac{r-1}{r(3-r)}(3n - (n+2)r)$, we suppose that $f \in L^\infty(Q_T^\varepsilon)^n$, $\partial_t f \in L^{r'}(Q_T^\varepsilon)^n$. Moreover we suppose that $f(0) = 0$. Then there exists ε_0 such that for $\varepsilon < \varepsilon_0$ there is a unique solution $v^\varepsilon \in L^\infty(0,T;V_r(\Omega^\varepsilon))$, $\partial_t v^\varepsilon \in L^2(0,T;V_r(\Omega^\varepsilon)) \cap L^\infty(0,T;L^2(\Omega^\varepsilon)^n)$ of (2.11)–(2.12). It satisfies the following a priori estimates*

$$\|v^\varepsilon\|_{L^\infty(0,T;L^2(\Omega^\varepsilon)^n)} \leq C\left\{\varepsilon^{\gamma+1} + \varepsilon^{\frac{r'}{2}(\gamma+1)}\right\} \qquad (2.33)$$

$$\|D(v^\varepsilon)\|_{L^r(0,T,L^r(\Omega^\varepsilon)^{n^2})} \leq C\left\{\varepsilon^{\gamma+1} + \varepsilon^{\frac{\gamma+1}{r-1}}\right\} \qquad (2.34)$$

$$\|v^\varepsilon\|_{L^r(0,T,L^r(\Omega^\varepsilon)^{n^2})} \leq C\varepsilon\left\{\varepsilon^{\gamma+1} + \varepsilon^{\frac{\gamma+1}{r-1}}\right\} \qquad (2.35)$$

$$\|\partial_t v^\varepsilon\|_{L^\infty(0,T,L^2(\Omega^\varepsilon)^n)} \leq C\left\{\varepsilon^{\gamma+1} + \varepsilon^{(\gamma+1)r'/2}\right\} \qquad (2.36)$$

$$\|D(v^\varepsilon)\|_{L^\infty(0,T,L^r(\Omega^\varepsilon)^{n^2})} \leq C\left\{\varepsilon^{\gamma+1} + \varepsilon^{\frac{\gamma+1}{r-1}}\right\} \qquad (2.37)$$

$$\|D(\partial_t v^\varepsilon)\|_{L^2(0,T,L^r(\Omega^\varepsilon)^{n^2})} \leq C\left\{\varepsilon^{\gamma+1} + \varepsilon^{\frac{\gamma+1}{r-1}}\right\}. \qquad (2.38)$$

In the previous theorem we have derived the a priori estimates for the velocity. For deriving the a priori estimates for the pressure we use the momentum equation. For $\varphi \in V_r(\Omega^\varepsilon)$ we have

$$\int_0^T \langle \nabla p^\varepsilon, \varphi \rangle_{\Omega^\varepsilon} = -\int_0^T \int_{\Omega_\varepsilon} \left(\eta_r(D(v^\varepsilon))D(v^\varepsilon)D(\varphi) + \partial_t v^\varepsilon \varphi + (v^\varepsilon \nabla)v^\varepsilon \varphi - \varepsilon^\gamma f\varphi\right).$$
$$(2.39)$$

We use (2.39) to derive the estimates for the pressure.

Proposition 2.7. *Let η_r be given by the formula (2.3) (Bird-Carreau's law). Let $\gamma + 1 > \dfrac{1-r}{3-r}\dfrac{(n+2)r - 3n}{r}$.*
Then we have

$$\|\nabla p^\varepsilon\|_{L^{r'}(0,T;W^{-1,q}(\Omega^\varepsilon)^n)} \leq C\varepsilon^{\gamma+1}, \quad \forall q \in \left(1, \frac{n}{n-1}\right). \qquad (2.40)$$

Proof : We consider only the difficult case $\gamma + 1 < 0$. For $\gamma + 1 \geq 0$ everything is analogous but simpler.

In order to prove (2.40) we estimate the terms in (2.39). Using (2.33)-(2.38) we get

$$\left| \int_0^T \int_{\Omega^\varepsilon} \eta_r(D(v^\varepsilon))D(v^\varepsilon)D(\varphi) \right| \leq C\varepsilon^{\gamma+1}\|D(\varphi)\|_{L^r(Q_T^\varepsilon)}$$

$$\varepsilon^\gamma \left| \int_0^T \int_{\Omega^\varepsilon} f\varphi \right| \leq C\varepsilon^{\gamma+1}\|D(\varphi)\|_{L^r(Q_T^\varepsilon)}$$

$$\left| \int_0^T \int_{\Omega^\varepsilon} \partial_t v^\varepsilon \varphi \right| \leq C\varepsilon^{\frac{\gamma+1}{r-1}}\|\varphi\|_{L^r(Q_T^\varepsilon)}$$

$$\left| \int_0^T \int_{\Omega^\varepsilon} (v^\varepsilon \nabla) v^\varepsilon \varphi \right| \leq \int_0^T \varepsilon^{\frac{(n+1)r-2n}{r}} \|D(v^\varepsilon(t))\|^2_{L^r(\Omega^\varepsilon)^{n^2}}\|\varphi\|_{L^\infty(\Omega^\varepsilon)^n}$$

$$\leq C\varepsilon^{\frac{(n+1)r-2n}{r}+2\frac{\gamma+1}{r-1}}\|\varphi\|_{L^r(0,T;L^\infty(\Omega^\varepsilon)^n)}$$

Since $\gamma + 1 > \dfrac{1-r}{3-r}\dfrac{(n+2)r-3n}{r}$ implies $\dfrac{(n+1)r-2n}{r} + 2\dfrac{\gamma+1}{r-1} \geq \gamma$ and $\frac{\gamma+1}{r-1} \geq \gamma$, we deduce easily (2.40). ∎

Now we are able to extend the pressure.

Proposition 2.8. *Let all assumptions of Proposition 2.7 hold and let us suppose $\int_{\Omega^\varepsilon} p^\varepsilon\, dx = 0$. Let $\tilde{p}^\varepsilon$ be the pressure extension given by (1.27). Then we have*

$$\|\tilde{p}^\varepsilon\|_{L^{r'}(0,T;L_0^q(\Omega))} + \|\nabla\tilde{p}^\varepsilon\|_{L^{r'}(0,T;W^{-1,q}(\Omega))^n} \leq C\varepsilon^\gamma, \quad \forall q \in (1, \frac{n}{n-1}). \quad (2.41)$$

In complete analogy with Propositions 1.7 and 1.8 and Corollary 1.9, we have

Proposition 2.9. *Let $\tilde{p}^\varepsilon$ be the extension of the pressure, given by (1.27), and let $\{w^\varepsilon\} \subset L^r(0,T;L_0^{\frac{q}{q-1}}(\Omega)^n)$ be an arbitrary sequence which converges weakly to 0. Then*

$$\int_0^T \int_\Omega \tilde{p}^\varepsilon w^\varepsilon \to 0 \quad \text{as } \varepsilon \to 0,$$

and $\{\tilde{p}^\varepsilon\}$ is relatively strongly compact in $L^{r'}(0,T;L_0^q(\Omega))$, $\forall q \in (1, \frac{n}{n-1})$.

2.3. Effective filtration laws

In this section we consider the behavior of Navier-Stokes system (2.3)– (2.8) in the limit $\varepsilon \to 0$. The stationary case, with the viscosity given by the power-law and Carreau's law, was considered in [BM1]. Here the difficulty comes from the non-stationarity and from the more complicated viscosity law. However, due to the appropriate uniform a priori estimates we shall pass to the limit as in the cases considered in [BM1].

It should be noted that for $\gamma > -1$ the limit law is linear and the result coincide with the one from [BM1]. From that reason, we give here the proofs only for the case $\gamma \le -1$.

The first and the simplest possibility is $\gamma > -1$. Let us formulate the auxiliary problem.

Let $(w^i, \pi^i) \in H^1(Y_F)^n \times L_0^2(Y_F)$, $(i = 1, \ldots, n)$ be the Y-periodic solution of the problem

$$-\triangle w^i + \nabla \pi^i = e^i \text{ in } Y_F, \quad \operatorname{div} w^i = 0 \text{ in } Y_F \quad \text{and } w^i = 0 \text{ on } \partial Y_F \setminus \partial Y \quad (2.42)$$

and let

$$K = (K_{ij})_{i,j=1,2}, \quad K_{ij} = \int_{Y_F} w_j^i(y) \, dy. \quad (2.43)$$

The matrix K (permeability tensor) is symmetric and positive definite.
Let $\{v, p\}$ be the solution of the (homogenized) problem

$$\operatorname{div} v = 0 \text{ in } \Omega, \quad v = \frac{K}{\eta_0}(f(x,t) - \nabla p) \text{ in } \Omega, \quad \{p, v\} \text{ is } L - \text{periodic}, \int_\Omega p = 0. \quad (2.44)$$

Then we have

Theorem 2.10. *Let $\gamma > -1$ and let $\{v^\varepsilon, p^\varepsilon\}$ be the weak solutions of the (2.3)–(2.8) and let $\{v, p\}$ be the solution of (2.44). Let v^ε be extended by zero to Ω and let $\tilde{p}^\varepsilon$ be given by (1.27). Let $q \in (1, \frac{n}{n-1})$. Then in the limit $\varepsilon \to 0$ we have*

$$\frac{1}{\varepsilon^{\gamma+2}} v^\varepsilon \rightharpoonup v \text{ weakly in } L^r(Q_T)^n \text{ and } \varepsilon^{-\gamma} \tilde{p}^\varepsilon \to p \text{ strongly in } L^{r'}(0, T; L_0^q(\Omega)).$$

Proof of Theorem 2.10 : It is along the same lines as the proofs of Theorems 2.12 and 2.13, which follow. ■

Remark 2.11 It is important to note that Theorem 2 and (2.44) imply that for $\gamma > -1$, Bird-Carreau's law (2.3) leads to Darcy's law.

Now let us consider the case of the Bird-Carreau law (2.2) with $\gamma = -1$. First, we formulate the homogenized problem.

For $t \in [0, T]$, find

$$\{v_0(t), p(t), \pi(t)\} \in L^r(\Omega; W^{1,r}(Y_F))^n \times L_0^{r'}(\Omega) \times L^{r'}(\Omega; L_0^{r'}(Y_F)),$$

L-periodic in x and such that

$$- \operatorname{div}_y\{\eta_r(D_y(v_0))D_y(v_0)\} + \nabla_y \pi = f(x,t) - \nabla_x p \text{ in } Y_F \quad (2.45)$$

$$\operatorname{div}_y v_0(x,y) = 0 \text{ in } Y_F; \quad (v_0, \pi) \text{ is } Y - \text{periodic}, \text{ (a.e.) on } \Omega; \quad (2.46)$$

$$v_0 = 0 \text{ on } \partial Y_F \setminus \partial Y, \quad \operatorname{div}_x \int_{Y_F} v_0 \, dy = 0 \text{ in } \Omega. \quad (2.47)$$

We have the following result:

Theorem 2.12. *Problem (2.45)–(2.47) has a unique solution. Moreover, $p(t) \in W^{1,r'}(\Omega)$ for every $t \in [0,T]$.*

Proof of Theorem 2.12 : Uniqueness of solutions in $L^r(\Omega; W^{1,r}(Y_F))^n \times L_0^{r'}(\Omega) \times L^{r'}(\Omega; L_0^{r'}(Y_F))$ is a consequence of the intrinistic monotone structure of the problem (2.45)-(2.47). For more details we refer to [MH].

Furthermore, using the equation (2.45) we easily obtain $\nabla p(t) \in L^{r'}(\Omega)^n$.

∎

Theorem 2.13. *Let $\gamma = -1$, let $\{v^\varepsilon, p^\varepsilon\}$ be the weak solutions of (2.3)–(2.8) and let (v_0, p, π) be the weak solution of (2.45)–(2.47). Furthermore, let v^ε be extended by zero to Ω and let $\tilde{p}^\varepsilon$ be given by (1.7). Then in the limit $\varepsilon \to 0$ we have*

$$\varepsilon^{-1} v^\varepsilon \to v_0 \quad \text{in the two-scale sense in } L^r$$

$$\varepsilon^{-1} v^\varepsilon(t) \rightharpoonup v = \int_{Y_F} v_0 \, dy \quad \text{weakly in } L^r(\Omega)^n, \quad \text{uniformly on } [0,T]$$

$$\nabla v^\varepsilon \to \nabla_y v_0 \quad \text{in the two-scale sense in } L^r$$

$$\varepsilon \tilde{p}^\varepsilon \to p \quad \text{strongly in } L^{r'}(0,T; L_0^q(\Omega))$$

Proof of Theorem 2.13 is given in Section 2.4.

Remark 2.14. It turns out that there is a critical value of γ ($\gamma = -1$) for which the linear regime of flow (2.44) changes to a nonlinear and nonlocal one, described by system (2.45) $-$ (2.47). Linearity holds only if $r = 2$.

The last case of Bird–Carreau's law (2.3) that we will consider is with $\gamma < -1$. It exhibits an interesting behavior: the filtration law is the same as for fluids with the viscosity obeying the power law. For the homogenization of such flows in the stationary case we refer to [BM1].

In fact it shows that for such range of the parameters the flow through porous media of a polymer is governed by a nonlinear "power-like" law, no matter whether empiricism employed to describe the fluid behavior in the pores was Bird–Carreau's law or the power law.

Finally, let us describe the nonlinear law occuring in the case $\gamma < -1, 1 < r \le 2$.

We formulate the homogenized problem:

Find $\{v_0, p, \pi\} \in L^r(\Omega; W^{1,r}(Y_F))^n \times L_0^{r'}(\Omega) \times L^{r'}(\Omega; L_0^{r'}(Y_F))$, L-periodic in x and such that

$$-\text{div}_y\{\eta_0 \lambda^{\frac{r}{2}-1}|D_y(v_0)|^{r-2} D_y(v_0)\} + \nabla_y \pi = f(x,t) - \nabla_x p \text{ in } Y_F \quad (2.48)$$

$$\text{div}_y v_0 = 0 \text{ in } Y_F; \quad (v_0, \pi) \text{ is } Y - \text{ periodic, (a.e.) on } \Omega; \quad (2.49)$$

$$v_0 = 0 \text{ on } \partial Y_F \setminus \partial Y, \quad \text{div}_x \int_{Y_F} v_0(x,y) \, dy = 0 \text{ in } \Omega. \quad (2.50)$$

We have the following results:

Theorem 2.15. *Problem (2.48)–(2.50) has a unique solution. Furthermore,*
$p \in W^{1,r'}(\Omega)$.

Theorem 2.16. *Let* $n + 1 - \dfrac{4}{3-r} - \dfrac{n}{r} < \gamma < -1$. *Let* $\{v^\varepsilon, p^\varepsilon\}$ *be the weak solutions of (2.3)-(2.8) and let* $\{v_0, p, \pi\}$ *be the weak solution of problem (2.48)–(2.50). Furthermore, let* v^ε *be extended by zero and let* $\tilde{p}^\varepsilon$ *be defined by (1.7). Then in the limit* $\varepsilon \to 0$ *we have*

$$\frac{1}{\varepsilon^{\gamma+2}} v^\varepsilon \to v_0 \quad \text{in the two-scale sense in } L^r$$

$$\frac{1}{\varepsilon^{\gamma+1}} \nabla v^\varepsilon \to \nabla_y v_0 \quad \text{in the two-scale sense in } L^r$$

$$\frac{1}{\varepsilon^{\gamma+2}} v^\varepsilon(t) \rightharpoonup v = \int_{Y_F} v_0 \, dy \quad \text{weakly in } L^r(\Omega)^n, \quad \text{uniformly on } [0,T]$$

$$\varepsilon^{-\gamma} \tilde{p}^\varepsilon \to p \text{ strongly in } L^{r'}(0,T; L_0^q(\Omega)), \quad q \in (1, \frac{n}{n-1}).$$

2.4. Proof of Theorem 2.13

In this chapter we consider the situation from Theorem 4, i.e. Bird-Carreau's law with $\gamma = -1$. It is important to notice that for $\gamma = -1$ we cannot conclude the convergence $\eta_r^\varepsilon \to \eta_0$. Consequently, the filtration law differs from Darcy's .

We start with the following result which is the straightforward consequence of the results from Propositions 1.12 and 2.9 :

Proposition 2.17. *Let* $\gamma = -1$ *and let* $\{v^\varepsilon, p^\varepsilon\}$ *be the solutions for problem (2.3)–(2.8). Then there exist subsequences of* $\{v^\varepsilon\}$ *and* $\{\tilde{p}^\varepsilon\}$ *(again denoted by the same symbols) and functions* $v_0^* \in L^r(Q_T \times Y)^n$, $p^* \in L_0^{r'}(Q_T)$ *such that in the limit* $\varepsilon \to 0$

$$\varepsilon^{-1} v^\varepsilon \to v_0^* \text{ in the two-scale sense in } L^r \tag{2.51}$$

$$\nabla v^\varepsilon \to \nabla_y v_0^* \text{ in the two-scale sense in } L^r \tag{2.52}$$

$$\varepsilon^{-1} v^\varepsilon(t) \rightharpoonup v^* = \int_Y v_0^*(x,y) \, dy \text{ weakly in } L^r(\Omega)^2 \text{ uniformly on } [0,T] \tag{2.53}$$

$$\varepsilon \tilde{p}^\varepsilon \to p^* \text{ in } L^{r'}(0,T; L_0^q(\Omega)) \tag{2.54}$$

The function v^* *satisfy the equations*

$$\operatorname{div} v^* = 0 \quad \text{in } \Omega \tag{2.55}$$

The properties of the two-scale limit v_0^* are more precisely described by the following Lemma:

Lemma 2.18. We have $v_0^* \in L^r\left(Q_T; W^{1,r}(Y_F)\right)^n$, $v_0^* = 0$ on $\Omega \times (\partial Y_F \setminus \partial Y)$ and $\operatorname{div}_y v_0^* = 0$.

Proof. See [BM1]. ∎

Our next goal is to derive the law giving the relationship between v_0^* and ∇p. Let the functional Φ_r be defined by

$$\Phi_r(x) = \frac{\eta_0}{r\lambda}\left\{\left[1 + \lambda|x|^2\right]^{r/2} - 1\right\},$$

for symmetric $n \times n$ matrices x. Note that $\nabla \Phi_r(x) = \eta_r(x)x$ and Φ_r is a strictly convex proper functional defined on symmetric matrices, $1 < r < +\infty$.

Proposition 2.19. The functions v_0^* and p^* defined, respectively, by (2.51) and (2.54), respectively, satisfy the equations

$$- \operatorname{div}_y\{\eta_r(D_y(v_0^*))D_y(v_0^*)\} + \nabla_y \pi = f(x,t) - \nabla_x p^*(x,t) \text{ in } Y_F \quad (2.56)$$
$$\operatorname{div}_y v_0^* = 0 \text{ in } Y_F, \quad (v_0^*, \pi) \text{ is } Y - \text{periodic}, \quad v_0^* = 0 \text{ on } \partial Y_F \setminus \partial Y \quad (2.57)$$

Proof. We write the equation (2.4) in the following equivalent form

$$\int_0^T \int_\Omega \Phi_r(\varepsilon D(z)) - \int_\Omega \Phi_r(D(v^\varepsilon)) \geq \varepsilon \int_0^T \langle \nabla p^\varepsilon, \frac{v^\varepsilon}{\varepsilon} - z\rangle_{\Omega^\varepsilon} + \int_0^T \int_\Omega \partial_t v^\varepsilon(v^\varepsilon - \varepsilon z)$$
$$- \int_0^T \int_\Omega f(\frac{v^\varepsilon}{\varepsilon} - z) + \int_0^T \int_\Omega (v^\varepsilon \nabla)v^\varepsilon(v^\varepsilon - \varepsilon z), \quad (2.58)$$

$\forall z \in L^r(0,T; W_0^{1,r}(\Omega^\varepsilon)^n)$.

Now let $\psi(x,y) \in C_0^\infty(\Omega; C_{\text{per}}^\infty(\bar{Y}_F)^2)$, $\psi(x,y)\big|_{\partial Y_F \setminus \partial Y} = 0$, $\operatorname{div}_y \psi(x,y) = 0$. We define $\psi^\varepsilon(x) = \psi(x, \frac{x}{\varepsilon})$. Due to the estimates in the proof of the pressure estimate (2.40), we see that there will be no contribution from the material derivative of v^ε. Let $\varphi \in C^\infty[0,T]$. Then, after inserting $z = \psi^\varepsilon \varphi(t)$ in (2.58) and using Propositions 1.13 and 1.14 we get

$$\int_0^T \int_\Omega \int_Y \Phi_r(D_y(\psi\varphi)) \, dx \, dy \, dt - \int_0^T \int_\Omega \int_Y \Phi_r(D_y(v_0^*)) \, dx \, dy \, dt$$
$$\geq - \int_0^T \int_\Omega \int_Y f(x,t)(v_0^* - \psi\varphi)) + \lim_{\varepsilon \to 0} \int_0^T \varphi\langle \varepsilon \tilde{p}^\varepsilon, \operatorname{div}_x \psi^\varepsilon\rangle_{\Omega^\varepsilon}$$

as $\varepsilon \to 0$. The elementary properties of the two-scale convergence imply

$$\varepsilon \int_0^T \int_\Omega \varphi \tilde{p}^\varepsilon \operatorname{div}_x \psi^\varepsilon \to \int_0^T \int_\Omega \int_Y \varphi p^* \operatorname{div}_x \psi(x,y) \, dx \, dy \text{ as } \varepsilon \to 0$$

and, for $\forall \varphi \in C^\infty[0,T]$ we get

$$\int_0^T \langle -\operatorname{div}_y\{\eta_r(D_y(v_0^*))D_y(v_0^*)\} + \nabla_x p^* - f(x,t), \psi(y)\rangle_{X',X} \, \varphi \, dt = 0 \quad (2.59)$$

$\forall \psi \in X = \{z \in W^{1,r}(Y_F)^n : z = 0 \text{ on } \partial Y_F \setminus \partial Y, z \text{ is } Y\text{-periodic and } \operatorname{div}_y z = 0 \text{ in } Y_F\}$. It is easy to see that (2.59) implies (2.56) by the variant of de Rham's formula in a periodic setting (see Temam [Te1]). (2.57) was established in Lemma 2.18. ∎

Now we are able to prove Theorem 2.13 :

Proof of Theorem 2.13 : The convergence for the subsequence is direct consequence of the preceding Propositions. Theorem 2.12 implies that problem (2.45)–(2.47) has a unique solution and therefore the limits do not depend on subsequences, i.e. $v = v_0^*$ and $p = p^*$. ∎

2.5 Proof of Theorem 2.16

In this section we prove Theorem 2.16.

Proposition 2.20. *Let* $n + 1 - \dfrac{4}{3-r} - \dfrac{n}{r} < \gamma < -1$. *Let* $\{v^\varepsilon, p^\varepsilon\}$ *be corresponding solutions of problem (2.3)-(2.8). Then there exist subsequences of* $\{v^\varepsilon\}$ *and* $\{\tilde{p}^\varepsilon\}$ *(again denoted by the same symbols) and functions* $v_0^* \in L^r(Q_T \times Y)^2$, $p^* \in L_0^r(Q_T)$ *such that*

$$\varepsilon^{-\gamma-2} \quad v^\varepsilon \to v_0^* \text{ in the two-scale sense in } L^r \tag{2.60}$$

$$\frac{1}{\varepsilon^{\gamma+1}} \quad \nabla v^\varepsilon \to \nabla_y v_0^* \text{ in the two-scale sense in } L^r \tag{2.61}$$

$$\frac{1}{\varepsilon^{\gamma+2}} v^\varepsilon(t) \to v^* = \int_{Y_F} v_0^* \, dy \text{ weakly in } L^r(\Omega)^n, \text{ uniformly on } [0, T] \tag{2.62}$$

$$\varepsilon^{-\gamma} \tilde{p}^\varepsilon \to p \text{ strongly in } L^{r'}(0, T; L_0^q(\Omega)), \quad q \in (1, \frac{n}{n-1}) \tag{2.63}$$

as $\varepsilon \to 0$. *The average* v^* *satisfies the equations*

$$\operatorname{div}_x v^* = 0 \text{ in } \Omega. \tag{2.64}$$

Analogously to Lemma 2.18 we have

Lemma 2.21 . $v_0^*(t) \in L^r(\Omega; W^{1,r}(Y_F)^n)$, $z = 0$ on $\partial Y_F \setminus \partial Y$ and $\operatorname{div}_y v_0^* = 0$ in Y_F.

Now we are going to establish the filtration law giving a relationship between v_0^* and ∇p^*.

Proposition 2.22. *Let* $n + 1 - \dfrac{4}{3-r} - \dfrac{n}{r} < \gamma < -1$. *Then* v_0^* *and* p^* *defined, by (2.60) and (2.63), respectively, satisfy the equations*

$$-\eta_0 \lambda^{r/2-1} \operatorname{div}\{|D_y(v_0^*)|^{r-2} D_y(v_0^*)\} + \nabla_y \pi = f(x, t) - \nabla_x p^* \text{ in } Y_F \tag{2.65}$$

$$\operatorname{div}_y v_0^* = 0 \text{ in } Y_F, \quad (v_0^*, \pi) \text{ is } Y\text{-periodic}, v_0^* = 0 \text{ on } \partial Y_F \setminus \partial Y \tag{2.66}$$

Proof : It is analogous to the proof of Proposition 2.19. The only difference is that

$$\int_0^T \int_\Omega \frac{\eta_0}{r\lambda} \left(\varepsilon^{-2\frac{\gamma+1}{r-1}} + \lambda |D(\psi(x, x/\varepsilon))|^2 \right)^{r/2} \to \int_0^T \int_\Omega \int_{Y_F} \eta_0 \lambda^{r/2-1} |D_y(\psi(x, y))|^r$$

$$\liminf{}_{\varepsilon \to 0} \int_0^T \int_\Omega \frac{\eta_0}{r\lambda} \left(\varepsilon^{-2\frac{\gamma+1}{r-1}} + \lambda |\varepsilon^{-\frac{\gamma+1}{r-1}} D(v^\varepsilon)|^2 \right)^{r/2}$$

$$\geq \int_0^T \int_\Omega \int_{Y_F} \eta_0 \lambda^{r/2-1} |D_y(v_0^*)|^r.$$

Hence, Bird-Carreau's law changes to the power law in the homogenization limit. ∎

Proof of Theorem 2.16: It is a direct consequence of the preceding propositions. ∎

2.6. Conclusion

In this chapter we applied the method of homogenization on modelling nonstationary non-Newtonian flow through porous media. Starting from the incompressible Navier-Stokes equations, with the viscosity depending on the shear rate through Bird-Carreau's law, we have obtained both nonlinear and linear relations between the filtration velocity and the effective pressure.

When the forcing term is small (i.e. when $\gamma > -1$) we found that the filtration velocity obeys Darcy's law. It involves the parameter η_0 as the "effective viscosity". The nonlinear structure of the flow is not visible any more.

When the forcing term is "moderate " (i.e. when $n+1-\frac{4}{3-r} - \frac{n}{r} < \gamma < -1$) the filtration velocity is the average of a velocity v_0. This velocity depends on the fast variable y and the slow variable x and we determine $v_0(x, y)$ and the effective pressure $p(x)$ by solving the system (2.45) - (2.47) (or (2.48)-(2.50), respectively). This system of PDE's does not involve the extremly complicated geometry of our porous medium any more and it should be used instead of the original flow equations. Let us however point out that it has an intrinistic nonlinear two-scale coupling, which poses some chalenges in the numerical simulation. This structural complexity comes from nonlinear viscosity law.

At the other hand, there are nonlinear filtration laws used in standard engineering treatments as e.g. the law from [WPW], obtained by modelling a porous medium as a collection of long capillary tubes through which the fully developed laminar flow occurs.

$$v = \left(\frac{K}{\mu_{eff}} [-\frac{\partial p}{\partial x}] \right)^{\frac{1}{r-1}} \tag{2.67}$$

If we suppose the flow only in the x_1 direction then the variables x and y in the nonlinear Stokes system (2.48) − (2.50) can be separated and we obtain

$$\int_{Y_F} v(x, y)\, dy = |\frac{dp}{dx_1}|^{r'-2} \cdot (-\frac{dp}{dx_1}) \int_{Y_F} w(y)\, dy,$$

where w is the solution of the nonlinear Stokes system (2.48) – (2.50) when the right hand side of (2.48) is $\vec{e}_1$. For more details we refer to [MH]. Derivations of the law (2.67) at the physical level of rigour are in [BSL] and [CMi].

However, it should be noticed that this argument holds only in the one dimensional case. Our laws for $n > 1$ are nonlocal and they cannot be reduced to the n-dimensional variants of (2.67), connecting the Darcy velocity v and some power of ∇p. It is reasonable to study the possible decoupling for small values of $f - \nabla p$. This leads to the study of the problem

$$- \operatorname{div}_y\big\{|D_y(v(\alpha))|^{r-2} D_y(v(\alpha))\big\} + \nabla_y \pi(\alpha) = \alpha \text{ in } Y_F \qquad (2.68)$$

$$\operatorname{div}_y v(\alpha) = 0 \text{ in } Y_F, \qquad (2.69)$$

$$(v(\alpha), \pi(\alpha)) \text{ is } Y\text{-periodic}, v(\alpha) = 0 \text{ on } \partial Y_F \setminus \partial Y \qquad (2.70)$$

Then we easily prove that $\nabla_\alpha v$ exists and

$$\|D(v(\alpha))\|_{L^r(Y_F)^{n^2}} \le C|\alpha|^{\frac{1}{r-1}}; \ \|\frac{\partial v(\alpha)}{\partial \alpha_i}\|_{L^r(Y_F)^{n^2}} \le C\|D(v(\alpha))\|_{L^r(Y_F)^{n^2}}^{2-r}.$$

Those estimates indicate that linearization around 0 is not possible. In general, for v_0 defined by (2.48)-(2.50), we have

$$v_0(x, y) = \eta_0^{1-r'} \lambda^{r'/2-1} |f(x) - \nabla_x p(x)|^{\frac{1}{r-1}} v_0\big(\frac{f(x) - \nabla_x p(x)}{|f(x) - \nabla_x p(x)|}, y\big) \qquad (2.71)$$

Hence (2.67) gives the correct dependance on the length of $f - \nabla p$ in the effective law, but the direction of the flow is not correctly described. Formula (2.71) also confirms the " dual " functional relationship between shear stress and shear rate usually found when modelling porous media as bundle of capillary tubes. Contrary to the case of the Darcy law, when homogenization only justified something already well-established, here homogenization was used in deriving an a priori unknown effective model!

When $n + 1 - \dfrac{4}{3 - r} - \dfrac{n}{r} > \gamma$ then the inertial term $(v^\varepsilon \nabla)v^\varepsilon$ can't be neglected any more. In particular, the expansions for the viscous energy term and for the inertial term are of the same order if $\gamma + 1 = 2\frac{1-r}{3-r}$.

In the case $r = 2$ there is a rigorous mathematical theory for the corresponding nonlinear and nonlocal two-scale system of partial differential equations established in [MaMi], which will be discussed in the next chapter. In the non-Newtonian case there is no theory covering the inertia effects.

There are also other open modelling problems as presence of thermal and memory effects for non-Newtonian flow.

3. On the non-linear seepage caused by the inertia effects

3.0 Introduction

The famous law of Darcy states that the specific discharge q (or the seepage velocity) is proportional to the gradient of the piezometric head φ. For an incompressible flow through a porous medium it reads

$$q = -\frac{k}{\mu}(\nabla p - \rho \vec{g}), \tag{3.1}$$

where k is the medium's permeability, μ is the dynamic viscosity of the fluid and ρ is the density of the fluid. The average velocity v is then the specific discharge q divided by the porosity m.

Derivation of Darcy's law (3.1) for a slow viscous incompressible flow through a periodic porous medium is explained in details in §1. However, as the seepage velocity q increases, the relationship between q and ∇p gradually becomes non-linear. One should make an analogy with the Reynolds number analysis for the Navier-Stokes system describing the flow through conduits. Then the dimensional analysis gives

$$-\frac{k}{\mu}\nabla p = qF(Re^{loc}, m), \tag{3.2}$$

where Re^{loc} denotes the Reynolds number for the seepage microflow. If d denotes the representative length of the porous matrix then $Re^{loc} = |q|\frac{\rho d}{\mu}$. In petroleum engineering literature $d = \sqrt{\frac{k}{m}}$, which is qualitatively in agreement with the scaling of the permeability tensor defined in §1. Experimental evidence indicates that Darcy's law (3.1) is valid for $Re^{loc} \leq 10$ (see Bear [BE]). As the seepage velocity increases then the following scenario is observed :

a) $10 < Re^{loc} \leq 100$. The region of a gradual transition. The viscous forces still dominate the inertial ones, but the inertia effects are observable. The filtration law is a non-linear perturbation of (3.1).

b) $100 \leq Re^{loc} \leq 150$. Darcy's law (3.1) is no longer valid and the inertial forces are dominant.

c) $150 \leq Re^{loc}$. Some authors consider that the flow becomes turbulent, but it is contested by the others.

For more details on observed properties of the rapid seepage flow we refer to classic textbooks in water resourses and petroleum engineering, as e.g. [BE], [BER] and [Sch] and references therein.

The experimental observations did not allow to find a universally accepted formula for F. However, by expanding function $F(Re^{loc}, m)$ into a power serie and keeping only the first two terms we arrive at the law proposed by P.

Forchheimer in 1901

$$-\frac{k}{\mu}\nabla p = q + \beta\frac{\rho\sqrt{k}}{\mu}q|q|. \tag{3.3}$$

The application of the law (3.3) (or (3.2), respectively) is restricted to the flow in the vicinity of outlets or to the flow in fractured media. In chemical engineering the law (3.3) is called Ergun's law.

Using the continuum approach, Hassanizadeh and Gray have found in [HG], the constitutive law (3.2) compatible with thermodynamics. In their theory, F is an arbitrary smooth function depending on the porosity, temperature, density and the velocity magnitude.

Looking at all those results from the literature, some natural questions arise:
1) Is it possible to derive the non-linear filtration law (3.2) (and, in particular, the Forchheimer-Ergun law (3.3)) from the first principles, as it was the case with Darcy's law?
2) Could one determine the function F (or the constants in the law (3.3), respectively) ?
3) Is it possible to give a mathematically rigorous justification of the laws (3.2) and (3.3) ?

In order to answer these questions, we should fix our model. The non-linear filtration law (3.2) describes the laminar effects of a fast flow through a porous medium and it is stationary. Hence it is natural to consider the stationary incompressible viscous flow through a porous medium. The flow regime is assumed to be laminar through the fluid part of porous medium, which is supposed to be a network of interconnected channels.

In order to avoid handling the effects of outer boundary layers, which aren't of the fundamental importance for studying our problem, we suppose that they are contained in the forcing term. Let the fluid viscosity be μ_0. We introduce the macroscopic characteristic length L_0, the characteristic velocity V_0, the characteristic pressure P_0 and the characteristic volume force F_0. Then characteristic numbers are defined as follows: $E_u = \dfrac{P_0}{\rho V_0}$ is the Euler number; $Re = \dfrac{V_0 L_0}{\mu_0}$ is the Reynolds number and the Froude number is $Fr = \dfrac{V_0^2}{F_0 L_0}$. As customary in modeling the filtration using homogenization, we use the fact that the porous medium has a microscopic length scale ℓ (e.g. a representative pore size) which is small compared to the characteristic length L_0. Therefore there is a small parameter $\varepsilon = \ell/L_0$ in the problem and we suppose that $Re = Re_\varepsilon$ and $Fr = Fr_\varepsilon$ behave as powers of ε.

With these conventions, the non-dimensional incompressible Navier-Stokes system is given by

$$-\frac{1}{Re_\varepsilon}\Delta v_\varepsilon + (v_\varepsilon\nabla)v_\varepsilon + E_u\nabla p_\varepsilon = \frac{1}{Fr_\varepsilon}f \text{ in } \Omega^\varepsilon \tag{3.4}$$

$$\text{div } v_\varepsilon = 0 \text{ in } \Omega^\varepsilon \tag{3.5}$$

$$v_\varepsilon = 0 \text{ on } \partial\Omega^\varepsilon , \tag{3.6}$$

where Ω^ε is the fluid part of the porous medium $\Omega =]0, L_0[^n$, v_ε is the velocity, p_ε is the pressure and f is the density of an external body force.

We suppose that the Ω^ε is a Lipschitz domain in $I\!\!R^n$, $n = 2, 3$, defined as in §1.2. Then its boundary consists of two disjointed parts $\partial\Omega$ (the outer boundary) and $S_\varepsilon = \partial\Omega^\varepsilon \backslash \partial\Omega$ (the inner boundary). It is assumed that $L_0/\varepsilon \in I\!\!N$. Furthermore, S_ε is a union of a finite, but large, number of boundaries of some open sets of strictly positive measures, strictly contained in Ω. Actually the geometry of Ω^ε is periodic with a period εY where $Y = (0, 1)^n$.

We make the following assumptions

$$f \in C_{per}^3(\Omega)^n \ , \ \mathrm{div}\, f = 0 \ \mathrm{in} \ \Omega \ . \tag{3.7}$$

Then for $n = 2, 3$ the classical theory gives the existence of at least one weak solution $(v_\varepsilon, p_\varepsilon) \in V_{per}(\Omega^\varepsilon) \times L_0^2(\Omega^\varepsilon)$ for the problem (3.4), (3.5) with the boundary conditions

$$v_\varepsilon = 0 \ \mathrm{on} \ S_\varepsilon \ , \ (v_\varepsilon, p_\varepsilon) \ \mathrm{is} \ L_0 - \mathrm{periodic} \tag{3.8}$$

and

$$V_{per}(\Omega^\varepsilon) = \{z \in H^1(\Omega^\varepsilon)^n \ : \ z = 0 \ \mathrm{on} \ S_\varepsilon \ , \ z \ \mathrm{is} \ L_0 - \mathrm{periodic}, \ \mathrm{div}\, z = 0 \ \mathrm{in} \ \Omega^\varepsilon\} \ .$$

Let us now suppose that $Re_\varepsilon = \frac{1}{\mu}\varepsilon^{-\gamma_1}$ and $Fr_\varepsilon = \varepsilon^{\gamma_2}$. Then the a priori estimate for velocities is immediate:

$$\|v_\varepsilon\|_{L^2(\Omega^\varepsilon)^n} \le C\varepsilon^{2-\gamma_1-\gamma_2} \ , \ \|\nabla v_\varepsilon\|_{L^2(\Omega^\varepsilon)^{n^2}} \le C\varepsilon^{1-\gamma_1-\gamma_2} \ . \tag{3.9}$$

We have supposed the non-dimensional form of the equation (3.4). Therefore we have $\gamma_2 = 2 - \gamma_1$ and we distinguish three typical cases. More discussion on the a priori estimate for the pressure and on the two-scale asymptotic expansion is in the next subsection.

- **Case 1:** $\gamma_1 < 1$. This case is well-understood. For $\gamma_1 \ll 1$ the inertia effects are negligible and it is customary to consider the Stokes equation in place of (3.4). The formal asymptotic expansion gives the well known Darcy law as the averaged momentum equation (3.4) and the permeability tensor K is calculated using an auxiliary problem in the periodic cell Y. For details we refer to §1. Even if we keep the inertial term, for $\gamma_1 < 1$ the homogenization is always the Darcy law. For the rigorous proof we refer to Mikelić [Mik1]. The main difference in comparison with the homogenization of the Stokes system is in the convergence of the pressures. However, when γ_1 approaches 1 from below the presence of the lower order terms starts to be non-negligible, as it was found out by the formal asymptotic expansions in [WL], [MA] and [RA]. This explains the experimental observations for $10 < Re^{loc} < 100$ and, as we are going to see, the appearance of inertia effects when $Re^{loc} \ge 100$. Two other papers related to this problem are

[FG] and [FGL]. Using the notion of the Reynolds number for the seepage microflow $Re_\varepsilon^{loc} = \varepsilon Re_\varepsilon = \varepsilon^{1-\gamma_1}$ the following filtration law is found:

$$v_D = \frac{K}{\mu}(f - \nabla p^0) + \frac{1}{\mu^3} Re_\varepsilon^{loc} \sum_{i,j} M^{ij}(f_i - \frac{\partial p^1}{\partial x_i})(f_j - \frac{\partial p^1}{\partial x_j}) + O((Re_\varepsilon^{loc})^2) .$$

(3.10)

Obviously, when $1 - \gamma_1$ is a small positive number the second order term in ∇p^1 should not be neglected and we can consider (3.10) as a nonlinear filtration law.

The rigorous derivation of the law (3.10) using the homogenization theory is in [BMM], where it was shown that difference between the velocities defined by the problem (3.4), (3.5) and (3.7) and the corresponding two-scale homogenized velocity behaves as $(Re_\varepsilon^{loc})^2$. Here, the filtration velocity v_D is the average over the unit cell of the two-scale homogenized velocity. We give the details of that proof in subsections which follow.

- **Case 2:** $\gamma_1 > 1$. It leads to the homogenized problems which are not well-posed. This is the turbulent case and some progress can be expected only if the thermodynamics is included.
- **Case 3:** $\gamma_1 = 1$. In this case the local Reynolds number is of order 1 and the validity of filtration law (3.10) is not clear any more. The formal asymptotic expansion is performed in [SP80] and [LiPe]. Starting from the system (3.4), (3.5) and (3.7) the following homogenized problem was obtained

$$\text{Find } v_0 = v_0(x,y), \quad \pi_1 = \pi_1(x,y) \quad \text{and} \quad p_0 = p_0(x) \quad \text{such that}$$
$$-\mu\Delta_y v_0 + (v_0\nabla_y)v_0 + \nabla_y\pi_1 = f(x) - \nabla_x p_0 \quad \text{in } \Omega \times Y_F,$$
$$\text{div}_y v_0 = 0 \text{ in } \Omega \times Y_F,$$
$$v_0 = 0 \text{ on } \Omega \times S , \qquad (v_0, \pi_1) \text{ is } Y\text{-periodic}$$
$$\text{div}_x\{\int_{Y_F} v_0(x,y)dy\} = 0 \text{ in } \Omega, \qquad \{v_0, \pi_1, p_0\} \text{ is } - \text{periodic. (3.11)}$$

Here Y_F is the fluid part of the unit cell $Y =]0,1[^n$, $n = 2,3$, and $S = \partial Y_F \setminus \partial Y$.

The well-posedness of the problem (3.11) was left as an open problem in the above references. It should be noted that in (3.11) we lack the derivatives with respect to x. Furthermore, the problem is non monotone which indicates that it does not confirm the law of Forchheimer and Ergun (3.3). We will discuss this point, which contradicts the engineering literature, in some details.

In the subsections which follow, we address two questions. Firstly, we discuss the problem (3.11) and prove the existence, uniqueness and regularity of solution under the assumptions that f is not too large. Secondly, we establish the convergence of the homogenization process as $\varepsilon \to 0$. More precisely we estimate appropriate norms for the differences between the solutions to the problem (3.4), (3.5), (3.7) and the problem (3.11) with $y = x/\varepsilon$, as $C\varepsilon$.

The results have already been published in [Mik2] and [MaMi]. [MaMi] contains also the results for the case of an arbitrary domain and we'll refer to it frequently. Before starting with the detailed study of the cases 1 and 3, we derive an *a priori* estimate for the pressure field p_ε :

Proposition 3.1. *Let* $\{v_\varepsilon, p_\varepsilon\}_{\varepsilon > 0}$ *be solutions for (3.4)-(3.6) and let* $\gamma_1 \leq 1$. *Then we have*

$$\|v_\varepsilon\|_{L^2(\Omega^\varepsilon)^n} \leq C, \quad \|\nabla v_\varepsilon\|_{L^2(\Omega^\varepsilon)^{n^2}} \leq C\varepsilon^{-1}. \tag{3.12}$$

Furthermore, let the pressure extension $\tilde{p}_\varepsilon$ *be given by (1.27). Then* $\tilde{p}_\varepsilon$ *satisfies the following a priori estimate*

$$\|\tilde{p}_\varepsilon\|_{L^q_0(\Omega)} \leq \frac{C}{\varepsilon} \left\{ \varepsilon^{\gamma_1} \|\nabla v_\varepsilon\|_{L^2(\Omega^\varepsilon)^{n^2}} + \varepsilon^{\gamma_1 - 1} + \varepsilon \|\nabla v_\varepsilon\|_{L^2(\Omega^\varepsilon)^{n^2}} \|v_\varepsilon\|_{L^2(\Omega^\varepsilon)^n} \right\}$$

$$\leq C\varepsilon^{\gamma_1 - 2} \left\{ 1 + \varepsilon Re_\varepsilon^{loc} \|\nabla v_\varepsilon\|_{L^2(\Omega^\varepsilon)^{n^2}} \|v_\varepsilon\|_{L^2(\Omega^\varepsilon)^n} \right\}, \quad \forall q \in (1, \frac{n}{n-1}). \tag{3.13}$$

Proof : Due to the results in § 1.4, we only have to estimate the contribution coming from the inertial term. For $r > n$, we have

$$\left| \int_{\Omega^\varepsilon} (v_\varepsilon \nabla) v_\varepsilon \varphi \right| \leq \|\nabla v_\varepsilon\|_{L^2(\Omega^\varepsilon)^{n^2}} \|v_\varepsilon\|_{L^2(\Omega^\varepsilon)^n} \|\varphi\|_{L^\infty(\Omega^\varepsilon)^n} \tag{3.14}$$

for all $\varphi \in W^{1,r}(\Omega^\varepsilon)^n$, such that $\varphi = 0$ on S_ε and they are L_0-periodic.
   Since

$$\langle \nabla \tilde{p}_\varepsilon, \varphi \rangle_\Omega = \langle \nabla p_\varepsilon, R_\varepsilon \varphi \rangle_{\Omega^\varepsilon}$$

we are going to estimate the quantity $\|R_\varepsilon \varphi\|_{L^\infty(\Omega^\varepsilon)^n}$. This requires an analogue of Proposition 1.4 for $g \in L^r_0(\Omega)$, $r > n$.
   We replace Δ in (1.16) by r'-Laplacean, $r' = r/(r-1)$, and get $F^\varepsilon = |\nabla u^\varepsilon|^{r'-2} \nabla u^\varepsilon \chi_{\Omega^\varepsilon} \in L^r(\Omega)^n$ with

$$\|F^\varepsilon\|_{L^r(\Omega)^n} \leq C\|g\|_{L^r(\Omega)}.$$

Inspection of the part b) of the proof of Proposition 1.14 gives

$$\|z^\varepsilon\|_{L^\infty(\Omega^\varepsilon)^n} \leq C\left\{ \|F^\varepsilon\|_{L^r(\Omega)^n} + \varepsilon \|g\|_{L^r(\Omega)} \right\}$$

Therefore (3.14) reads

$$\left| \int_{\Omega^\varepsilon} (v_\varepsilon \nabla) v_\varepsilon R_\varepsilon \varphi \right| \leq C\|\nabla v_\varepsilon\|_{L^2(\Omega^\varepsilon)^{n^2}} \|v_\varepsilon\|_{L^2(\Omega^\varepsilon)^n} \left\{ \|\nabla u^\varepsilon\|_{L^r(\Omega)^n} \right.$$

$$\left. + \varepsilon \| \operatorname{div}(R_\varepsilon \varphi)\|_{L^r(\Omega)} \right\}, \tag{3.15}$$

where u^ε is a solution for (1.6) with Δ replaced by the r'-Laplacean. (3.15) and estimates from § 1.4 imply (3.13). ∎

Remark 3.2. *Estimate (3.13) shows clearly how the inertial term becomes important when* $\gamma_1 \to 1$.

Corollary 3.3. *Sequence $\{\tilde{p}_\varepsilon\}$ is strongly relatively compact in $L_0^q(\Omega)$, $\forall q \in$ $(1, n/(n-1))$.*

Remark 3.4. *Now the result from [Mik1] is clear. For $\gamma_1 < 1$ the inertial term vanishes in the limit $\varepsilon \to 0$ and we get the Darcy law, as in the homogenization of the Stokes equation. More precisely, for $\gamma_1 < 1$ we have*

$$
\begin{cases}
v_\varepsilon \rightharpoonup q^0 & \text{in } L^2(\Omega)^n & \text{weakly} \\
\varepsilon^{2-\gamma_1}\tilde{p}_\varepsilon \to p^0 & \text{in } L_0^q(\Omega), & 1 < q < n/(n-1)
\end{cases}
\tag{3.16}
$$

where $q^0 = \int_{Y_F} v^0 \, dy$ and $\{v^0, p^0\}$ is the solution for the problem (1.49)-(1.50).

The plan of the chapter is the following:

After the introductory Section 3.0, Section 3.1 is devoted to the case $\gamma_1 < 1$. The existence uniqueness and regularity for the problem (3.11) is in Section 3.2. Furthermore, a version of (3.10) for $\gamma_1 = 1$ (i.e. for the local Reynolds number of order 1) is established and the non-monotonicity is discussed. Convergence of the homogenization process in the case $\gamma_1 = 1$ is proved in Section 3.3 and in Section 3.4 we present a conclusion of our results.

3.1 Weak non-linear corrections to Darcy's law

In this subsection we study the subcritical case $\gamma_1 < 1$. As it can be expected the filtration law depends on the range of γ_1; on one hand, for $\gamma_1 < 1$, the homogenization of the system (3.4)-(3.5) leads, taking the limit as $\varepsilon \to 0$, to the linear Darcy's law as in [Mik1] but on the other hand, as already mentioned, in the critical case $\gamma_1 = 1$, the fully nonlinear system of equations (3.11) is found in the limit.

However, for γ_1 sufficiently close to 1, the linear Darcy's law will not be accurate enough to approximate the filtration law and the neglected terms in the asymptotic expansion of order ε^{-1} become significant compared to ε^{γ_1-2}. In order to see the apparition of this discrepancy in the linear Darcy's law, when γ_1 is close to 1, we compute the correctors appearing in the homogenization process for the system (3.4)-(3.5) when $\gamma_1 < 1$ and estimate the error between $\{v_\varepsilon, \varepsilon^{2-\gamma_1}p_\varepsilon\}$ and the homogenized limit plus these correctors. These limits and correctors could also be obtained from the zeroth and first order terms in the asymptotic expansion in powers of the **Reynolds number for the seepage microflow** $Re_\varepsilon^{loc} = \varepsilon^{1-\gamma_1}$. As already mentioned, the formal multiscale expansion was constructed in [WL]. The special cases $\gamma_1 = 1/2$ and $3/4$ were investigated in [RA] and [MA], respectively. For a very clear exposition of the experimental results, the asymptotic expansion and numerical simulation we refer to the recent paper [FGL]. The conclusion of [FGL] is that for a fairly large class of periodic porous media the Forchheimer's correction to Darcy's law is in fact cubic, i.e. in (3.2) $F(z, m) = A(m) + B(m)z^2$.

It is easy to see that for $\gamma_1 \leq 1/2$ and taking into account only the first term of the expansion gives an error estimate of order ε, like in the linear case.

Furthermore, we expect that being close to the critical value $\gamma_1 = 1$ produces non-linear effects of a polynomial type. For this reason, we restrict our investigation to the weak nonlinear effects, i.e. we will keep γ_1 in the range $0 < \gamma_1 < 1$.

According to the scaling of data, as defined in the introduction, i.e. ε^{γ_1} being the order of the viscosity, $\varepsilon^{2-\gamma_1}$ being the order of the velocity v_ε in the pores and ε being the size of the pores, we seek an asymptotic expansion in powers of the local Reynolds number $Re_\varepsilon^{loc} = \varepsilon^{1-\gamma_1}$ for $\{v_\varepsilon, p_\varepsilon\}$ solution of (3.4)-(3.6). Due to the range of γ_1 we have $0 < 1 - \gamma_1 < 1$. For simplicity, we set $E_u = 1$.

Motivated by (3.13), we set for γ_1 *sufficiently close* to 1, the following asymptotic expansion :

$$v_\varepsilon(x) = v^0(x,y) + Re_\varepsilon^{loc} v^1(x,y) + (Re_\varepsilon^{loc})^2 v^2(x,y) + \cdots$$
$$+ \varepsilon\{v^{0,1}(x,y) + Re_\varepsilon^{loc} v^{1,1}(x,y) + \cdots\} + \cdots \qquad (3.17)$$
$$\varepsilon^{2-\gamma_1} p_\varepsilon(x) = p^0(x,y) + Re_\varepsilon^{loc} p^1(x,y) + (Re_\varepsilon^{loc})^2 p^2(x,y) + \cdots$$
$$+ \varepsilon\{p^{0,1}(x,y) + Re_\varepsilon^{loc} p^{1,1}(x,y) + (Re_\varepsilon^{loc})^2 p^{2,1}(x,y) + \cdots\} + \cdots , \qquad (3.18)$$

where $y = x/\varepsilon$.

After collecting equal powers of ε we obtain, as in §1.5 , a sequence of the problems in $Y_F \times \Omega$. First we have
$O(\varepsilon^{\gamma_1 - 3})$:

$$\nabla_y p^0 = 0 \ , \quad \text{i.e.} \quad p^0 = p^0(x) \qquad (3.19)$$

$O(\varepsilon^{-2})$:

$$\nabla_y p^1 = 0 \ , \quad \text{i.e.} \quad p^1 = p^1(x) \qquad (3.20)$$

$O(\varepsilon^{\gamma_1 - 2})$:

$$\begin{cases} -\mu\Delta_y v^0 + \nabla_y p^{0,1} = & f - \nabla_x p^0 \ \text{in} \ \Omega \times Y_F \\ \text{div}_y v^0 = 0 \ \text{in} \ \Omega \times Y_F, & v^0 = 0 \ \text{on} \ \Omega \times (\partial Y_F \setminus \partial Y) \\ \{v^0, p^{0,1}\} \ \text{is} \ Y - \text{periodic}, & \text{div}_x\{\int_{Y_F} v^0 \, dy\} = 0 \ \text{in} \ \Omega \\ \{\int_{Y_F} v^0, p^0\} & \text{is} \ L_0 - \text{periodic}. \end{cases} \qquad (3.21)$$

$O(\varepsilon^{-1})$:

$$\begin{cases} -\mu\Delta_y v^1 + \nabla_y p^{1,1} = & -(v^0 \nabla_y)v^0 - \nabla_x p^1 \ \text{in} \ \Omega \times Y_F \\ \text{div}_y v^1 = 0 \ \text{in} \ \Omega \times Y_F, & v^1 = 0 \ \text{on} \ \Omega \times (\partial Y_F \setminus \partial Y) \\ \{v^1, p^{1,1}\} \ \text{is} \ Y - \text{periodic}, & \text{div}_x\{\int_{Y_F} v^1\} = 0 \ \text{in} \ \Omega \\ \{\int_{Y_F} v^1, p^1\} & \text{is} \ L_0 - \text{periodic}. \end{cases} \qquad (3.22)$$

If we go further in identifying terms in the expansions (3.17)-(3.18) and keeping γ_1 close to 1, we will find higher order effects (like cubic, etc.).

Remark 3.5. *Problems (3.20) and (3.21) are standard Stokes problems in Y_F and the regularity of the solutions v^0 and v^1 follows from the regularity of the geometry and of the data.*

In analogy with (1.62)-(1.63), separation of scales leads to the following formulas for v^0 and $p^{0,1}$.

$$v^0(x,y) = \frac{1}{\mu} \sum_{i=1}^{n} w^i(y)[f_i(x) - \frac{\partial p^0}{\partial x_i}(x)] \tag{3.23}$$

$$p^{0,1}(x,y) = \sum_{i=1}^{n} \pi^i(y)[f_i(x) - \frac{\partial p^0}{\partial x_i}(x)] \ , \tag{3.24}$$

where $\{w^i, \pi^i\}$ are given by (1.57) and the permeability matrix K by (1.58). In addition p^0 is defined by (1.65).

Similarly, we have v^1 and $p^{1,1}$ given by :

$$\mu v^1(x,y) = \frac{1}{\mu^2} \sum_{i,j=1}^{n} \omega^{ij}(y)[f_i(x) - \frac{\partial p^0}{\partial x_i}(x)] \, [f_j(x) - \frac{\partial p^0}{\partial x_j}(x)]$$

$$- \sum_{i=1}^{n} w^i(y)\frac{\partial p^1}{\partial x_i}(x) \tag{3.25}$$

$$p^{1,1}(x,y) = \frac{1}{\mu^2} \sum_{i,j=1}^{n} \Lambda^{ij}(y)[f_i(x) - \frac{\partial p^0}{\partial x_i}(x)] \, [f_j(x) - \frac{\partial p^0}{\partial x_j}(x)]$$

$$- \sum_{i=1}^{n} \pi^i(y)\frac{\partial p^1}{\partial x_i}(x) \ , \tag{3.26}$$

where $\{\omega^{ij}, \Lambda^{ij}\}$ is the Y-periodic solution of the following auxiliary Stokes problem:

$$\begin{cases} -\Delta_y \omega^{ij} + \nabla_y \Lambda^{ij} = -\frac{1}{2}[(w^i \nabla_y)w^j + (w^j \nabla_y)w^i] & \text{in } Y_F \\ \text{div}_y \omega^{ij} = 0 & \text{in } Y_F \\ \omega^{ij} = 0 \text{ on } (\partial Y_F \setminus \partial Y) \ , & \int_{Y_F} \Lambda^{ij} = 0. \end{cases} \tag{3.27}$$

In addition $\{q^1 = \int_{Y_F} v^1 \, dy, p^1\}$ is the solution of

$$\text{div}_x q^1 = 0 \text{ in } \Omega \tag{3.28}$$

$$\mu q_k^1 = \frac{1}{\mu^2} \sum_{i,j=1}^{n} M_k^{ij}\{f_i - \frac{\partial p^0}{\partial x_i}\}\{f_j - \frac{\partial p^0}{\partial x_j}\} - \sum_{j=1}^{n} K_{kj}\frac{\partial p^1}{\partial x_j} \tag{3.29}$$

$$\{q^1, p^1\} \text{ is } \Omega - \text{periodic}, \int_{\Omega} p^1 = 0 \ , \tag{3.30}$$

where M is the third order tensor, defined by

$$M_k^{ij} = \int_{Y_F} \omega_k^{ij}(y)dy, \ i,j,k = 1,\dots,n. \tag{3.31}$$

Remark 3.6. We note that without p^1 in the right hand side of (3.26), as in [WL], it would not be possible to satisfy (3.25). This is the reason why we introduced the pressure p^1 in the expansion (1.19).

Now we introduce the effective seepage velocity $\mathcal{Q}^1$ and the effective pressure $\mathcal{P}^1$ by

$$\mathcal{Q}^1 = q^0 + Re_\varepsilon^{loc} q^1, \qquad \mathcal{P}^1 = p^0 + Re_\varepsilon^{loc} p^1 \qquad (3.32)$$

and with these notations we have

$$\mathcal{Q}^1 = \frac{1}{\mu} K(f - \nabla \mathcal{P}^1) + \frac{1}{\mu^3} Re_\varepsilon^{loc} \sum_{i,j=1}^n M^{ij}(f_i - \frac{\partial \mathcal{P}^1}{\partial x_i})(f_j - \frac{\partial \mathcal{P}^1}{\partial x_j}) + O((Re_\varepsilon^{loc})^2).$$

$$(3.33)$$

The quadratic expression entering in equation (3.33) starts to be important when Re_ε^{loc} is not too small, i.e. when γ_1 is close to 1.

Remark 3.7. If $1 - 1/(2(k+1)) < \gamma_1 < 1$, $k \in \mathbb{N}$; introducing in the same way as above

$$\mathcal{Q}^{k+1} = \sum_{\ell=0}^{k+1} (Re_\varepsilon^{loc})^\ell q^\ell, \qquad \mathcal{P}^{k+1} = \sum_{\ell=0}^{k+1} (Re_\varepsilon^{loc})^\ell p^\ell, \qquad (3.34)$$

leads to

$$\mathcal{Q}^{k+1} = \frac{1}{\mu} K(f - \nabla \mathcal{P}^{k+1}) +$$

$$\sum_{\ell=1}^{k+1} \frac{1}{\mu^{2\ell+1}} (Re_\varepsilon^{loc})^\ell \sum_{i_1,\dots,i_{\ell+1}=1}^n M^{i_1 \cdots i_{\ell+1}} \prod_{m=1}^{\ell+1} (f_{i_m} - \frac{\partial \mathcal{P}^{k+1}}{\partial x_{i_m}})$$

$$+ O((Re_\varepsilon^{loc})^{k+1}). \qquad (3.35)$$

Remark 3.8. Laws (3.33) and (3.35) formally justify a polynomial relationship between the seepage velocity and the pressure gradient. However, this relationship is in fact determined through a set of linear filtration problems and this justifies the name weak non-linear corrections.

Due to the construction, we can express $f - \nabla \mathcal{P}^1$ as a non-linear function of $\mathcal{Q}^1$ and (3.33) gives the following variant of Forchheimer's law

$$\frac{1}{\mu} K(f - \nabla \mathcal{P}^1) = \mathcal{Q}^1 - \frac{1}{\mu} Re_\varepsilon^{loc} M(K^{-1}\mathcal{Q}^1)(K^{-1}\mathcal{Q}^1) + O((Re_\varepsilon^{loc})^2). \qquad (3.36)$$

The obtained non-linear constitutive laws are always invertible, which explain why they are taken to be monotone in applications. However, the relationship should be analytic in $\mathcal{Q}^1$ and this exlude the classical form (3.3) of Forchheimer law. One should replace $|\mathcal{Q}^1|$ by $\sqrt{\delta + |\mathcal{Q}^1|^2}$.

In the remaining part of the subsection we prove the convergence. It is possible to study the two-scale convergence of $(Re_\varepsilon^{loc})^{-1}(v_\varepsilon - v^{0,\varepsilon})$ and justify rigorously the multiscale expansion developped above. However, it is of interest to have an error estimate which could be generalized to corrections of any order.

We follow the strategy of § 1.8. Let $\{v^{0,\varepsilon}, p^{1,\varepsilon}\}$ be defined by (1.77), but with $p^1(x,y)$ replaced by $p^{0,1}(x,y)$, and Q^ε by (1.78)-(1.79). Then we introduce $U \in H^1_{per}(\Omega; H^2_{per}(Y_F)^n)$ satisfying (1.78) with $\operatorname{div}_x v^0(x,y)$ replaced by $\operatorname{div}_x v^1(x,y)$ and

$$U^\varepsilon(x) = \varepsilon U(x, \frac{x}{\varepsilon}), x \in \Omega^\varepsilon. \tag{3.37}$$

Having introduced the corrections for the compressibility effects caused by v^0 and v^1, we are in situation to annonce our convergence result

Theorem 3.9. *Let $\{v_\varepsilon, p_\varepsilon\}$ be a solution of (3.4)-(3.6) with $\gamma_1 \in]0,1[$, and let $\{v^0, p^0\}$ be given by (3.21), $\{v^1, p^1\}$ by (3.22), Q^ε by (1.79) and U^ε by (3.37), respectively. Then we have the following error estimate for the velocity:*

$$\|\mu v_\varepsilon - \{v^0(.,\frac{\cdot}{\varepsilon}) + Re^{loc}_\varepsilon v^1(.,\frac{\cdot}{\varepsilon}) +$$

$$\varepsilon[Q(.,\frac{\cdot}{\varepsilon}) + Re^{loc}_\varepsilon U(.,\frac{\cdot}{\varepsilon})]\}\|_{H_{per}(\Omega, \, div \,)} \le C \max\{(Re^{loc}_\varepsilon)^2, \varepsilon\}. \tag{3.38}$$

Furthermore, let $\tilde{p}_\varepsilon$ be the pressure extension defined by (1.27). Then the following estimate is valid

$$\|\varepsilon^{2-\gamma_1}\tilde{p}_\varepsilon - \{p^0 + Re^{loc}_\varepsilon p^1\}\|_{L^s(\Omega)} \le C \max\{(Re^{loc}_\varepsilon)^2, \varepsilon\}, \ \forall s \in]1, \frac{n}{n-1}[. \tag{3.39}$$

Remark 3.10. *If $0 \le \gamma_1 \le 3/4$ then $\max\{(Re^{loc}_\varepsilon)^2, \varepsilon\} = \varepsilon$, but if $3/4 \le \gamma_1 < 1$ then $\max\{(Re^{loc}_\varepsilon)^2, \varepsilon\} = (Re^{loc}_\varepsilon)^2 = \varepsilon^{2(1-\gamma_1)}$.*

Remark 3.11. *The error estimate (3.38) implies the L^2-error estimate. By neglecting the terms Q^ε and U^ε we obtain*

$$\|\mu v_\varepsilon - v^0(.,\frac{\cdot}{\varepsilon}) - Re^{loc}_\varepsilon v^1(.,\frac{\cdot}{\varepsilon})\|_{L^2(\Omega)^n} \le C \max\{(Re^{loc}_\varepsilon)^2, \varepsilon\}$$

The terms Q^ε and U^ε are used only to control the divergence and to get the pressure estimate.

Remark 3.12. *Adding any element from the kernel of the divergence operator in (1.78), to Q^ε or to U^ε, does not affect the result of Theorem 3.9 ; since such elements are divergence free and have a L^2-norm bounded by $C\varepsilon$.*

Before proving Theorem 3.9, we give some preliminary results in the three following lemmas.

Lemma 3.13. *Let v^0, p^0 and $p^{0,1}$ be defined by (3.21) and Q by (1.78). Then for*

$$\Psi^1_\varepsilon(x) = \Delta\{v^0(x, \frac{x}{\varepsilon}) + \varepsilon Q(x, \frac{x}{\varepsilon})\} - \varepsilon^{-2}\nabla\{\varepsilon p^{0,1}(x, \frac{x}{\varepsilon}) + p^0(x)\} + \varepsilon^{-2}f(x), \tag{3.40}$$

we have

$$\|\Psi^1_\varepsilon\|_{(W^\varepsilon)'} \le C \quad . \tag{3.41}$$

The proof of Lemma 3.13 follows from the calculations fron § 1.8.

Lemma 3.14. Let v^1 and $p^{1,1}$ be defined by (3.22), U by (3.37) and let w^i be the solution of (1.57). Let Ψ_ε^2 be defined by

$$\Psi_\varepsilon^2(x) = \Delta\{v^1(x,\frac{x}{\varepsilon}) + \varepsilon U(x,\frac{x}{\varepsilon})\} - \varepsilon^{-2}\nabla\{\varepsilon p^{1,1}(x,\frac{x}{\varepsilon}) + p^1(x)\} -$$

$$\varepsilon^{-1}\{(v^0(x,\frac{x}{\varepsilon})\nabla)v^0(x,\frac{x}{\varepsilon}) + \sum_{j=1}^{n} w^j(\frac{x}{\varepsilon})(v^0(x,\frac{x}{\varepsilon})\nabla)[f_j(x) - \frac{\partial p^0}{\partial x_j}(x)]\}. \quad (3.42)$$

Then

$$\|\Psi_\varepsilon^2\|_{(W^\varepsilon)'} \leq C \ . \quad (3.43)$$

Proof of Lemma 3.14 is lengthy and technical, but does not represent real difficulties. For detailed proof we refer to [BMM].

Lemma 3.15. Let

$$X_\varepsilon(x) = \mu v_\varepsilon - \{v^0(x,\frac{x}{\varepsilon}) + \varepsilon Q(x,\frac{x}{\varepsilon}) + Re_\varepsilon^{loc}[v^1(x,\frac{x}{\varepsilon}) + \varepsilon U(x,\frac{x}{\varepsilon})]\}. \quad (3.44)$$

Then we have

$$\|\operatorname{div} X_\varepsilon\|_{L^q(\Omega)} \leq C\varepsilon, \ \forall q \in [1,+\infty]. \quad (3.45)$$

This result follows directly from the definitions of Q and U.

Now we write an Oseen system for X_ε. Following [BMM], we find out that it reads

$$-\varepsilon^2\Delta X_\varepsilon + \varepsilon Re_\varepsilon^{loc}\{(v^\varepsilon\nabla)X_\varepsilon + (X_\varepsilon\nabla)v^{0,\varepsilon}\} + \nabla\Pi^\varepsilon = \varepsilon^2(\Psi_\varepsilon^1 + Re_\varepsilon^{loc}\Psi_\varepsilon^2)$$

$$+\varepsilon Re_\varepsilon^{loc}\{\sum_{j=1}^{n} w^j(\frac{x}{\varepsilon})(v^{0,\varepsilon}\nabla)(f_j - \frac{\partial p^0}{\partial x_j}) - [(v_\varepsilon\nabla)(Re_\varepsilon^{loc}v^1(x,\frac{x}{\varepsilon}) + \varepsilon Q^\varepsilon + \varepsilon Re_\varepsilon^{loc}U^\varepsilon)$$

$$+((Re_\varepsilon^{loc}v^1(x,\frac{x}{\varepsilon}) + \varepsilon Q^\varepsilon + \varepsilon Re_\varepsilon^{loc}U^\varepsilon)\nabla)v^{0,\varepsilon}]\}, \quad (3.46)$$

where

$$\Pi^\varepsilon(x) = \varepsilon^{2-\gamma_1}p_\varepsilon(x) - \{p^0(x) + Re_\varepsilon^{loc}p^1(x) + \varepsilon[p^{0,1}(x,\frac{x}{\varepsilon}) + Re_\varepsilon^{loc}p^{1,1}(x,\frac{x}{\varepsilon})]\}. \quad (3.47)$$

Our last auxiliary step before proving Theorem 3.9. consists of elimination of the pressure field Π^ε. We have the following estimate

Proposition 3.16. Let $\widetilde{\Pi^\varepsilon}$ be the extension of Π^ε defined by (1.27). Then we have

$$\|\widetilde{\Pi^\varepsilon}\|_{L_0^s(\Omega)} \leq C\{(Re_\varepsilon^{loc})^2 + \varepsilon(1 + \|\nabla X_\varepsilon\|_{L^2(\Omega^\varepsilon)^{n^2}})\}, \ \forall s \in (1,\frac{n}{n-1}). \quad (3.48)$$

Proof of Proposition 3.16 is analogue to the proof of Proposition 3.1. We refer to [BMM] for details. Now we are ready for proving Theorem 3.9.

Proof of Theorem 3.9. Using X_ε as a test function in (3.46) we obtain

$$\varepsilon^2\|\nabla X_\varepsilon\|^2_{L^2(\Omega^\varepsilon)^{n^2}} + \varepsilon Re^{loc}_\varepsilon \int_{\Omega^\varepsilon} (X_\varepsilon \nabla) v^{0,\varepsilon} X_\varepsilon = \int_\Omega \widetilde{\Pi}^\varepsilon \, \mathrm{div} \, X_\varepsilon +$$

$$\varepsilon^2 \langle \Psi^1_\varepsilon, X_\varepsilon \rangle_{\Omega^\varepsilon} + \varepsilon^2 Re^{loc}_\varepsilon \langle \Psi^2_\varepsilon, X_\varepsilon \rangle_{\Omega^\varepsilon} + \varepsilon Re^{loc}_\varepsilon \sum_{j=1}^n \int_{\Omega^\varepsilon} w^j(\frac{x}{\varepsilon}) X_\varepsilon (v^{0,\varepsilon} \nabla)(f_j - \frac{\partial p^0}{\partial x_j})$$

$$-\varepsilon (Re^{loc}_\varepsilon)\{ \int_{\Omega^\varepsilon} (v_\varepsilon \nabla)(Re^{loc}_\varepsilon v^1(x, \frac{x}{\varepsilon}) + \varepsilon Q^\varepsilon + \varepsilon Re^{loc}_\varepsilon U^\varepsilon) X_\varepsilon +$$

$$\int_{\Omega^\varepsilon} ((Re^{loc}_\varepsilon v^1(x, \frac{x}{\varepsilon}) + \varepsilon Q^\varepsilon + \varepsilon Re^{loc}_\varepsilon U^\varepsilon) \nabla) v^{0,\varepsilon} X_\varepsilon \}. \tag{3.49}$$

For the lower order terms we have the following estimates:

$$\varepsilon^2 |\langle \Psi^1_\varepsilon, X_\varepsilon \rangle_{\Omega^\varepsilon}| \le C\varepsilon^2 \|\nabla X_\varepsilon\|_{L^2(\Omega^\varepsilon)^{n^2}} \tag{3.50}$$

$$\varepsilon^2 Re^{loc}_\varepsilon |\langle \Psi^2_\varepsilon, X_\varepsilon \rangle_{\Omega^\varepsilon}| \le C\varepsilon^2 Re^{loc}_\varepsilon \|\nabla X_\varepsilon\|_{L^2(\Omega^\varepsilon)^{n^2}} \tag{3.51}$$

from (3.41) and (3.43), and

$$\left| \int_{\Omega^\varepsilon} (v^\varepsilon \nabla)(Re^{loc}_\varepsilon v^1(x, \frac{x}{\varepsilon}) + \varepsilon Q^\varepsilon + \varepsilon Re^{loc}_\varepsilon U^\varepsilon) X_\varepsilon \right| \le C\|\nabla X_\varepsilon\|_{L^2(\Omega^\varepsilon)^{n^2}} \tag{3.52}$$

$$\left| \int_{\Omega^\varepsilon} ((Re^{loc}_\varepsilon v^1(x, \frac{x}{\varepsilon}) + \varepsilon Q^\varepsilon + \varepsilon Re^{loc}_\varepsilon U^\varepsilon) \nabla) v^{0,\varepsilon} X_\varepsilon \right| \le C\|\nabla X_\varepsilon\|_{L^2(\Omega^\varepsilon)^{n^2}} \tag{3.53}$$

$$\varepsilon Re^{loc}_\varepsilon \sum_{j=1}^n \left| \int_{\Omega^\varepsilon} w^j(x, \frac{x}{\varepsilon}) X_\varepsilon (v^0_\varepsilon \nabla)(f_j - \frac{\partial p^0}{\partial x_j}) \right| \le$$

$$C\varepsilon^2 Re^{loc}_\varepsilon \|\nabla X_\varepsilon\|_{L^2(\Omega^\varepsilon)^{n^2}}. \tag{3.54}$$

From the Proposition 3.16 and Lemma 3.15 we have:

$$\left| \int_\Omega \widetilde{\Pi}^\varepsilon \, \mathrm{div} \, X_\varepsilon \right| \le C\{ \varepsilon (Re^{loc}_\varepsilon)^2 + \varepsilon^2 [1 + \|\nabla X_\varepsilon\|_{L^2(\Omega^\varepsilon)^{n^2}}] \}. \tag{3.55}$$

Finally we have for the second term on the left-hand side in (3.49):

$$\varepsilon Re^{loc}_\varepsilon \left| \int_{\Omega^\varepsilon} (X_\varepsilon \nabla) v^{0,\varepsilon} X_\varepsilon \right| \le \varepsilon^2 Re^{loc}_\varepsilon \|\nabla X_\varepsilon\|^2_{L^2(\Omega^\varepsilon)^{n^2}}. \tag{3.56}$$

Adding up all the previous estimates leads to

$$\|\nabla X_\varepsilon\|_{L^2(\Omega^\varepsilon)^{n^2}} \le C\left\{ 1 + \frac{(Re^{loc}_\varepsilon)^2}{\varepsilon} \right\}. \tag{3.57}$$

Using Poincaré's inequality in Ω^ε we finally get

$$\|X_\varepsilon\|_{L^2(\Omega^\varepsilon)^{n^2}} \le C\{ \varepsilon + (Re^{loc}_\varepsilon)^2 \}.$$

Since for $3/4 < \gamma_1 < 1, \; 1 > 2(1 - \gamma_1)$ we finally have

$$\|X_\varepsilon\|_{L^2(\Omega^\varepsilon)^{n^2}} \leq C \begin{cases} (Re_\varepsilon^{loc})^2 & \text{for } 3/4 < \gamma_1 < 1 \\ \varepsilon & \text{for } 0 < \gamma_1 \leq 3/4 \end{cases}$$

and (3.38) follows. Pressure estimate (3.39) is now the consequence of (3.45) and (3.57). ∎

3.2. Study of the homogenized problem in the fully non-linear case

In this subsection we start studying the case $\gamma_1 = 1$. Effective flow in this case is non-linear and cannot be obtained as a perturbation of Darcy's law, but it is not turbulent.

The a priori estimates for velocities v_ε and pressures p_ε were derived in § 3.0 and they read

$$\|v_\varepsilon\|_{L^2(\Omega^\varepsilon)^n} \leq C, \;\; \|\nabla v_\varepsilon\|_{L^2(\Omega^\varepsilon)^{n^2}} \leq \frac{C}{\varepsilon}; \;\; \|\tilde{p}_\varepsilon\|_{L^q_0(\Omega)} \leq \frac{C}{\varepsilon}, \; \forall q \in (1, \frac{n}{n-1}). \text{(3.58)}$$

For $\gamma_1 = 1$, we have $Re_\varepsilon^{loc} = O(1)$. Consequently, expansions (3.17)-(3.18) does not make sense any more and we use the expansions (1.37)-(1.38) adapted to the a priori estimate (3.58). We set

$$\begin{cases} v_\varepsilon(x) = v_0(x,y) + \varepsilon v^1(x,y) + \varepsilon^2 v^2(x,y) + \dots, & y = \frac{x}{\varepsilon} \\ p_\varepsilon(x) = \varepsilon^{-1} p_0(x) + \pi_1(x,y) + \varepsilon p^2(x,y) + \dots, & y = \frac{x}{\varepsilon} \end{cases} \tag{3.59}$$

and after straightfoward calculations obtain the system (3.11). It is called the **Navier-Stokes system with two pressures** (see [LiPe] and [SP80]).

The system (3.11) is different from the classical incompressible Navier-Stokes system. The difficulty comes from the fact that we lack derivatives in respect of x and only L^2-estimate on the velocity with respect to x are straightforward. Moreover, the inertial term is not continuous with respect to the weak L^2-convergence and some compactness-type results with respect to x are necessary. The difficulty is overcome in the case of the special geometry in [Mik2] using the differences in x. Then it was possible to obtain an a priori estimate in $H^\alpha(\Omega; H^1(Y_F))$, $\alpha > 1$ for velocity (for *small* f) and Tychonoff's fixed point theorem implied the existence. Using regularity and *smallness* for f it is also possible to prove uniqueness.

In [MaMi] a different strategy, which works for general domains, was used. Namely, it was proved that the seepage velocity $q^0 = \int_{Y_F} v_0 \, dy$ was a monotone and coercive as a function of the right hand side on a neighborhood of zero.

For an exhausitive discussion about the existence and uniqueness problems for (3.11) we advise the reader to consult [MaMi]. Here we give only a very simple existence and uniqueness result using the Implicit Function Theorem.

Let the functional space $\mathcal{W}(Y_F)$ be defined by (1.51). We introduce next the Banach space

$$Z_{per} = \{\psi \in C^{1,\gamma}_{per}(\Omega; \mathcal{W}(Y_F)) \mid \text{div}_x \int_{Y_F} \psi(x,y)dy = 0 \text{ in } \Omega \} \qquad (3.60)$$

with $0 < \gamma < 1$. Furthermore, we introduce the nonlinear mapping G by

$$G(\{v, \tilde{\pi}, \tilde{p}\}, g) = -\mu\Delta_y v(x,y) + (v(x,y)\nabla_y)v(x,y)+$$
$$\nabla_y \tilde{\pi}(x,y) + \nabla_x \tilde{p}(x) - g(x) \qquad (3.61)$$

defined for $\{v, \tilde{\pi}, \tilde{p}\} \in \Upsilon_1 = Z_{per} \times C^{1,\gamma}_{per}(\Omega; L^2_0(Y_F)) \times C^{2,\gamma}_{per}(\Omega)$ and $g \in X = C^{1,\gamma}_{per}(\Omega)^n$. Then G is a continuous mapping between $\Upsilon_1 \times X$ and $\Upsilon_2 = C^{1,\gamma}_{per}(\Omega; \mathcal{W}(Y_F)')$. We have the following result

Theorem 3.17. *There exists a neighborhood $\mathcal{N}$ of 0 in $C^{1,\gamma}_{per}(\Omega)^n$, $0 < \gamma < 1$, such that the problem (3.11) is uniquely solvable in $Z_{per} \times C^{1,\gamma}_{per}(\Omega; L^2_0(Y_F)) \times (C^{2,\gamma}_{per}(\Omega) \cap L^2_0(\Omega))$ for any $f \in \mathcal{N}$.*

Proof. We check the assumptions of the Implicit Function Theorem. Firstly, $G(0,0) = 0$ and the mapping G is continuously differentiable at $(0,0)$. Its partial Fréchet derivative with respect to $u = \{v, \tilde{\pi}, \tilde{p}\}$ at $(0,0)$ reads

$$< G_u(0,0), \{z, z_1, z_2\} > = -\mu\Delta_y z(x,y) + \nabla_y z_1(x,y) + \nabla_x z_2(x), \qquad (3.62)$$

$\forall \{z, z_1, z_2\} \in \Upsilon_1$. It remains to prove that $G_u(0,0)$ is an invertible mapping, $G_u(0,0) : \Upsilon_1 \to \Upsilon_2$.

Let $h \in \Upsilon_2 = C^{1,\gamma}_{per}(\Omega; \mathcal{W}(Y_F)')$. Then we consider the problem
Find $\{z, z_1, z_2\}$, 1-periodic in y and L_0-periodic in x, such that

$$\begin{cases} -\mu\Delta_y z(x,y) + \nabla_y z_1(x,y) + \nabla_x z_2(x) = h(x,y) & \text{in } \Omega \times Y_F, \\ \text{div}_y z(x,y) = 0 & \text{in } \Omega \times Y_F, \\ z = 0 \text{ on } \Omega \times S, \\ \text{div}_x\{\int_{Y_F} z(x,y)dy\} = 0 & \text{in } \Omega, \end{cases} \qquad (3.63)$$

Let the functional space V be given by (1.52). Then we have the following variational formulation for (3.63):

$$\begin{cases} \text{Find } z \in V \text{ such that for every } \varphi \in V \text{ we have} \\ \mu \int_\Omega \int_{Y_F} \nabla_y z(x,y)\nabla_y \varphi(x,y)dy\, dx = \int_\Omega < h(x,\cdot), \varphi(x,\cdot) >_{W',W} dx. \end{cases} \qquad (3.64)$$

Obviously, (3.64) admits a unique solution $z \in V$ for any $h \in L^2(\Omega; \mathcal{W}(Y_F)')$. In addition $h \in C^{1,\gamma}_{per}(\Omega; \mathcal{W}(Y_F)')$ implies $z \in Z_{per}$.

Corresponding considerations from Chapter 2 of [LiPe] give the equivalence of (3.64)) and (3.63) and the existence of unique $z_1 \in C^{1,\gamma}_{per}(\Omega; L^2_0(Y_F))$ and $z_2 \in C^{2,\gamma}_{per}(\Omega) \cap L^2_0(\Omega)$ satisfying (3.63).

Therefore we are in a situation to apply the Implicit Function Theorem and the proof is completed. ∎

Remark 3.18. *Clearly, it is possible to replace $C^{1,\gamma}$ with any other Banach algebra with respect to x. The results from Theorem 3.17 extend directly to the choice of $C^{0,\gamma}$, L^∞, $C^{0,1}$ or $C^{k,\gamma}$ as regularity with respect to x.*

We do not know much about the neighborhood $\mathcal{N}$ and it is the weak point of the above result. In general $\mathcal{N}$ depends on the pore geometry (i.e. on Y_F) and on the porous medium Ω on way which can be fairly complicated.

It is of some interest to know more about $\mathcal{N}$ and it is proved in [MaMi] that $\mathcal{N}$ depends only on the porosity $|Y_F|$, the permeability K and the domain Ω. Furthermore, an explicit bound on f assuring the solvability is given. Such improvement requires a different existence proof, involving the monotonicity properties of the seepage velocity.

More precisely, in the analogy with the linear case the auxiliary problem is studied

$$\begin{cases} -\mu\Delta w(\alpha) + (w(\alpha)\nabla)w(\alpha) + \nabla\pi(\alpha) = \alpha \text{ in } Y_F, & \text{div } w(\alpha) = 0 \text{ in } Y_F, \\ \{w(\alpha), \pi(\alpha)\} \text{ is } Y\text{-periodic}, & w(\alpha) = 0 \text{ on } S. \end{cases}$$
(3.65)

Now we introduce a $\mathbb{R}^n$-valued function $\mathcal{U}(\alpha)$ by

$$\mathcal{U}(\alpha) = \int_{Y_F} w(\alpha)\, dy. \tag{3.66}$$

Obviously, $\mathcal{U}$ is well-defined only if the problem (3.65) has a unique solution. In [MaMi] it is proved that $\mathcal{U}$ is smooth and monotone and the effective pressure field p is defined through the following well posed quasi-linear problem in Ω :

$$\begin{cases} \text{div } \mathcal{U}(f - \nabla p_0) = 0 & \text{in } \Omega \\ \quad\quad p_0 \text{ is } \quad L_0 - \text{periodic .} \end{cases}$$

However, at first glance there are no relationship between the homogenized problem (3.11) and filtration laws (3.2) and (3.3). We conlude this subsection by explaining results from [MaMi] establishing such relationship. The idea is to establish the analiticity of $\mathcal{U}$ in a neighborhood of zero. Let $\sigma_0 = (1 - |Y_F|)^3/(\sqrt{|Y_F|}C(n))$ with $C(n) = n^{3n/2}\omega_n^{3-3/n}(3n/(4-n))^{3/2}$. Then we have the following result from [MaMi] :

Theorem 3.19. *The function $\mathcal{U}$ is (real) analytic in B_ρ, where*

$$\rho = \sigma_0\{2\cdot 3^n + \mu - 4 - 2\cdot\sqrt{(3^n - 2)\cdot(3^n + \mu - 2)}\}. \tag{3.67}$$

Furthermore,

$$|D_\alpha^m \mathcal{U}(\alpha)| \leq C\frac{m!}{\pi(m)^{3/2}}\left(\frac{4(3^n - 2)\sigma_0}{(\mu\sigma_0 - |\alpha|)^2}\right)^{|m|}, \quad m \in \mathbb{N}_0^n,$$

where $\pi(m)$ denotes the product of all nonzero components of m.

Corollary 3.20. *The expansions*

$$\mathcal{U}(\alpha) = \sum_{m \in \mathbb{N}^n} \frac{1}{m!}(D_\alpha^m \mathcal{U})(0)\alpha^m \tag{3.68}$$

and (3.35), with $Re_\varepsilon^{loc} = 1$, coincide in B_ρ and (3.68) can be viewed as the limit $Re_\varepsilon^{loc} \to 1$ of the asymptotic expansion (3.35). In addition, the seepage velocity $q^0 = \mathcal{U}(f - \nabla p_0)$ can be approximated by polynomials in $f - \nabla p_0$, for $|f - \nabla p_0| < \mu\sigma_0$, and the quadratic approximation reads

$$q^0 = \frac{1}{\mu}K(f - \nabla p_0) + \frac{1}{\mu^3}\sum_{i,j=1}^n M^{ij}(f_i - \frac{\partial p_0}{\partial x_i})(f_j - \frac{\partial p_0}{\partial x_j}) + O((|f - \nabla p_0|^3). \tag{3.69}$$

where M_k^{ij} are given by (3.31).

Let us now establish a quadratic filtration law of the form (3.2) or (3.3). It can be considered as a non-classical Dupuit - Ergun - Forchheimer's law obtained directly from the conservation of momentum equation.

Corollary 3.21. *Let $|f - \nabla p_0| < \rho$, where ρ is given by (3.67). Then we have the following relation between the pressure gradient and the filtration velocity*

$$\frac{K}{\mu}(f - \nabla p_0) = q^0 - \frac{1}{\mu}\sum_{1 \leq j,k \leq n} \mathcal{H}^{jk}q_j^0 q_k^0 + O(|q^0|^3), \tag{3.70}$$

where the components of the vector $\mathcal{H}^{jk}$ are defined by

$$\mathcal{H}_i^{jk} = \sum_{1 \leq l,r \leq n} M_i^{lr}(K^{-1})_{lj}(K^{-1})_{rk} , \ i = 1,\ldots,n, \tag{3.71}$$

Remark 3.22. *As in [WL] we notice that $M_i^{\ell p} + M_p^{\ell i} + M_\ell^{ip} = 0$ implying that $M_i^{ii} = 0$.*

Remark 3.23. *It was proved in [FG] and [FGL] that in case of an isotropic porous medium $T^{\ell p} = 0$ (and consequently $\mathcal{H}^{\ell p} = 0$). In fact, it is easy to see that if $Y_F - (\frac{1}{2}, \frac{1}{2})$ is invariant under transformations $y_j \to -y_j$, $j = 1, \ldots, n$, then*

$$w(\alpha)(y) = -w(-\alpha)(-y) \ , \ \ \pi(\alpha)(y) = \pi(-\alpha)(-y).$$

Therefore the function $\mathcal{U}$ is odd, i.e.

$$\mathcal{U}(-\alpha) = -\mathcal{U}(\alpha) \ .$$

As the consequence, in that case not only $T^{\ell p} = 0$ but our expansion (3.67) contains only odd powers of α.

We shall compare the nonlinear filtration law (3.70) with the existing engineering literature in § 3.4

3.3 Convergence of the homogenization procedure in the fully non-linear case

In this subsection we prove the convergence of the homogenization procedure for $\gamma_1 = 1$. We mimic the proof from the case $\gamma_1 < 1$, but there will some additionnal difficulties coming from the inertial term. We use the solution to the problem (3.11) constructed in § 3.2, i.e. we deal with the solution $\{v_0, \pi_1, p_0\} \in W_{per}^{2,r}(\Omega; \tilde{Z}) \times W_{per}^{2,r}(\Omega; W_{per}^{1,r}(Y_F) \cap L_0^2(Y_F)) \times (W_{per}^{3,r}(\Omega) \cap L_0^2(\Omega))$, $r > n$, for (3.11), where $\tilde{Z}$ is the subspace of $\mathcal{W}(Y_F)$ given by

$$\tilde{Z} = \{\varphi \in \mathcal{W}(Y_F) \mid \varphi \in W^{1,\infty}(Y_F)^n\}. \tag{3.72}$$

Now set

$$v_0^\varepsilon(x) = \begin{cases} v_0(x, \frac{x}{\varepsilon}), & x \in \Omega^\varepsilon \\ 0, & x \in \Omega \setminus \Omega^\varepsilon \end{cases} \tag{3.73}$$

and

$$\pi_1^\varepsilon(x) = \pi_1(x, \frac{x}{\varepsilon}), \ x \in \Omega^\varepsilon. \tag{3.74}$$

As mentioned before, $\operatorname{div} v_0^\varepsilon$ introduces important compressibility effects. We follow the construction from § 1.8 and take $Q \in W_{per}^{1,r}(\Omega; W_{per}^{1,r}(Y_F)^n)$, $r > n$, satisfying (1.78) with $\operatorname{div}_x v^0$ replaced by $\operatorname{div}_x v_0$.

Now we set

$$Q^\varepsilon(x) = \begin{cases} \varepsilon Q(x, \frac{x}{\varepsilon}), & x \in \Omega^\varepsilon \\ 0, & x \in \Omega \setminus \Omega^\varepsilon \end{cases} \tag{3.75}$$

and state the result :

Theorem 3.24. *Let* $f \in W_{per}^{2,r}(\Omega)^n$ *,* $r > n$ *,* $\operatorname{div} f = 0$ *be such that*

$$\|f - \nabla p_0\|_{L^\infty(\bar{\Omega})^n} \leq \frac{\mu^2}{2} \frac{(1 - |Y_F|)^3}{|Y_F|^{1/2} n^{3n/2-2}} \omega_n^{3/n-3} \left(\frac{4-n}{3n}\right)^{3/2} \tag{3.76}$$

Then we have

$$\|v_\varepsilon - v_0^\varepsilon + Q^\varepsilon\|_{H_{per}(\Omega, \operatorname{div})} \leq C\varepsilon \tag{3.77}$$

where v_ε *is a solution for (3.4)-(3.6), with* $\gamma_1 = 1$, *extended by zero to* Ω. *Furthermore, let* $\tilde{\Pi}^\varepsilon$ *be the extension of* $\Pi^\varepsilon = \varepsilon p_\varepsilon - p_0 - \varepsilon \pi_1^\varepsilon$ *given by (1.27). Then we have the following estimate*

$$\|\tilde{\Pi}^\varepsilon\|_{L_0^s(\Omega)} \leq C\varepsilon, \ \forall s \in]1, \frac{n}{n-1}[\ . \tag{3.78}$$

Now, as in § 1.8, we find the equations for v_0^ε , $v_0^\varepsilon - Q^\varepsilon$ and $v_\varepsilon - v_0^\varepsilon + Q^\varepsilon$, respectively. Clearly $\{v_0^\varepsilon, \pi_1^\varepsilon\}$ satisfies the stationary Navier-Stokes system

$$\begin{cases} -\mu\varepsilon^2 \Delta v_0^\varepsilon + \varepsilon(v_0^\varepsilon \nabla)v_0^\varepsilon + \nabla(p_0 + \varepsilon\pi_1^\varepsilon) = f - \Psi_1^\varepsilon \ \text{in} \ \Omega^\varepsilon \\ \operatorname{div} v_0^\varepsilon = \operatorname{div}_x v_0^\varepsilon \ \text{in} \ \Omega^\varepsilon \\ v_0^\varepsilon = 0 \ \text{on} \ \partial\Omega^\varepsilon \setminus \partial\Omega \ , \ \{v_0^\varepsilon, \varepsilon\pi_1^\varepsilon + p_0\} \ \text{is} \ L_0 - \text{periodic} \end{cases} \tag{3.79}$$

where the subscript x denotes the differentiation only with respect to the slow variable and

$$\Psi_{1,j}^\varepsilon = \mu\varepsilon^2 \Delta_x v_{0,j}^\varepsilon - 2\mu\varepsilon^2 \mathrm{div}(\nabla_x v_{0,j}^\varepsilon) + \varepsilon(v_0^\varepsilon \nabla_x)v_{0,j}^\varepsilon + \varepsilon\frac{\partial \pi_1^\varepsilon}{\partial x_j}. \tag{3.80}$$

It is easy to verify that

$$\begin{cases} \|\frac{\partial^2 v_0^\varepsilon}{\partial x_i \partial x_j}\|_{L^2(\Omega^\varepsilon)^n} \le C; \quad \|\nabla_x v_0^\varepsilon\|_{L^r(\Omega^\varepsilon)^{n^2}} \le C \\ \|v_0^\varepsilon\|_{L^\infty(\Omega^\varepsilon)^n} \le C; \quad \|\nabla v_0^\varepsilon\|_{L^r(\Omega^\varepsilon)^{n^2}} \le C\varepsilon \\ \|\pi_1^\varepsilon\|_{L^\infty(\Omega^\varepsilon)} \le C; \quad \|\nabla_x \pi_1^\varepsilon\|_{L^2(\Omega^\varepsilon)^n} \le C \ , \ r > n \end{cases} \tag{3.81}$$

and consequently

$$|<\Psi_1^\varepsilon,\varphi>_{\Omega^\varepsilon}| \le C\varepsilon^2 \|\nabla\varphi\|_{L^2(\Omega^\varepsilon)^{n^2}} \ , \ \forall\varphi \in W^\varepsilon. \tag{3.82}$$

At this stage we have to correct the compressibility effects and derive an appropriate Navier-Stokes system for $v_0^\varepsilon - Q^\varepsilon$. It reads

$$\begin{cases} -\mu\varepsilon^2 \Delta(v_0^\varepsilon - Q^\varepsilon) + \varepsilon((v_0^\varepsilon - Q^\varepsilon)\nabla)(v_0^\varepsilon - Q^\varepsilon) + \nabla(p_0 + \varepsilon\pi_1^\varepsilon) \\ \qquad\qquad = f + \Psi^\varepsilon \ \text{ in } \Omega^\varepsilon \\ \mathrm{div}\,(v_0^\varepsilon - Q^\varepsilon) = -\mathrm{div}_x Q^\varepsilon \ \text{ in } \Omega^\varepsilon \\ v_0^\varepsilon - Q^\varepsilon = 0 \ \text{ on } \partial\Omega^\varepsilon \backslash \partial\Omega, \ \{v_0^\varepsilon - Q^\varepsilon, \varepsilon\pi_1^\varepsilon + p_0\} \text{ is } L_0 - \text{periodic} \end{cases} \tag{3.83}$$

with

$$\Psi^\varepsilon = -\Psi_1^\varepsilon - \mu\varepsilon^2 \Delta Q^\varepsilon + \varepsilon(Q^\varepsilon\nabla)Q^\varepsilon + \varepsilon(Q^\varepsilon\nabla)v_0^\varepsilon + \varepsilon(v_0^\varepsilon\nabla)Q^\varepsilon \ \text{ in } \Omega^\varepsilon. \tag{3.84}$$

In view of (1.78) we obtain

$$\begin{cases} \|Q^\varepsilon\|_{L^\infty(\Omega)^n} \le C\varepsilon \ , \quad \|\nabla Q^\varepsilon\|_{L^r(\Omega)^{n^2}} \le C \ , \ r > n \ , \\ \qquad\qquad \|\nabla_x Q^\varepsilon\|_{L^r(\Omega)^{n^2}} \le C\varepsilon \end{cases} \tag{3.85}$$

and consequently

$$|<\Psi^\varepsilon,\varphi>_{\Omega^\varepsilon}| \le C\varepsilon^2 \|\nabla\varphi\|_{L^2(\Omega^\varepsilon)^{n^2}}, \ \forall\varphi \in W^\varepsilon. \tag{3.86}$$

Now we get a Navier-Stokes system in Ω^ε for $w^\varepsilon = v_\varepsilon - v_0^\varepsilon + Q^\varepsilon$. In view of (3.83)-(3.84) we have the following equations

$$\begin{cases} -\mu\varepsilon^2 \Delta w^\varepsilon + \varepsilon(v_\varepsilon \nabla)w^\varepsilon + \varepsilon(w^\varepsilon\nabla)(v_0^\varepsilon - Q^\varepsilon) + \nabla\Pi^\varepsilon = -\Psi^\varepsilon \ \text{ in } \Omega^\varepsilon \\ \qquad\qquad\qquad \mathrm{div}\,w^\varepsilon = \mathrm{div}_x Q^\varepsilon \ \text{ in } \Omega^\varepsilon \\ \qquad w^\varepsilon = 0 \ \text{ on } \partial\Omega\backslash\partial\Omega \ , \ \{w^\varepsilon, \Pi^\varepsilon\} \text{ is } L_0 - \text{periodic} \end{cases} \tag{3.87}$$

where $\Pi^\varepsilon = \varepsilon p_\varepsilon - p_0 - \varepsilon\pi_1^\varepsilon$.

As before, we observe immediately that the forcing term Ψ^ε is estimated in a satisfactory way by (3.86). It remains to estimate the pressure term using ∇w^ε. Following the construction from [Mik2] we have

Proposition 3.25. *Let $\widetilde{\Pi^{\varepsilon}}$ be the extension of Π^{ε} given by*

$$\|\widetilde{\Pi^{\varepsilon}}\|_{L_0^s(\Omega)} \leq C\varepsilon\{1 + \|\nabla w^{\varepsilon}\|_{L^2(\Omega^{\varepsilon})^{n^2}}\}, \ \forall s \in]1, \frac{n}{n-1}[. \tag{3.88}$$

Proof. The first equation in (3.87) gives

$$< \nabla \Pi^{\varepsilon}, \varphi >_{\Omega^{\varepsilon}} = -\mu\varepsilon^2 \int_{\Omega^{\varepsilon}} \nabla w^{\varepsilon} \nabla \varphi + \varepsilon \int_{\Omega^{\varepsilon}} \text{div } w^{\varepsilon} (v_0^{\varepsilon} - Q^{\varepsilon})\varphi +$$

$$+\varepsilon \int_{\Omega^{\varepsilon}} w^{\varepsilon} \otimes (v_0^{\varepsilon} - Q^{\varepsilon})\nabla\varphi - < \Psi^{\varepsilon}, \varphi >_{\Omega^{\varepsilon}} -\varepsilon \int_{\Omega^{\varepsilon}} (v_{\varepsilon}\nabla)w^{\varepsilon}\varphi \tag{3.89}$$

for any $\varphi \in W_r^{\varepsilon} = \{W^{1,r}(\Omega^{\varepsilon})^n \mid ; z = 0 \text{ on } \partial\Omega^{\varepsilon} \setminus \partial\Omega , z \text{ is } L_0 - \text{periodic }\}, r > n$

Owing to the estimates (3.86) and (3.58) we get

$$| < \nabla\Pi^{\varepsilon}, \varphi >_{\Omega^{\varepsilon}} | \leq C\varepsilon^2\{1 + \|\nabla w^{\varepsilon}\|_{L^2(\Omega^{\varepsilon})^{n^2}}\}\|\nabla\varphi\|_{L^2(\Omega^{\varepsilon})^n} +$$

$$C\varepsilon\|\nabla w^{\varepsilon}\|_{L^2(\Omega^{\varepsilon})^{n^2}}\|\varphi\|_{L^{\infty}(\Omega^{\varepsilon})^n}, \ \forall \varphi \in W_r^{\varepsilon}, \tag{3.90}$$

for any $r \in]n, +\infty[$.

Now we use the identity

$$\langle \nabla\widetilde{\Pi^{\varepsilon}}, \varphi \rangle_{\Omega} = \langle \nabla\Pi^{\varepsilon}, R_{\varepsilon}\varphi \rangle_{\Omega^{\varepsilon}}$$

and owing to the results from §1.4 and to the proof of Proposition 3.1, we get (3.88). ∎

Proof of Theorem 3.24. We test (3.87) with w^{ε}. Owing to (3.86) and (3.88) we have

$$\mu\varepsilon^2 \int_{\Omega^{\varepsilon}} |\nabla w^{\varepsilon}|^2 + \varepsilon \int_{\Omega^{\varepsilon}} (w^{\varepsilon}\nabla)(v_0^{\varepsilon} - Q^{\varepsilon})w^{\varepsilon} = \int_{\Omega^{\varepsilon}} \widetilde{\Pi^{\varepsilon}} \text{div } w^{\varepsilon} -$$

$$< \Psi^{\varepsilon}, w^{\varepsilon} >_{\Omega^{\varepsilon}} \leq \leq C\varepsilon^2(1 + \|\nabla w^{\varepsilon}\|_{L^2(\Omega^{\varepsilon})^{n^2}}). \tag{3.91}$$

It remains to estimate the second term at the left hand side. We have

$$\varepsilon \left| \int_{\Omega^{\varepsilon}} (w^{\varepsilon}\nabla)(v_0^{\varepsilon} - Q^{\varepsilon})w^{\varepsilon} \right| \leq \varepsilon \left| \int_{\Omega^{\varepsilon}} (w^{\varepsilon}\nabla)v_0^{\varepsilon}w^{\varepsilon} \right| + \varepsilon \left| \int_{\Omega^{\varepsilon}} (w^{\varepsilon}\nabla)Q^{\varepsilon}w^{\varepsilon} \right|. \tag{3.92}$$

Since ∇Q^{ε} is uniformly bounded in $L^r(\Omega)^{n^2}$, $r > n$, we obtain by interpolation

$$\varepsilon \left| \int_{\Omega^{\varepsilon}} (w^{\varepsilon}\nabla)Q^{\varepsilon}w^{\varepsilon} \right| \leq \varepsilon\|\nabla Q^{\varepsilon}\|_{L^r(\Omega)^{n^2}}\|w^{\varepsilon}\|_{L^{\frac{2r}{r-1}}(\Omega)^n}^2. \tag{3.93}$$

Owing to $r > n$, Poincaré's inequality and Sobolev's injection, we have

$$\|w^{\varepsilon}\|_{L^{\frac{2r}{r-1}}(\Omega)^n} \leq C\varepsilon^{\theta}\|\nabla w^{\varepsilon}\|_{L^2(\Omega)^{n^2}}$$

with $\theta = 1 - 3/(2r) > 1/2$ for $n = 3$ and $\theta = 1 - 1/r + O(\delta) > 1/2$ for $n = 2$. Therefore we have

$$\varepsilon \left| \int_{\Omega^\varepsilon} (w^\varepsilon \nabla) Q^\varepsilon w^\varepsilon \right| \le C\varepsilon^\gamma \|\nabla w^\varepsilon\|^2_{L^2(\Omega^\varepsilon)^{n2}}, \tag{3.94}$$

with $\gamma > 2$. The first term on the right hand side of (3.92) is more complicated to estimate. Let

$$(Y_F)^k_\varepsilon = \varepsilon(k + Y_F) \subset \Omega .$$

Then after simple change of variables we obtain

$$\int_{(Y_F)^k_\varepsilon} \|\nabla v^\varepsilon_0\|^2 dx \le \varepsilon^{n-2} \sup_{x \in \Omega} \int_{Y_F} |\nabla_y v_0(x,y)|^2 dy + C\varepsilon^{n-1}. \tag{3.95}$$

Back to the Navier-Stokes system with two pressures (3.11), we test it with v_0 and obtain

$$\mu \int_{Y_F} |\nabla_y v_0|^2 \, dy = (f(x) - \nabla_x p_0(x)) \int_{Y_F} v_0 \, dy$$
$$\le |f(x) - \nabla_x p_0(x)| |Y_F|^{1/2} \|v_0\|_{L^2(Y_F)^n}. \tag{3.96}$$

From the other side, it is well-known that

$$\|\varphi\|_{L^2(Y_F)} \le \frac{n^{n/2}}{1 - |Y_F|} \omega_n^{1-1/n} \|\nabla \varphi\|_{L^2(Y_F)^n}, \ \forall \varphi \in H^1(Y_F), \ \varphi = 0 \ \text{on} \ \partial Y_F \setminus \partial Y,$$

where $\omega_n = 2\pi^{n/2}/(n\Gamma(n/2))$ is the volume of the unit ball in $\mathbb{R}^n$ (see e.g. [GT]). Consequently, (3.95)-(3.96) gives

$$\|\nabla v^\varepsilon_0\|_{L^2((Y_F)^k_\varepsilon)} \le \varepsilon^{n/2-1} \frac{|f(x) - \nabla_x p_0(x)|}{\mu} \frac{|Y_F|^{1/2} n^{n/2}}{1 - |Y_F|} \omega_n^{1-1/n} + C\varepsilon^{n/2 - 1/2}. \tag{3.97}$$

Using (3.97), and the precise Sobolev injection inequality (see e.g. [GT])

$$\|\varphi\|_{L^4(Y_F)} \le \frac{n^{n/2-1}}{1 - |Y_F|} \omega_n^{1-1/n} \left(\frac{3n}{4-n}\right)^{3/4} \|\nabla \varphi\|_{L^2(Y_F)^n},$$
$$\forall \varphi \in H^1(Y_F), \ \varphi = 0 \ \text{on} \ \partial Y_F \setminus \partial Y,$$

the estimate of the first term in (3.92) reads

$$\varepsilon \left| \int_{\Omega^\varepsilon} (w^\varepsilon \nabla) v^\varepsilon_0 w^\varepsilon \right| \le \varepsilon \sum_k \left| \int_{(Y_F)^k_\varepsilon} (w^\varepsilon \nabla) v^\varepsilon_0 w^\varepsilon \right| \le$$

$$\varepsilon^{n/2} \sum_k \frac{\|f - \nabla_x p_0\|_{L^\infty(\Omega)^n}}{\mu} \frac{|Y_F|^{1/2} n^{n/2}}{1 - |Y_F|} \omega_n^{1-1/n} \|w^\varepsilon\|^2_{L^4((Y_F)^k_\varepsilon)} \le$$

$$\frac{\|f - \nabla_x p_0\|_{L^\infty(\Omega)^n}}{\mu} \frac{|Y_F|^{1/2} n^{3n/2-2}}{(1 - |Y_F|)^3} \omega_n^{3-3/n} \left(\frac{3n}{4-n}\right)^{3/2} \varepsilon^{n/2} \times$$

$$\sum_k \varepsilon^{2- n/2} \|\nabla w^\varepsilon\|^2_{L^2((Y_F)^k_\varepsilon)^{n2}} \le$$

$$\frac{\|f - \nabla_x p_0\|_{L^\infty(\Omega)^n}}{\mu} \frac{|Y_F|^{1/2} n^{3n/2-2}}{(1 - |Y_F|)^3} \omega_n^{3-3/n} \left(\frac{3n}{4-n}\right)^{3/2} \varepsilon^2 \|\nabla w^\varepsilon\|^2_{L^2(\Omega^\varepsilon)^{n2}} \tag{3.98}$$

After inserting (3.94) and (3.98) into (3.91) and using the condition (3.76) we obtain

$$\|\nabla w^\varepsilon\|_{L^2(\Omega)^{n^2}} \leq C$$

which implies the theorem. ∎

Remark 3.26. *For an explicit estimate on f assuring (3.76) we refer to [MaMi].*

3.4 Conclusion

In this chapter we derived the non-linear filtration law (3.2) starting from the stationary incompressible Navier-Stokes equations in a periodic porous medium, with the local Reynolds number of order 1 or of an order not much smaller than 1. For $Re_\varepsilon^{loc} = O(1)$ this relationship is nonlocal and described by the stationary incompressible Navier-Stokes system in the fluid part of the unit cell of the porous medium (the representative elementary volume) with $f - \nabla p$ as the external force. This is the effective constitutive law derived in [LiPe] and [SP80]. We proved solvability of the homogenized problem (3.11) for f not too large and found out a constitutive law of the form

$$q^0(x) = \int_{Y_F} v_0(x,y)\, dy = \mathcal{U}\big(f(x) - \nabla p(x)\big), \qquad (3.99)$$

where q^0 is the filtration velocity and $\mathcal{U}$ is a nonlinear function defined by (3.65)-(3.66) on some ball around zero.

At first glance, the nonlinear filtration law (3.99) doesn't equal the Forchheimer's law (3.3). Furthermore, we have other nonlinear filtration laws as e.g. the Sokolovskij-Rose law (see [Sch]).

At the other hand, for the local Reynolds number $Re_\varepsilon^{loc} = \varepsilon^{1-\gamma_1}$, $\gamma_1 \to 1$, we established the formula (3.36). It determinates the function F in (3.2) up to order $(Re_\varepsilon^{loc})^2$ and, using (3.35), it is can be easily generalized to any order of accurancy. Therefore the results from § 3.1 allow to determine F in the range of gradual transition and the expression for F is *rigorously justified*.

Furthermore, we established the analyticity of $\mathcal{U}$ in a ball around zero. It allowed to find the relation between $\mathcal{U}$ and the weakly nonlinear filtration laws (3.33) and (3.35), corresponding to the local Reynolds number proportional to some positive power of ε (the sub-critical case). More precisely, we proved that one obtains $\mathcal{U}$ in the limit $Re_\varepsilon^{loc} \to 1$. Consequently, it is possible to approximate the nonlinear filtration law (3.99) by the quadratic law (3.70)

$$\frac{K}{\mu}(f - \nabla p_0) = q^0 - \frac{1}{\mu}\sum_{1 \leq j,k \leq n} \mathcal{H}^{jk} q_j^0 q_k^0 + O(|q^0|^3).$$

Forchheimer effects are now described through the quantities $\mathcal{H}_i^{jk}$, which confirms the observation by [RuMa] that they occur because microscopic inertial effects alter the velocity and pressure fields. For large f, the monotonicity properties seem to be lost.

The classical Forchheimer's law (3.3) contradicts the analyticity of $\mathcal{U}$ in a neighborhood of zero. It is likely that one should replace $|q|$ by $\sqrt{\delta + |q|^2}$, $\delta > 0$.

Using the continuum approach it was heuristically shown in [HG] that the growth of microscopic viscous forces (drag forces) at high velocities give rise to nonlinear effects. Then, employing the constitutive theory, they developed a general nonlinear relationship of the form

$$f(x) - \nabla p(x) = R(|v|)$$

where R depends also on the porosity, density and temperature. After the series expansion of R in terms of $|v|$, they got the following variant of (3.3):

$$f(x) - \nabla p(x) = \big(a + b|v(x)| + c|v(x)|^2\big)v(x),$$

The analyticity of $\mathcal{U}$ around zero suggests $b = 0$.

It is important to note that $\mathcal{H}^{jk} = 0$ for isotropic media (see e.g. [FG] and [FGL]). Then, the important second-order term disappears from (3.70) and it was one of the reasons for introducing the term $b|v|v$. However, the quadratic filtration law (3.70) holds true only for a limited range of $|v|$ and for large values the Navier-Stokes system with two pressures (3.11) should be used.

4. On the effective boundary conditions at a naturally permeable boundary

4.0 Introduction

We have derived the law of Darcy, describing a creeping flow through a porous medium, in § 1. For its derivation the periodicity of the porous medium was required. The periodicity condition can be relaxed to a kind of statistical homogeneity and ergodicity, but clearly such assumptions break down close to the boundaries. Deviations from Darcy's law are expected only in thin layers near the interfaces, however they can significantly change the structure of coefficients and even the effective constitutive law.

The simplest possible problem is to find relationship between the seepage velocity and the pressure gradient for an incompressible viscous flow through a domain consisting of two different periodic porous media separated by an interface. The argument from § 1 is local and we get the Darcy law in every porous piece. However, due to the different geometric structures, the permeability matrix are different. In order to couple the flow we need conditions at the interface. From the incompressibility condition we conclude immediately the continuity of the normal components of the seepage velocities. Another physically natural interface condition is the continuity of the effective pressure field. However, it is usually imposed without discussion. For a rigorous derivation we refer to [JM5], where the pressure continuity was obtained after using the corresponding boundary layers from [JM3]. It is interesting to note the deterioration of the error estimate. In general the approximation of the pressure is of order $O(\varepsilon^{1/8})$, contrary to the order $O(\varepsilon)$ obtained in § 1.8 in the absence of the boundaries. The same deterioration is present when introducing the velocity corrector. It is due to the jump of the normal derivatives of the effective pressure at the boundary. For discussion of these interface conditions, sometimes called the *refraction at a boundary between two porous bodies* , at physical level of rigour, we refer to [DA].

The next interesting problem is the effective behavior of the seepage velocity at an impervious boundary. The incompressibility condition and the no-slip boundary condition imply that the normal component of the seepage velocity is zero. It is interesting to find out what is the error estimate for the approximations of the pressure and velocity fields. As already mentioned, the case of a general bounded domain was considered in [MPM] and an error estimate was obtained. Presence of an arbitrary boundary introduces serious technical difficulties and construction of an approximation of order $O(\varepsilon)$ is possible only in the case of special geometries. The case of a flat impervious boundary was considered in [JM2]. Using the boundary layers constructed in [JM3], an approximation of order $O(\varepsilon)$ was constructed and it was shown that the tangential component of the seepage velocity is of order $O(1)$.

The above described cases are obvious from modelling point of view. The problem of finding effective boundary conditions at a naturally permeable wall is much more complicated. Namely, if the flow region contains a porous medium, a channel with a free flow and an interface between them, then one wishes to have an effective model. As before we are considering a slow viscous and incompressible flow. Clearly, the effective flow through a porous medium will be described by the Darcy law. In the channel, the free fluid flow remains governed by the Stokes system (or by the Navier-Stokes system if the inertia effects in the free fluid are important). We note that this means that one should couple two systems of PDE's, one being a second order system for the velocity and a first order equation for the pressure, respectively, and the other being a scalar second order equation for the pressure and a first order system for the seepage velocity. The coupling conditions should be imposed at the interface. One coupling condition is very simple. It is a consequence of the incompressibility and says that we have the continuity of the normal mass flux. This is not enough for determinating the effective flow and one should specify more conditions. Classically, the tangential velocity of the free fluid velocity was set to zero at the interface. This condition corresponds to an impervious boundary and could not be justified, neither from mathematical nor modelling or experimental point of view.

G.S. Beavers and D.D. Joseph concluded experimentally in [BJ] that the difference, between the slip velocity of the free fluid and the tangential component of the seepage velocity at the interface, was proportional to the shear stress from the free fluid. This law was justified at a physical level of rigour by P.G. Saffman in [SAF], where it was observed that the seepage velocity contribution could be neglected. He used a statistical approach to extend Darcy's law to non-homogeneous porous media. However, it should be noted that his argument is not entirely satisfactory since he made an *ad hoc* hypothesis about the representation of the averaged interfacial forces as a linear integral functional of the velocity, with an unknown kernel. A similar argument is developped in [DA], where Slattery's linear relationship between the pressure gradient and a combination of derivatives of the seepage velocity was assumed. Neither paper [SAF] nor [DA] contain construction of the boundary layers describing the flow behavior close to the interface. The Saffman's modification of the law by Beavers and Joseph is widely accepted and it reads

$$\sqrt{k^\varepsilon}\frac{\partial v_\tau}{\partial \nu} = \alpha v_\tau + O(k^\varepsilon). \tag{4.1}$$

Here α is a dimensionless parameter depending on the geometrical structure of the porous medium, ε is the characteristic pore size and $k^\varepsilon = \varepsilon^2 k$ is the (scalar) permeability. ν denotes the unit normal vector at the interface and v_τ is the slip velocity of the free fluid in the channel.

In the papers [ESP] and [LSP], H.Ene, Th. Levy and E. Sanchez-Palencia have undertaken the effort to find the effective interface laws by a formal asymptotic argument. They have considered two essentially different cases. The case of the flow in a cavity, lying inside of a porous matrix, was considered in [LSP]. By comparing the orders of the magnitude of characteristic quantities, it was found out that the effective pressure should be constant at the interface. This conclusion was rigorously justified in [JM3], after constructing the appropriate boundary layers.

The second case corresponds to the flow considered by Beavers and Joseph. In the paper [ESP] the continuity of the effective pressure was deduced, but without a rigorous argument or an asymptotic expansion. From modelling point of view this interface law is acceptable. It can be considered as an alternative to (4.1), however it should be noted that the well-posedness of the averaged problem is not clear.

In this section we are going to justify the law (4.1) by the technique developped in [JM1] for Laplace's operator and then in [JM3] for the Stokes system. We suppose the conditions of the experiment from [BJ], i.e. we consider the 2D laminar stationary incompressible viscous flow over a porous bed. The flow is governed by the pressure drop $p_b - p_0$ over the bed of length b. The mathematical justification of the law (4.1) for the Navier-Stokes system and the boundary conditions for the pressure on the inlet and outlet boundaries is in [JM6]. Since the inertia effects and the outer boundary layers effects, due to the choice of the pressure boundary conditions, are not of the fundamental importance for the study of the interface boundary conditions, we'll make some non-essential simplifications. First, we neglect the inertial term. We note that anyhow we are note able to find the boundary behavior for the turbulent free flow. The nonlinear stability results for the laminar Navier-Stokes system are in [JM6]. Second, we suppose that the boundary is sufficiently long and one can suppose the periodic boundary conditions at inlet/outlet boundary. The flow is then governed by a force coming from the pressure drop and equal to $\dfrac{p_b - p_0}{b} e_1$.

Let us also mention the derivation of the the effective laws for flows through sieves and filters. We mention only the papers [CO] and [JM4]. The paper [JM4] is on the effective equations for a viscous incompressible flow through a filter of a finite thickness and it also uses the boundary layers developed in [JM3].

This section is organized as follows: In §4.1 we construct the necessary boundary layer and in §4.2 the law (4.1).

4.1 Navier's boundary layer

As observed in hydrology, the phenomena relevant to the boundary occur in a thin layer surrounding the interface between a porous medium and a free flow. Theory of the boundary layers at the interface between a perforated and a non-perforated domain, was first developped in [JM1] for Laplace's operator with Dirichlet boundary conditions. The corresponding theory for the Stokes system is in [JM3] and in this subsection we are going to present a self-contained construction of the main boundary layer, used for determining the coefficient α in (4.1) and, also, for a rigorous justification of the law by Beavers and Joseph. Since the law by Beavers and Joseph is an exemple of the Navier slip condition, we call it Navier's boundary layer.

Let $Y = (0,1)^2$ and Z^* a Lipschitz domain strictly contained in Y. We introduce the pore space Y_F by $Y_F = Y \setminus \bar{Z}^*$, $S = (0,1) \times \{0\}$, $Z^+ = (0,1) \times$

$(0, +\infty)$ and the semi-infinite porous slab $Z^- = \cup_{k=1}^{\infty}(Y_F - \{0, k\})$. The flow region is then $Z_{BL} = Z^+ \cup S \cup Z^-$.

We consider the following problem:

Find $\{\beta^{bl}, \omega^{bl}\}$ with square-integrable gradients satisfying

$$-\Delta_y \beta^{bl} + \nabla_y \omega^{bl} = 0 \qquad \text{in } Z^+ \cup Z^- \qquad (4.2)$$

$$\text{div}_y \beta^{bl} = 0 \qquad \text{in } Z^+ \cup Z^- \qquad (4.3)$$

$$[\beta^{bl}]_S(\cdot, 0) = 0 \qquad \text{on } S \qquad (4.4)$$

$$[\{\nabla_y \beta^{bl} - \omega^{bl} I\} e_2]_S(\cdot, 0) = e_1 \quad \text{on } S \qquad (4.5)$$

$$\beta^{bl} = 0 \quad \text{on } \cup_{k=1}^{\infty}(\partial Z^* - \{0, k\}), \qquad \{\beta^{bl}, \omega^{bl}\} \text{ is } 1 - \text{periodic in } y_1 \ (4.6)$$

Let $V = \{z \in L^2_{loc}(Z_{BL})^2 : \nabla_y z \in L^2(Z_{BL})^4; z \in L^2(Z^-)^2; z = 0 \text{ on } \cup_{k=1}^{\infty}(\partial Z^* - \{0, k\}); \text{div}_y z = 0 \text{ in } Z_{BL} \text{ and } z \text{ is } 1\text{-periodic in } y_1\}$. Then, by Lax-Milgram lemma, there is a unique $\beta^{bl} \in V$ satisfying

$$\int_{Z_{BL}} \nabla \beta^{bl} \nabla \varphi \, dy = -\int_S \varphi_1 \, dy, \qquad \forall \varphi \in V. \qquad (4.7)$$

Using De Rham's theorem we obtain a function $\omega^{bl} \in L^2_{loc}(Z^+ \cup Z^-)$, unique up to a constant and satisfying (4.2). By the elliptic theory, $\{\beta^{bl}, \omega^{bl}\} \in C^{\infty}(Z^+ \cup Z^-)^3$ and $\forall R > 0$, $\{\beta^{bl}, \omega^{bl}\} \in H^2\left((Z^+ \cap \{0 < y_2 < R\}) \cup (Z^- \cap \{-R < y_2 < 0\})\right)^2 \times H^1\left((Z^+ \cap \{0 < y_2 < R\}) \cup (Z^- \cap \{-R < y_2 < 0\})\right)$. Following [JM3] we can eliminate the shear stress jump by modifying β^{bl} and for a neighbourhood $\mathcal{O}$ of S, which does not include any solid boundary, we have $\beta^{bl} - ((y_2 - y_2^2/2)e^{-y_2} H(y_2), 0) \in W^{2,q}(\mathcal{O})^2$ and $\omega^{bl} \in W^{1,q}(\mathcal{O})$, $\forall q \in [1, \infty)$.

The goal of this subsection is to prove that the system (4.2)-(4.6) describes a boundary layer, i.e. that β^{bl} and ω^{bl} stabilize exponentially towards constants, when $|y_2| \to \infty$.

Since we are studying an incompressible flow, it is useful to prove properties of the conserved averages.

Lemma 4.1 ([JM3]). *Any solution* $\{\beta^{bl}, \omega^{bl}\}$ *satisfies*

$$\int_0^1 \beta_2^{bl}(y_1, b) \, dy_1 = 0, \qquad \forall b \in \mathbb{R} \qquad (4.8)$$

$$\int_0^1 \omega^{bl}(y_1, b_1) \, dy_1 = \int_0^1 \omega^{bl}(y_1, b_2) \, dy_1, \qquad \forall b_1 > b_2 \geq 0 \qquad (4.9)$$

$$\int_0^1 \beta_1^{bl}(y_1, b_1) \, dy_1 = \int_0^1 \beta_1^{bl}(y_1, b_2) \, dy_1, \qquad \forall b_1 > b_2 \geq 0 \qquad (4.10)$$

$$\int_0^1 \beta_1^{bl}(y_1, 0) \, dy_1 = -\int_{Z_{BL}} |\nabla \beta^{bl}(y)|^2 \, dy \qquad (4.11)$$

Proof : (4.8) is obtained by integrating the incompressibility condition (4.3) between 0 and 1 with respect to y_1, for $y_2 = b$.

In order to prove (4.9) we integrate the second component of the momentum equation (4.2) over the rectangle $(0, 1) \times (b_2, b_1)$. We have

$$\int_0^1 \left(\frac{\partial \beta_2^{bl}}{\partial y_2} - \omega^{bl} \right) (y_1, b_1) \, dy_1 = \int_0^1 \left(\frac{\partial \beta_2^{bl}}{\partial y_2} - \omega^{bl} \right) (y_1, b_2) \, dy_1,$$

and, after using (4.8), we obtain (4.9).

Integration of the first component of (4.2)

$$\text{div } \{\nabla \beta_1^{bl} - \omega^{bl} e_1\} = 0$$

between 0 and 1 with respect to dy_1 gives $\int_0^1 \beta_1^{bl}(y_1, y_2) \, dy_1 = Ay_2 + B$, for some constants A, B. Since β^{bl} is a smooth function for $y_2 \geq 0$ and $\nabla \beta^{bl} \in L^2(Z_{BL})^4$, we have $A = 0$ and (4.10) is proved.

(4.11) is the equation (4.7) tested with $\varphi = \beta^{bl}$. ∎

In the next step we determine the decay in Z^+ by reduction to the Laplace operator. We note that in such situation decay can be obtained by using Tartar's lemma from [LiPo]. However, the results from Lemma 4.1 allow us a very simple proof in this particular case.

Proposition 4.2. Let $\xi^{bl} = \text{curl } \beta^{bl} = \frac{\partial \beta_1^{bl}}{\partial y_2} - \frac{\partial \beta_2^{bl}}{\partial y_1}$. Then for every $a \in (0, 1)$ and $y \in Z^+ \cap \{y_2 > a\}$ we have

$$\begin{cases} | \, \xi^{bl}(y_1, y_2) \, | \leq C(a) e^{-2\pi y_2} \\[2mm] | \, D^\alpha \xi^{bl}(y_1, y_2) \, | \leq C(\alpha, a) e^{-2\pi y_2}. \end{cases} \tag{4.12}$$

Proof : After applying the operator curl to the equation (4.2) in $Z^+ \cap \{y_2 > a\}$, we find out that ξ^{bl} is a harmonic function. It is 1-periodic in y_1 and has C^∞-regularity at $\{y_2 = a\}$. Furthermore, $\xi^{bl} \in H^1(Z^+ \cap \{y_2 > a\})$ and $\int_0^1 \xi^{bl}(y_1, b) \, dy_1 = 0$, $\forall b > a$. Let us note that $\int_0^1 |\psi(y_1) - \int_0^1 \psi \, dz|^2 \, dy_1 \leq 4\pi^2 \int_0^1 |\frac{d\psi}{dy_1}|^2 \, dy_1$. Then the unique solution $\xi^{bl} \in W_a^+ = \{z \in H^1(Z^+ \cap \{y_2 > a\}) : z$ is 1-periodic in y_1 and $\int_0^1 z(y_1, b) \, dy_1 = 0$ for all $b \geq a\}$ for the variational problem

$$\int_{Z^+ \cap \{y_2 > a\}} \nabla \xi^{bl} \nabla \varphi \, dy = 0, \qquad \forall \varphi \in W_a^+, \; \varphi = 0 \text{ on } \{y_2 = a\} \tag{4.13}$$

is given by

$$\xi^{bl}(y_1, y_2) = \sum_{k=1}^{+\infty} \left(C_{1,k} \sin(2\pi k y_1) + C_{2,k} \cos(2\pi k y_1) \right) e^{-2\pi k y_2}, \tag{4.14}$$

where $C_{1,k}$ and $C_{2,k}$, $k \in \mathbb{N}$, are determined by the trace of $\xi^{bl} = \text{curl } \beta^{bl}$ at $\{y_2 = a\}$. This implies the first part of (4.12). The estimate for the derivatives is analogous. ∎

Knowing the behavior of curl β^{bl} and div β^{bl} in Z^+, we are able to establish a boundary layer estimate for β^{bl}.

Proposition 4.3. *Let*

$$C_1^{bl} = \int_0^1 \beta_1^{bl}(y_1, 0) dy_1. \tag{4.15}$$

Then for every $y_2 \geq 0$ and $y_1 \in (0,1)$

$$|\beta^{bl}(y_1, y_2) - (C_1^{bl}, 0)| \leq C e^{-\delta y_2}, \qquad \forall \delta < 2\pi. \tag{4.16}$$

Proof: As in [JM3], we note that in Z^+

$$\begin{cases} -\Delta \beta_1^{bl} = F_1 = -\dfrac{\partial \xi^{bl}}{\partial y_2} \quad ; \\[3mm] -\Delta \beta_2^{bl} = F_2 = \dfrac{\partial \xi^{bl}}{\partial y_1}. \end{cases} \tag{4.17}$$

By Lemma 4.1, the averages of β^{bl} over the sections $\{y_2 = b\}$ do not depend on y_2. Hence (4.17) is also satisfied by $\tilde{\beta}^{bl} = \beta^{bl} - (C_1^{bl}, 0) \in W_0^+$, which is the unique solution for

$$\int_{Z^+} \nabla \tilde{\beta}^{bl} \nabla \varphi \, dy = \int_{Z^+} F\varphi \, dy, \qquad \forall \varphi \in W_0^+, \ \varphi = 0 \ \text{on} \ \{y_2 = 0\} \tag{4.18}$$

Using the formula (4.14) for ξ^{bl} and its derivatives, we obtain

$$\begin{cases} \tilde{\beta}_1^{bl}(y) = \displaystyle\sum_{k=1}^{+\infty} \left((D_{1,k}^1 + \tfrac{C_{1,k} y_2}{2}) \sin(2\pi k y_1) + \right. \\[3mm] \left. (D_{2,k}^1 + \tfrac{C_{2,k} y_2}{2}) \cos(2\pi k y_1) \right) e^{-2\pi k y_2}, \\[4mm] \tilde{\beta}_2^{bl}(y) = \displaystyle\sum_{k=1}^{+\infty} \left((D_{1,k}^2 - \tfrac{C_{2,k} y_2}{2}) \sin(2\pi k y_1) + \right. \\[3mm] \left. (D_{2,k}^2 + \tfrac{C_{1,k} y_2}{2}) \cos(2\pi k y_1) \right) e^{-2\pi k y_2} \end{cases}$$

and the proposition is proved. ∎

Corollary 4.4. *Let*

$$C_\omega^{bl} = \int_0^1 \omega^{bl}(y_1, 0) \, dy_1. \tag{4.19}$$

Then for every $y_2 \geq 0$ and $y_1 \in (0,1)$

$$|\omega^{bl}(y_1, y_2) - C_\omega^{bl}| \leq e^{-2\pi y_2} \tag{4.20}$$

In the last step we study the decay of β^{bl} and ω^{bl} in Z^-.

We start with an estimate on ω^{bl} over the cell $Z_k = Z^- \cap (]0, 1[\times]k, k+1[)$.

Proposition 4.5. Let β^{bl} and ω^{bl} be defined by (4.2)-(4.6) and let r_k be given by

$$r_k = \frac{1}{|Y_F|} \int_{Z_k} \omega^{bl}(y) \, dy \tag{4.21}$$

Then we have

$$\|\omega^{bl} - r_k\|_{L^2(Z_k)} \leq C\|\nabla\beta^{bl}\|_{L^2(Z_k)^4} \tag{4.22}$$

and

$$|r_{k+1} - r_k| \leq C\|\nabla\beta^{bl}\|_{L^2(Z_k \cup Z_{k+1} \cup (]0,1[\times\{k+1\}))^4}. \tag{4.23}$$

Proof: We define the function space V_k by

$$V_k = \{z \in H^1(Z_k)^2 : z = 0 \text{ on } \partial Z_k \setminus ((\{0\} \cup \{1\})\times]k, k+1[); z \text{ is } y_1 - \text{periodic}\}$$

and start from the weak form of (4.2)-(4.6):

$$\int_{Z_k} \nabla\beta^{bl}\nabla\varphi - \int_{Z_k} \omega^{bl} \, \text{div}\varphi = 0, \quad \forall\varphi \in V_k. \tag{4.24}$$

(4.24) can be written as

$$\int_{Z_k} \nabla\beta^{bl}\nabla\varphi - \int_{Z_k} (\omega^{bl} - r_k)\text{div}\varphi = 0, \quad \forall\varphi \in V_k. \tag{4.25}$$

Due to the surjectivity of div, there exists a $\varphi_k \in V_k$ being a solution to

$$\text{div } \varphi_k = \omega^{bl} - r_k \quad \text{in } Z_k \tag{4.26}$$

and satisfying

$$\|\nabla\varphi_k\|_{L^2(Z_k)^4} \leq C\|\omega^{bl} - r_k\|_{L^2(Z_k)}, \tag{4.27}$$

where C does not depend on k. We refer to [JM3] for an explicit construction. Inserting φ_k as a test function for (4.25) gives

$$\|\omega^{bl} - r_k\|_{L^2(Z_k)} \leq C\|\nabla\beta^{bl}\|_{L^2(Z_k)^4}$$

where C does not depend on k.

It remains to discuss behavior of the averages $\{r_k\}$ as $k \to -\infty$. In order to obtain it we set $Z_{k,k+1} = Z_k \cup Z_{k+1} \cup (]0,1[\times\{k+1\})$ and choose $\varphi_{k,k+1}$ which satisfies

$$\text{div } \varphi_{k,k+1} = \begin{cases} 1 & \text{in } Z_k \\ -1 & \text{in } Z_{k+1} \end{cases}$$

being 1-periodic in y_1 and zero on $\partial Z_{k,k+1} \setminus ((\{0\} \cup \{1\})\times]k, k+2[)$. After inserting $\varphi_{k,k+1}$ into the analogue of (4.25) defined on $Z_{k,k+1}$ we get

$$-\int_{Z_k} \omega^{bl} + \int_{Z_{k+1}} \omega^{bl} - \int_{Z_{k,k+1}} \nabla\beta^{bl}\nabla\varphi_{k,k+1} = 0$$

and finally (4.23). ■

At this stage we turn to the variational equation for $\{\beta^{bl}, \omega^{bl}\}$. Since it holds only for test functions with compact support we prove the following auxiliary result:

Proposition 4.6. Let $\sigma_k \in C_0^\infty(\mathbb{R}_-)$ be such that $\sigma_k = 0$ for $y \geq k+1$, $\sigma_k \geq 0$ and $\sigma_k = 1$ for $y \leq k$, $k \in \mathbb{N}_-$. Let $\{\beta^{bl}, \omega^{bl}\}$ be a solution for (4.2)-(4.6). Then

$$\int_{Z^-} |\nabla\beta^{bl}|^2 \, \sigma_k = \int_{Z_k} (\omega^{bl} - r_k)\beta^{bl}\nabla\sigma_k - \int_{Z^-} \nabla\beta^{bl}\beta^{bl}\nabla\sigma_k. \tag{4.28}$$

Proof: We start with the weak form of (4.2)-(4.6):

$$\int_{Z_{BL}} \nabla\beta^{bl}\nabla\varphi - \int_{Z_{BL}} \omega^{bl} \, \mathrm{div}\varphi = 0, \quad \forall\varphi \in C_0^\infty(Z_{BL})^2, \tag{4.29}$$

such that $\varphi = 0$ on $\Pi = \partial Z^- \setminus (\,(\{0\} \cup \{1\}0 \times (-\infty, 0) \cup S)$ and φ is 1-periodic in y_1. Now we choose $\varphi = \beta^{bl}\sigma_{k,l}$, where $\sigma_{k,l} = \sigma_k \cdot (1 - \sigma_l)$, $l \leq k - 1$. Then we get

$$\int_{Z^-} |\nabla\beta^{bl}|^2 \, \sigma_{k,l} = \int_{Z^-} \omega^{bl}\beta^{bl}\nabla\sigma_{k,l} - \int_{Z^-} \nabla\beta^{bl}\beta^{bl}\nabla\sigma_{k,l}. \tag{4.30}$$

The idea is to pass to the limit $l \to -\infty$ for fixed k. Obviously we only need to consider the first term on the right hand side, containing ω^{bl}. We write it as

$$\int_{Z^-} \omega^{bl}\beta^{bl}\nabla\sigma_{k,l} = \int_{Z_k \cup Z_l} (\omega^{bl} - r_k)\beta^{bl}\nabla\sigma_{k,l} = \int_{Z_k} (\omega^{bl} - r_k)\beta^{bl}\nabla\sigma_{k,l} +$$
$$\int_{Z_l} (\omega^{bl} - r_l)\beta^{bl}\nabla\sigma_{k,l} + (r_l - r_k)\int_{Z_l} \beta^{bl}\nabla\sigma_{k,l}. \tag{4.31}$$

Now

$$\left| (r_l - r_k)\int_{Z_l} \beta^{bl}\nabla\sigma_{k,l} \right| \leq C\|\nabla\beta^{bl}\|_{L^2(Z_l)^4} \to 0 \qquad \text{as } l \to -\infty$$

and

$$\left| \int_{Z_l} (\omega^{bl} - r_l)\beta^{bl}\nabla\sigma_{k,l} \right| \leq C\|\nabla\beta^{bl}\|_{L^2(Z_l)^4}^2 \to 0 \qquad \text{as } l \to -\infty,$$

using Poincaré's inequality.

Therefore,

$$\lim_{l \to -\infty} \int_{Z^-} \omega^{bl}\beta^{bl}\nabla\sigma_{k,l} = \int_{Z_k} (\omega^{bl} - r_k)\beta^{bl}\nabla\sigma_{k,l}$$

and (4.28) is proved. ∎

Now we are ready to prove the exponential decay for our Stokes system. Let $Z^-(k) = Z^- \cap (]0, 1[\times] - \infty, k[)$. We have

Proposition 4.7. Let β^{bl} and ω^{bl} be defined by (4.2)-(4.6). Then there exist positive constants C and γ_0, independents of k, such that

$$\int_{Z^-(k)} |\nabla\beta^{bl}|^2 \, dy_1 dy_2 \leq Ce^{\gamma_0 k} \tag{4.32}$$

for every negative integer k.

Proof: Using (4.28), (4.22) and definition of σ_k we get

$$\int_{Z^-} |\nabla\beta^{bl}|^2 \, \sigma_k \leq C\|\nabla\beta^{bl}\|^2_{L^2(Z_k)^4}, \quad \forall \delta > 0 \tag{4.33}$$

Therefore, we have

$$(1+\overline{C})\int_{Z^-(k)} |\nabla\beta^{bl}|^2 \, dy_1 dy_2 \leq \overline{C}\int_{Z^-(k+1)} |\nabla\beta^{bl}|^2 \, dy_1 dy_2.$$

Finally,

$$a_k \leq \gamma a_{k+1}, \qquad k \in I\!\!N_- \tag{4.34}$$

where

$$a_k = \|\nabla\beta^{bl}\|^2_{L^2(Z^-(k))}, \; \gamma = \overline{C}/(1+\overline{C}) < 1.$$

(4.34) implies (4.32) with $\gamma_0 = -\ln\gamma$ and $C = \int_{Z^-} |\nabla\beta^{bl}|^2 \, dy$. ∎

Corollary 4.8. Let us suppose the assumptions of Proposition 4.7. Then there exist constants κ_∞, given by

$$\kappa_\infty = \lim_{k\to-\infty} \frac{1}{|Y_F|} \int_{Z_k} \omega^{bl}(y) \, dy \tag{4.35}$$

and C_1, independent of k, such that $\forall k \in I\!\!N_-$ we have

$$\|\omega^{bl} - \kappa_\infty\|^2_{L^2(Z^-(k))} \leq C_1 e^{\gamma_0 k}. \tag{4.36}$$

Proof: Using the proof of Proposition 4.6 we obtain

$$\begin{cases} \|\omega^{bl} - r_k\|_{L^2(Z_k)} \leq C\|\nabla\beta^{bl}\|_{L^2(Z_k)^4} \\[2mm] |r_{k+1} - r_k| \leq C\|\nabla\beta^{bl}\|_{L^2(Z_{k,k+1})^4} \end{cases}$$

Hence the sequence $\{r_k\}$ has the limit κ_∞. Finally

$$\|\omega^{bl} - \kappa_\infty\|^2_{L^2(Z^-(k))} \leq 2\sum_{l=-\infty}^{k} \|\omega^{bl} - r_l\|^2_{L^2(Z_l)} + 2\,|Y_F|\sum_{l=-\infty}^{k} |\kappa_\infty - r_l|^2$$

and after summation we obtain (4.36). ∎

Since ω^{bl} is unique up to a constant, we fix it by setting $\kappa_\infty = 0$. Hence in the text which follows C_ω^{bl} has the meaning of the pressure drop between $y_2 = +\infty$ and $y_2 = -\infty$.

Remark 4.9. *If the geometry of Z^- is axysymmetric with respect to reflections around the axis $y_1 = 1/2$, then $C_\omega^{bl} = 0$. For the proof we refer to [JMN]. In [JMN] a detailed numerical analysis of the problem (4.2)-(4.6) is given. Through numerical experiments it is shown that for a general geometry of Z^-, $C_\omega^{bl} \neq 0$.*

4.2 Justification of the law by Beavers and Joseph

We consider the laminar viscous two-dimensional incompressible flow through a domain Ω consisting of the porous medium $\Omega_2 = (0,b) \times (-L,0)$, the channel $\Omega_1 = (0,b) \times (0,h)$ and the permeable interface $\Sigma = (0,b) \times \{0\}$ between them. We assume that the structure of the porous medium is periodic and generated by translations of a cell $Y^\varepsilon = \varepsilon Y$, where Y is the standard cell, $Y = (0,1)^2$, containing an open Lipshitzian set Z^*, strictly included in Y. Let $Y_F = Y \setminus \overline{Z^*}$ and let χ be the characteristic function of Y_F, extended by periodicity to $\mathbb{R}^2$. We set $\chi^\varepsilon(x) = \chi(\frac{x}{\varepsilon})$, $x \in \mathbb{R}^2$, and define Ω_2^ε by $\Omega_2^\varepsilon = \{x \mid x \in \Omega_2, \chi^\varepsilon(x) = 1\}$. Furthermore, $\Omega^\varepsilon = \Omega_1 \cup \Sigma \cup \Omega_2^\varepsilon$ is the fluid part of Ω. It is supposed that $(b/\varepsilon, L/\varepsilon) \in \mathbb{N}^2$.

Therefore, our porous medium is supposed to consist of a large number of periodically distributed channels of characteristic length ε, being small compared with a characteristic length of the macroscopic domain.

The flow is supposed to be slow and governed by the following equations

$$-\mu \Delta v^\varepsilon + \nabla p^\varepsilon = -\frac{p_b - p_0}{b} e_1 \qquad \text{in } \Omega^\varepsilon, \tag{4.37}$$

$$\operatorname{div} v^\varepsilon = 0 \qquad \text{in } \Omega^\varepsilon, \tag{4.38}$$

$$v^\varepsilon = 0 \qquad \text{on } \partial\Omega^\varepsilon \setminus \partial\Omega \quad \text{and on} \quad (0,b) \times (\{-L\} \cup \{h\}), \tag{4.39}$$

$$\{v^\varepsilon, p^\varepsilon\} \qquad \text{is } b-\text{periodic in } x_1 \tag{4.40}$$

where $\mu > 0$ is the viscosity and p_0 and p_b are given constants. $\varepsilon > 0$ is the characteristic pore size, v^ε is the velocity and p^ε is the pressure field. Problem (4.37)-(4.40) has a unique solution $\{v^\varepsilon, p^\varepsilon\} \in H^1(\Omega^\varepsilon)^2 \times L_0^2(\Omega^\varepsilon)$.

Now one would like to study of the effective behavior of the velocities v^ε and pressures p^ε as $\varepsilon \to 0$. We follow the decomposition approach from [JM6]. Firstly we observe that the classic Poiseuille flow in Ω_1, satisfying the no-slip conditions at Σ, is given by

$$\begin{cases} v^0 = \left(\dfrac{p_b - p_0}{2b\mu} x_2(x_2 - h), 0 \right) & \text{for } 0 \leq x_2 \leq h \, ; \\[2mm] \pi^0 = 0 & \text{for } 0 \leq x_1 \leq b \end{cases} \tag{4.41}$$

We extend this solution to Ω_2 by setting $v^0 = 0$ for $-L \leq x_2 \leq 0$ and keeping the same form of π^0. Now, the idea is to construct the solution to (4.37)-(4.40) as a small perturbation to the Poiseuille flow (4.41). First, we need the following simple auxiliary result:

Lemma 4.10. Let $\varphi \in H^1(\Omega_2^\varepsilon)$ be such that $\varphi = 0$ on $\partial\Omega_2^\varepsilon \setminus \partial\Omega_2$. Then we have

$$\|\varphi\|_{L^2(\Omega_2^\varepsilon)} \le C\varepsilon\|\nabla\varphi\|_{L^2(\Omega_2^\varepsilon)^2} \tag{4.42}$$

$$\|\varphi\|_{L^2(\Sigma)} \le C\varepsilon^{1/2}\|\nabla\varphi\|_{L^2(\Omega_2^\varepsilon)^2} \tag{4.43}$$

$$\|\varphi(0,\cdot)\|_{H^{1/2}(-L,0)} \le C\|\nabla\varphi\|_{L^2(\Omega_2^\varepsilon)^2} \tag{4.44}$$

Proof: The estimate (4.42) is well-known (see § 1).

Let us prove (4.43). Let $\tilde\varphi(y) = \varphi(\varepsilon y), y \in Y_F + k$. Then $\tilde\varphi \in H^1(Y_F + k)$, $\forall k$, and $\varphi = 0$ on $S_k = \partial(Y_F + k) \setminus \partial Y$. Therefore by the trace theorem and the Poincaré's inequality

$$\int_S |\tilde\varphi(y_1,0)|^2 \, dy_1 \le C \int_{Y_F} |\nabla_y\tilde\varphi|^2 \, dy.$$

Change of variables gives

$$\int_{\varepsilon S_k} |\varphi(x_1,0)|^2 \, \varepsilon^{-1}dx_1 \le C \int_{\varepsilon(Y_F+k)} \varepsilon^2 |\nabla_x\varphi|^2 \, \varepsilon^{-2}dx.$$

Now after summation over k we get

$$\left(\int_\Sigma |\varphi(x_1,0)|^2 \, dx_1\right)^{1/2} \le C\varepsilon^{1/2}\left(\int_{\Omega_2^\varepsilon} |\nabla_x\varphi(x)|^2 \, dx\right)^{1/2}$$

and (4.43) is proved.

It remains to prove (4.44). We note that the $H^{1/2}$-norm on the boundary of Y is given by

$$\|v\|^2_{H^{1/2}(0,1)} = \int_0^1 \int_0^1 \frac{|v(0,y_2') - v(0,z_2')|^2}{|y_2' - z_2'|^2} \, dy_2' \, dz_2' \le C \int_Y |\nabla_y v|^2 \, dy$$

Now, after the change of variables $x = \varepsilon y$, we obtain

$$\|\varphi(0,\cdot)\|^2_{H^{1/2}(\varepsilon((0,1)+k))} \le C\|\nabla\varphi\|^2_{L^2(\varepsilon(Y_F+k))^2}, \quad \forall k \in \mathbb{N}.$$

After summation over k one obtains (4.44). ∎

Our next step is to consider the following Stokes system in Ω_1

$$\begin{cases} -\Delta B + \nabla\beta = G_1 + \operatorname{div} G_2 & \text{in } \Omega_1 ; \\ \operatorname{div} B = \Theta & \text{in } \Omega_1 ; \\ B = \xi \text{ on } \Sigma \cup (0,b) \times \{h\}, & \{B,\beta\} \text{ is b-periodic in } x_1. \end{cases} \tag{4.45}$$

and we are interested in solvability of (4.45) for

$$\begin{cases} \xi \in L^2(\Sigma \cup (0,b) \times \{h\})^2, \ \Theta \in L^2_0(\Omega_1), \ (|\,G_1\,| + |\,G_2\,|) \in L^2(\Omega_1) \ \text{ and} \\ \int_{\Omega_1} \Theta \ dx - \int_\Sigma \xi_2 \ dx_1 + \int_{(0,b) \times \{h\}} \xi_2 \ dx_1 = 0. \end{cases}$$

$$(4.46)$$

Our goal is to estimate B using only the $L^2(\Sigma \cup (0,b) \times \{h\})$-norm of ξ and standard norms for G and Θ. Therefore we introduce the notion of the very weak solution for (4.45).

In general such very weak solutions (weaker then variational ones) are obtained by transposition. The particular case of the very weak solution for the Stokes system in bounded domains was considered in [CO]. We adapt the transposition approach from [CO] to our case the periodic boundary conditions in x_1-variable.

We start with considering the transposed problem:

Let $W_3 = \{z \in H^1(\Omega_1) : z \text{ is } b\text{-periodic in } x_1 \text{ and } \int_{\Omega_1} h = 0\}$. We take functions $\{g, h\}$ such that

$$g \in L^2(\Omega_1)^2, \ u \in W_3 \qquad\qquad (4.47)$$

and consider the problem

$$\begin{cases} -\Delta \Phi + \nabla \pi = g & \text{in } \Omega_1 \ ; \\ \text{div } \Phi = u & \text{in } \Omega_1 \ ; \qquad (4.48) \\ \Phi = 0 \text{ on } \Sigma \cup (0,b) \times \{h\}, \quad \{\Phi, \pi\} \text{ is } b\text{-periodic in } x_1 \ . \end{cases}$$

Obviously, there exists a solution $\{\Phi, \pi\} \in H^2(\Omega_1)^2 \times H^1(\Omega_1)$ for (4.48), such that Φ and $\nabla \pi$ are uniquely determined and

$$\|\Phi\|_{H^2(\Omega_1)^2} + \|\nabla \pi\|_{L^2(\Omega_1)^2} \le C\{\|g\|_{L^2(\Omega_1)^2} + \|u\|_{W_3}\}.$$

Before writing the very weak form of (4.45) we have to eliminate Θ. it is eliminated through the following auxilliary problem:

$$\begin{cases} \Delta w = \Theta - \frac{1}{|\Omega_1|} \int_{\Omega_1} \Theta \ dx & \text{in } \Omega_1 \ ; \\ \dfrac{\partial w}{\partial \nu} = 0 & \text{on } \Sigma \cup (0,b) \times \{h\} \ ; \qquad (4.49) \\ w \text{ is } b\text{-periodic in } x_1 \end{cases}$$

which has a solution $w \in H^2(\Omega_1)$, unique up to a constant, such that

$$\|\nabla w\|_{H^1(\Omega_1)^2} \le C\|\Theta\|_{L^2(\Omega_1)}$$

Let $\{g, h\}$ be b-periodic functions in x_1 satisfying (4.47). Then, after a simple computation, we have

$$\int_{\Omega_1} (B - \nabla w)g - \langle \beta - \Theta, u \rangle_{W_3', W_3} = \int_{\Sigma} \{\nabla \Phi - \pi I\} e_2(\xi - \nabla w)$$

$$- \int_{(0,b) \times \{h\}} \{\nabla \Phi - \pi I\} e_2(\xi - \nabla w) + \int_{\Omega_1} G_1 \Phi - \int_{\Omega_1} G_2 \nabla \Phi$$

where w is defined by (4.49) and $\{\Phi, \pi\}$ is the solution for (4.48).

Now in analogy with [CO] we introduce the notion of the very weak solution:

Definition 4.11. $\{V, Q\}$ is the very weak solution for the problem (4.45) if $V \in L^2(\Omega_1)^2$, $Q \in W_3'$ and

$$\forall g \in L^2(\Omega_1)^2, \forall u \in W_3$$

$$\int_{\Omega_1} (V - \nabla w)g - \langle Q - \Theta, u \rangle_{W_3', W_3} = \int_{\Sigma} \{\nabla \Phi - \pi I\} e_2(\xi - \nabla w) -$$

$$\int_{(0,b) \times \{h\}} \{\nabla \Phi - \pi I\} e_2(\xi - \nabla w) + \int_{\Omega_1} G_1 \Phi - \int_{\Omega_1} G_2 \nabla \Phi, \qquad (4.50)$$

where $\{\Phi, \pi\}$ is defined by (4.48) and w by (4.49).

We have the following result

Lemma 4.12. There exists a unique very weak solution $\{V, Q\}$ for (4.50).

Proof : It is a direct consequence of Riesz' representation theorem. ∎

After all those auxiliary results we easily obtain the a priori estimate for B in $L^2(\Omega_1)^2$, which was our goal:

Proposition 4.13. Let (B, β) be a solution for (4.45), with G_1, G_2, Θ and ξ satisfying (4.46). Then

$$\|B\|_{L^2(\Omega_1)^2} \le C\{\|\Theta\|_{L^2(\Omega_1)}$$
$$+ \|G_1\|_{L^2(\Omega_1)^2} + \|G_2\|_{L^2(\Omega_1)^4} + \|\xi\|_{L^2(\Sigma \cup (0,b) \times \{h\})}\}. \qquad (4.51)$$

Proof : Because of Lemma 4.12 $\{B, \beta\}$ is the very weak solution for (4.50). We choose $h = 0$ and $g = B$ as test functions in (4.50). Then (4.51) follows immediately from the equation (4.50). ∎

We have the following result

Proposition 4.14. *Let $\{v^\varepsilon, p^\varepsilon\}$ be the solution for (4.37)-(4.39) and $\{v^0, \pi^0\}$ defined by (4.41). Then we have*

$$\|\nabla(v^\varepsilon - v^0)\|_{L^2(\Omega^\varepsilon)^4} \leq C\sqrt{\varepsilon} \tag{4.52}$$

$$\|v^\varepsilon\|_{L^2(\Omega_2^\varepsilon)^2} \leq C\varepsilon\sqrt{\varepsilon} \tag{4.53}$$

$$\|v^\varepsilon\|_{L^2(\Sigma)} \leq C\varepsilon \tag{4.54}$$

$$\|v^\varepsilon - v^0\|_{L^2(\Omega_1)^2} \leq C\varepsilon \tag{4.55}$$

$$\|p^\varepsilon - p^0\|_{L^2(\Omega_1)} \leq C\sqrt{\varepsilon} \tag{4.56}$$

Proof : We search v^ε in the form $v^\varepsilon = v^0 + w^\varepsilon$. Let

$$\mathcal{Z}^\varepsilon = \Big\{ z \in H^1(\Omega^\varepsilon)^2 : z = 0 \text{ on } (\partial\Omega^\varepsilon \setminus \partial\Omega) \cup (0,b) \times (\{-L\} \cup \{h\}),$$

$$z \text{ is } b-\text{periodic in } x_1 \Big\}$$

Then we are looking for $w^\varepsilon \in \mathcal{W}^\varepsilon = \{\varphi \in \mathcal{Z}^\varepsilon : \text{ div } \varphi = 0 \text{ in } \Omega^\varepsilon\}$ such that

$$\mu \int_{\Omega^\varepsilon} \nabla w^\varepsilon \nabla \varphi = \mu \frac{\partial v_1^0}{\partial x_2}(0) \int_\Sigma \varphi_1 - \frac{p_b - p_0}{b} \int_{\Omega_2^\varepsilon} \varphi_1, \qquad \forall \varphi \in \mathcal{W}^\varepsilon. \tag{4.57}$$

Now we test (4.57) with $\varphi = w^\varepsilon$ and by (4.42)-(4.43) we get (4.52).

The inequalities (4.53) and (4.54) are also direct consequences of Poincaré's inequality (4.42) and the trace inequality (4.43) in Ω_2^ε, respectively.

In order to get the estimates (4.55)-(4.56) we note that $w^\varepsilon = v^\varepsilon - v^0$ and $\pi^\varepsilon = p^\varepsilon - \pi^0$ satisfy the system

$$\begin{cases} -\Delta w^\varepsilon + \nabla \pi^\varepsilon = 0 & \text{in } \Omega_1 , \\[2mm] \text{div } w^\varepsilon = 0 \text{ in } \Omega_1, \qquad \{w^\varepsilon, \pi^\varepsilon\} \text{ is b-periodic in } x_1, \\[2mm] w^\varepsilon = \xi^\varepsilon \text{ on } \Sigma \cup (0,b) \times \{h\}, \qquad w^\varepsilon = 0 \text{ on } (0,b) \times \{h\}, \end{cases} \tag{4.58}$$

with $\|\xi^\varepsilon\|_{L^2(\Sigma \cup (0,b) \times \{h\})^2} \leq C\varepsilon$, by (4.54).

By using the very weak variational formulation (4.50) with $\Theta = 0$, $w = 0$, $G_1 = 0$ and $G_2 = 0$, and the estimate (4.51) we get (4.55).

The estimate (4.56) follows from the first equation in (4.58) and Nečas' inequality in Ω_1. ∎

Therefore, we have obtained the uniform a priori estimates for $\{v^\varepsilon, p^\varepsilon\}$. Moreover, we have found that Poiseuille's flow in Ω_1 is an $O(\varepsilon)$ L^2-approximation for v^ε. Beavers and Joseph's law should correspond to the next order velocity correction.

The leading contribution for the estimate (4.52) was the the interface integral term $\int_\Sigma \varphi_1$. Following the approach from [JM3], we eliminate it by using the boundary layer-type functions

$$\beta^{bl,\varepsilon}(x) = \varepsilon\beta^{bl}(\frac{x}{\varepsilon}) \quad \text{and} \quad \omega^{bl,\varepsilon}(x) = \omega^{bl}(\frac{x}{\varepsilon}), \quad x \in \Omega^\varepsilon, \quad (4.59)$$

We extend $\beta^{bl,\varepsilon}$ by zero to $\Omega \setminus \Omega^\varepsilon$. Then we have

$$\begin{cases} \|\beta^{bl,\varepsilon} - \varepsilon(C_1^{bl},0)H(x_2)\|_{L^q(\Omega)^2} = C\varepsilon^{1+1/q}, \quad \forall q \geq 1 \\ \|\omega^{bl,\varepsilon} - C_\omega^{bl}H(x_2)\|_{L^q(\Omega^\varepsilon)} + \|\nabla\beta^{bl,\varepsilon}\|_{L^q(\Omega_1 \cup \Sigma \cup \Omega_2)^4} = C\varepsilon^{1/q}, \quad \forall q \geq 1 \end{cases}$$
$$(4.60)$$

$$\begin{cases} \|\omega^{bl,\varepsilon}(0,\cdot) - C_\omega^{bl}H(\cdot)\|_{H^{-1/2}(\mathbb{R})} + \sqrt{\varepsilon}\|\omega^{bl,\varepsilon}(0,\cdot) - C_\omega^{bl}H(\cdot)\|_{L^2(\mathbb{R})} = C\varepsilon \quad ; \\ \\ \varepsilon^{-1/2}\|\beta^{bl,\varepsilon}(0,\cdot) - \varepsilon(C_1^{bl},0)H(\cdot)\|_{L^2(\mathbb{R})^2} + \|\frac{\partial\beta^{bl,\varepsilon}}{\partial x_2}(0,\cdot)\|_{H^{-1/2}(\mathbb{R})^2} = C\varepsilon. \end{cases}$$
$$(4.61)$$

Finally,

$$-\Delta\beta^{bl,\varepsilon} + \nabla\omega^{bl,\varepsilon} = 0 \quad \text{in } \Omega_1 \cup \Omega_2^\varepsilon \quad (4.62)$$
$$\text{div } \beta^{bl,\varepsilon} = 0 \quad \text{in } \Omega^\varepsilon \quad (4.63)$$
$$[\beta^{bl,\varepsilon}]_{\Sigma\cup(0,b)\times\{h\}}(\cdot,0) = 0 \quad \text{on } \Sigma \cup (0,b) \times \{h\} \quad (4.64)$$
$$[\{\nabla\beta^{bl,\varepsilon} - \omega^{bl,\varepsilon}I\}e_2]_{\Sigma\cup(0,b)\times\{h\}}(\cdot,0) = e_1 \quad \text{on } \Sigma \cup (0,b) \times \{h\}. \quad (4.65)$$

As in [JM3] stabilization of $\beta^{bl,\varepsilon}$ towards a nonzero constant velocity $\varepsilon(C_1^{bl},0)$, at the upper boundary, generates a counterflow. It has the form of 2D Couette flow $d = (1 - \frac{x_2}{h})e_1$.

Now, we would like to prove that the following quantities are $o(\varepsilon)$ for the velocity and $O(1)$ for the pressure:

$$\mathcal{U}^\varepsilon(x) = v^\varepsilon - v^0 + \left(\beta^{bl,\varepsilon} - \varepsilon(C_1^{bl},0)H(x_2)\right)\frac{\partial v_1^0}{\partial x_2}(0) +$$
$$\varepsilon C_1^{bl}\frac{\partial v_1^0}{\partial x_2}(0)H(x_2)(1 - \frac{x_2}{h})e_1 \quad (4.66)$$

$$\mathcal{P}^\varepsilon = p^\varepsilon - p^0 H(-x_2) + (\omega^{bl,\varepsilon} - H(x_2)C_\omega^{bl})\mu\frac{\partial v_1^0}{\partial x_2}(0) \quad (4.67)$$

where $p^0 \in H^1(\Omega_2)$.

On $(0,b) \times \{h\}$ $\mathcal{U}^\varepsilon$ is not zero but it satisfies the inequality $|\mathcal{U}^\varepsilon(x)| \leq C\exp\{-C_0/\varepsilon\}$ for some positive constants C and C_0. Consequently, we consider it being zero without loosing generality.

We have the following result

Proposition 4.15. Let $\mathcal{U}^\varepsilon$ be given by (4.66) and $\mathcal{P}^\varepsilon$ by (4.67). Then $\mathcal{U}^\varepsilon \in \mathcal{Z}^\varepsilon$ and $\operatorname{div} \mathcal{U}^\varepsilon = 0$ in Ω^ε. Furthermore, we have the following estimate

$$|\mu \int_{\Omega^\varepsilon} \nabla \mathcal{U}^\varepsilon \nabla \varphi - \int_{\Omega^\varepsilon} \mathcal{P}^\varepsilon \operatorname{div}\varphi| \le C\varepsilon^{3/2}\|\nabla\varphi\|_{L^2(\Omega^\varepsilon)^4}+$$

$$|-\mu C_\omega^{bl}\frac{\partial v_1^0}{\partial x_2}(0)\int_{\Sigma \cup (0,b)\times\{h\}}\varphi_2 + \int_{\Omega_2^\varepsilon}(-\frac{p_b-p_0}{b}\varphi_1 + p^0 \operatorname{div}\varphi)|, \quad \forall \varphi \in \mathcal{Z}^\varepsilon \text{ (4.68)}$$

Proof : Firstly, we note that for $\varphi \in \mathcal{Z}^\varepsilon$ the equation (4.57) reads

$$\mu \int_{\Omega^\varepsilon} \nabla(v^\varepsilon - v^0)\nabla\varphi - \int_{\Omega^\varepsilon} p^\varepsilon \operatorname{div} \varphi = \mu\frac{\partial v_1^0}{\partial x_2}(0)\int_\Sigma \varphi_1 - \int_{\Omega_2^\varepsilon}\frac{p_b-p_0}{b}\varphi_1. \quad (4.69)$$

Next, the Couette velocity $d = (1 - \frac{x_2}{h})e_1$ satisfies

$$\int_{\Omega_1} \mu \nabla d \nabla\varphi = -\mu\int_\Sigma \frac{\partial d_1}{\partial x_2}\varphi_1, \qquad \forall \varphi \in \mathcal{Z}^\varepsilon \quad (4.70)$$

and, moreover, for $\{\beta^{bl,\varepsilon}, \omega^{bl,\varepsilon} - C_\omega^{bl} H(x_2)\}$ we have

$$\int_{\Omega^\varepsilon} (\nabla(\beta^{bl,\varepsilon} - \varepsilon(C_1^{bl},0)H(x_2))\nabla\varphi - (\omega^{bl,\varepsilon} - C_\omega^{bl}H(x_2)) \operatorname{div} \varphi) =$$

$$- \int_\Sigma (\varphi_1 + C_\omega^{bl}\varphi_2), \forall \varphi \in \mathcal{Z}^\varepsilon. \quad (4.71)$$

Because of the estimates (4.60)-(41), we have for $\varphi \in \mathcal{Z}^\varepsilon$

$$|\varepsilon C_1^{bl} \int_\Sigma \mu\frac{\partial d_1}{\partial x_2}\varphi_1| \le C\varepsilon^{3/2}\|\nabla\varphi\|_{L^2(\Omega_2^\varepsilon)^4} \quad (4.72)$$

Now the variational equations (4.69)-(4.71), the definition of $\mathcal{U}^\varepsilon$ and $\mathcal{P}^\varepsilon$ and the estimate (4.72) give the estimate (4.68). ∎

Obviously, the estimate (4.68) is useful only if the pressure field p^0 is choosen in an appropriate way. By analogy with the problem (1.65), one can try to take as p^0 the solution, b-periodic in x_1 for the problem

$$\operatorname{div}(K\nabla p^0) = 0 \qquad \text{in } \Omega_2, \quad (4.73)$$

$$p^0(x_1,-0) = \mu C_\omega^{bl}\frac{\partial v_1^0}{\partial x_2}(0) \qquad \text{on } \Sigma, \quad (4.74)$$

$$K\nabla p^0\, e_2 = 0 \qquad \text{on } (0,b)\times\{-L\}, \quad (4.75)$$

where the permeability tensor K is defined the auxiliary problem

$$\begin{cases} -\Delta_y w^j + \nabla_y \pi^j = e_j & \text{in } Y_F ; \\[2mm] \operatorname{div}_y w^j = 0 \quad \text{in } Y_F, \qquad \int_{Y_F}\pi^j = 0 ; \\[2mm] w^j = 0 \quad \text{on } \partial Z^*, \qquad \{w^j,\pi^j\} \text{ is 1-periodic,} \end{cases} \quad (4.76)$$

by $K_{ij} = \int_{Y_F} w_i^j(y) \, dy$. Obviously, $p^0 = \mu C_\omega^{bl} \frac{\partial v_1^0}{\partial x_2}(0)$.

Consequently, for every $\varphi \in Z^\varepsilon$ we have

$$|-\mu C_\omega^{bl} \frac{\partial v_1^0}{\partial x_2}(0) \int_\Sigma \varphi_2 + \int_{\Omega_2^\varepsilon}(-\frac{p_b - p_0}{b}\varphi_1 + p^0 \text{ div } \varphi)| \le C\varepsilon\|\nabla\varphi\|_{L^2(\Omega_2^\varepsilon)^4}. \quad (4.77)$$

Thus, we have obtained the following result

Corollary 4.16. *Let* $p^0 \doteq \mu C_\omega^{bl} \frac{\partial v_1^0}{\partial x_2}(0)$. *Then for* $\{\mathcal{U}^\varepsilon, \mathcal{P}^\varepsilon\}$ *given by (4.66)-(4.67), we have*

$$|\mu \int_{\Omega^\varepsilon} \nabla \mathcal{U}^\varepsilon \nabla\varphi - \int_{\Omega^\varepsilon} \mathcal{P}^\varepsilon \text{div}\varphi| \le C\varepsilon\|\nabla\varphi\|_{L^2(\Omega^\varepsilon)^4}, \quad \forall \varphi \in Z^\varepsilon. \quad (4.78)$$

At this stage we follow the ideas from [JM3], take $\varphi = \mathcal{U}^\varepsilon$ as the test function and get the required higher order a priori estimate. After introducing all auxiliary functions we are in position to prove our main result

Theorem 4.17. *Let*

$$\mathcal{U}^\varepsilon(x) = v^\varepsilon - v^0 + \beta^{bl,\varepsilon}\frac{\partial v_1^0}{\partial x_2}(0) - \varepsilon C_1^{bl}\frac{\partial v_1^0}{\partial x_2}(0)H(x_2)\frac{x_2}{h}e_1 \quad (4.79)$$

$$\mathcal{P}^\varepsilon = p^\varepsilon + \left(\omega^{bl,\varepsilon} - C_\omega^{bl}\right)\mu\frac{\partial v_1^0}{\partial x_2}(0), \quad (4.80)$$

where $\{v^0, \pi^0\}$ *is defined by (4.41) and* $\{\beta^{bl,\varepsilon}, \omega^{bl,\varepsilon}\}$ *by (4.59).*

Then we have the following estimates

$$\|\nabla\mathcal{U}^\varepsilon\|_{L^2(\Omega^\varepsilon)^4} \le C\varepsilon \quad (4.81)$$

$$\|\mathcal{U}^\varepsilon\|_{L^2(\Omega_2^\varepsilon)^2} \le C\varepsilon^2 \quad (4.82)$$

$$\|\mathcal{U}^\varepsilon\|_{L^2(\Sigma)^2} \le C\varepsilon^{3/2} \quad (4.83)$$

$$\|\mathcal{U}^\varepsilon\|_{L^2(\Omega_1)^2} \le C\varepsilon^{3/2} \quad (4.84)$$

$$\|\mathcal{P}^\varepsilon\|_{L^2(\Omega_1)} \le C\varepsilon \quad (4.85)$$

Proof: We test the estimate (4.78) with $\varphi = \mathcal{U}^\varepsilon$. With this choice we get the estimate (4.81). Poincaré's inequality (4.42) applied to (4.81) gives (4.82). (4.83) is a consequence of (4.43) and (4.85) follows from (4.81).

It remains to prove (4.84). We note that $\{\mathcal{U}^\varepsilon, \mathcal{P}^\varepsilon\}$ satisfies the following Stokes system in Ω_1:

$$\begin{cases} -\mu\Delta\mathcal{U}^\varepsilon + \nabla\mathcal{P}^\varepsilon = 0 & \text{in } \Omega_1 , \\[2mm] \text{div } \mathcal{U}^\varepsilon = 0 & \text{in } \Omega_1 , \\[2mm] \mathcal{U}^\varepsilon = \xi^\varepsilon \text{ on } \Sigma, & \|\xi^\varepsilon\|_{L^2(\Sigma)^2} \le C\varepsilon^{3/2}, \\[2mm] \mathcal{U}^\varepsilon = 0 \text{ on } (0,b) \times \{h\}, & \{\mathcal{U}^\varepsilon, \mathcal{P}^\varepsilon\} \text{ is b-periodic} . \end{cases} \quad (4.86)$$

Now (4.51) and (4.83) imply (4.84). ∎

The estimates (4.81) - (4.85) allow to justify Saffman's modification (8) of the Beavers and Joseph law (4.1).

We start with a result related to the behavior of the velocity field v^ε at the interface Σ with the porous body. Let $H^{1/2}_{per}(\Sigma) = [H^1_{per}(\Sigma), L^2(\Sigma)]_{1/2}$. Then we have

Theorem 4.18. *Let v^ε be the velocity field determined by (4.37)-(4.40) and let the boundary layer tangential velocity at infinity C^{bl}_1 be given by (4.15). Then we have*

$$\|v^\varepsilon_1 + \varepsilon C^{bl}_1 \frac{\partial v^\varepsilon_1}{\partial x_2}\|_{\left(H^{1/2}_{per}(\Sigma)\right)'} \leq C\varepsilon^{3/2}. \tag{4.87}$$

Proof: Using definition of the correction $\mathcal{U}^\varepsilon$, we get

$$\|v^\varepsilon_1 + \varepsilon C^{bl}_1 \frac{\partial v^\varepsilon_1}{\partial x_2}\|_{\left(H^{1/2}_{per}(\Sigma)\right)'} \leq C\varepsilon^2 + \|\mathcal{U}^\varepsilon_1 + \varepsilon C^{bl}_1 \frac{\partial \mathcal{U}^\varepsilon_1}{\partial x_2}\|_{\left(H^{1/2}_{per}(\Sigma)\right)'}$$

$$+ C\|\beta^{bl,\varepsilon}_1(0,\cdot) - \varepsilon C^{bl}_1\|_{\left(H^{1/2}_{per}(\Sigma)\right)'} + C\varepsilon\|\frac{\partial \beta^{bl,\varepsilon}_1}{\partial x_2}(0,\cdot)\|_{\left(H^{1/2}_{per}(\Sigma)\right)'}. \tag{4.88}$$

It should be noted that $\left(H^{1/2}_{per}(\Sigma)\right)' = [L^2(\Sigma), H^{-1}_{per}(\Sigma)]_{1/2}$ and that $\beta^{bl,\varepsilon}_1(0,\cdot) - \varepsilon C^{bl}_1$ and $\frac{\partial \beta^{bl,\varepsilon}_1}{\partial x_2}(0,\cdot)$ are ε-periodic functions with zero mean.

Consequently, by a simple duality argument we obtain

$$\|\beta^{bl,\varepsilon}_1(0,\cdot) - \varepsilon C^{bl}_1\|_{H^{-1}_{per}(\Sigma)} + \varepsilon\|\frac{\partial \beta^{bl,\varepsilon}_1}{\partial x_2}(0,\cdot)\|_{H^{-1}_{per}(\Sigma)} \leq C\varepsilon^2$$

and, after interpolation,

$$\|\beta^{bl,\varepsilon}_1(0,\cdot) - \varepsilon C^{bl}_1\|_{\left(H^{1/2}_{per}(\Sigma)\right)'} + \varepsilon\|\frac{\partial \beta^{bl,\varepsilon}_1}{\partial x_2}(0,\cdot)\|_{\left(H^{1/2}_{per}(\Sigma)\right)'} \leq C\varepsilon^{3/2}. \tag{4.89}$$

It remains to estimate the first term on the right hand side of the inequality (4.88). The difficulty comes from the derivatives of $\mathcal{U}^\varepsilon_1$. Since we have no information on the H^2-norm of $\mathcal{U}^\varepsilon$, the only possibility is to use the generalized Green formula.

Firstly, we have

$$\text{div}\,(\mu\nabla\mathcal{U}^\varepsilon_1 - \mathcal{P}^\varepsilon e_1) = 0$$

Now, by the generalized Green formula,

$$\|\frac{\partial \mathcal{U}^\varepsilon_1}{\partial x_2}\|_{\left(H^{1/2}_{per}(\Sigma)\right)'} \leq$$

$$C\left\{\|\mu\nabla\mathcal{U}^\varepsilon_1 - \mathcal{P}^\varepsilon e_1\|_{L^2(\Omega_1)^2} + \|\,\text{div}\,(\mu\nabla\mathcal{U}^\varepsilon_1 - \mathcal{P}^\varepsilon e_1)\|_{L^2(\Omega_1)}\right\} \leq C\varepsilon \tag{4.90}$$

Since (4.83) implies $\|\mathcal{U}_1^\varepsilon\|_{\left(H_{per}^{1/2}(\Sigma)\right)'} \leq C\varepsilon^{3/2}$, after inserting (4.89) and (4.90) into (4.88), we obtain the estimate (4.87). ∎

Now we introduce the effective flow equations in Ω_1 through the following boundary value problem:

Find a velocity field u^{eff} and a pressure field p^{eff} such that

$$-\mu\Delta u^{eff} + \nabla p^{eff} = -\frac{p_b - p_0}{b}e_1 \quad \text{in } \Omega_1, \tag{4.91}$$

$$\text{div } u^{eff} = 0 \quad \text{in } \Omega_1, \tag{4.92}$$

$$u^{eff} = 0 \quad \text{on } (0, b) \times \{h\}, \tag{4.93}$$

$$u^{eff} \text{ and } p^{eff} \quad \text{are } b - \text{periodic} \quad , \tag{4.94}$$

$$u_2^{eff} = 0 \quad \text{and} \quad u_1^{eff} + \varepsilon C_1^{bl}\frac{\partial u_1^{eff}}{\partial x_2} = 0 \quad \text{on } \Sigma. \tag{4.95}$$

Problem (4.91)-(4.95) has a unique solution

$$\begin{cases} u^{eff} = \left(\frac{p_b - p_0}{2b\mu}\left(x_2 - \frac{\varepsilon C_1^{bl}h}{h - \varepsilon C_1^{bl}}\right)(x_2 - h), 0\right) & \text{for } 0 \leq x_2 \leq h; \\ p^{eff} = 0 & \text{for } 0 \leq x_1 \leq b \end{cases} \tag{4.96}$$

The effective mass flow rate through the channel is then

$$M^{eff} = b\int_0^h u_1^{eff}(x_2)\, dx_2 = -\frac{p_b - p_0}{12\mu}h^3\frac{h - 4\varepsilon C_1^{bl}}{h - \varepsilon C_1^{bl}}, \tag{4.97}$$

where $C_1^{bl} < 0$.

Let us estimate the error made when replacing $\{v^\varepsilon, p^\varepsilon, M^\varepsilon\}$ by $\{u^{eff}, p^{eff}, M^{eff}\}$. We have

Proposition 4.19. *We have*

$$\|\nabla(v^\varepsilon - u^{eff})\|_{L^1(\Omega_1)^4} \leq C\varepsilon, \tag{4.98}$$

$$\|v^\varepsilon - u^{eff}\|_{L^2(\Omega_1)^2} \leq C\varepsilon^{3/2}, \tag{4.99}$$

$$|M^\varepsilon - M^{eff}| \leq C\varepsilon^{3/2}. \tag{4.100}$$

Proof: We have

$$v^\varepsilon - u^{eff} = \mathcal{U}^\varepsilon + v^0 - u^{eff} - \varepsilon C_1^{bl}\frac{\partial v_1^0}{\partial x_2}(0)(1 - \frac{x_2}{h})e_1 - \left(\beta^{bl,\varepsilon} - \varepsilon C_1^{bl}e_1\right)\frac{\partial v_1^0}{\partial x_2}(0) \text{ in } \Omega_1.$$

After a simple calculation we find the identity

$$v^0 - u^{eff} - \varepsilon C_1^{bl}\frac{\partial v_1^0}{\partial x_2}(0)(1 - \frac{x_2}{h})e_1 = \frac{p_b - p_0}{2b\mu}(x_2 - h)\frac{(\varepsilon C_1^{bl})^2}{h - \varepsilon C_1^{bl}}.$$

Now (4.98) and (4.99) follow from the estimates on $\mathcal{U}^\varepsilon$ and $\beta^{bl,\varepsilon}$.

It remains to prove (4.100), We use the above identities and get

$$|M^\varepsilon - M^{eff}| \le |\int_0^h \mathcal{U}_1^\varepsilon(0, x_2) \, dx_2| + C\varepsilon^2.$$

Now let $0 < \ell < b$ and $\varphi = \varphi(x_1) \in C^1[0, \ell]$, $\psi(0) = 1$ and $\psi(\ell) = 0$. Then we have

$$-\mathcal{U}_1^\varepsilon(0, x_2) = \int_0^\ell \frac{\partial}{\partial x_1} (\mathcal{U}_1^\varepsilon(x)\psi(x_1)) \, dx_1 =$$

$$\int_0^\ell \left(-\frac{\partial}{\partial x_2} (\mathcal{U}_2^\varepsilon \psi(x_1)) + \mathcal{U}_1^\varepsilon(x) \frac{d\psi}{dx_1}\right) \, dx_1$$

and $|\int_0^h \mathcal{U}_1^\varepsilon(0, x_2) \, dx_2| \le C\varepsilon^{3/2}$. This proves (4.100). ∎

Our interface is a mathematical one and it doesn't exists as a physical boundary. It is clear that we can take any straight line at the distance $O(\varepsilon)$ from the rigid parts as an interface. Hence it remains to prove that the law by Beavers and Joseph doesn't depend on the position of the interface. We have the following auxiliary result

Lemma 4.20. *Let $a < 0$ and let $\beta^{a,bl}$ be the solution for (4.2)-(4.6) with S replaced by $S_a = (0,1) \times \{a\}$, Z^+ by $Z_a^+ = (0,1) \times (a, +\infty)$ and $Z_a^- = Z_{BL} \setminus (S_a \cup Z_a^+)$. Then we have*

$$C_1^{a,bl} = C_1^{bl} - a. \tag{4.101}$$

Proof : By (4.10)

$$C_1^{bl} = \int_0^1 \beta_1^{bl}(y_1, c) \, dy_1, \qquad \forall c \ge 0$$

Let d_* be the distance between $Z^* - (0,1)$ and S and let $-d_* \le c_1 < 0 < c_2$. Integration of the first component of (4.2)

$$\text{div } \{\nabla \beta_1^{bl} - \omega^{bl} e_1\} = 0$$

over (c_1, c_2) gives

$$\int_0^1 \left\{ \frac{\partial \beta_1^{bl}}{\partial y_2}(y_1, c_2) - \frac{\partial \beta_1^{bl}}{\partial y_2}(y_1, a+0) + \frac{\partial \beta_1^{bl}}{\partial y_2}(y_1, a-0) - \frac{\partial \beta_1^{bl}}{\partial y_2}(y_1, c_1) \right\} \, dy_1 = 0.$$

Hence

$$\frac{d}{dy_2} \int_0^1 \beta_1^{bl}(y_1, y_2) = -1, \qquad \text{for} \quad c_1 < y_2 < 0$$

and

$$\int_0^1 \beta_1^{a,bl}(y_1, y_2) = -y_2 + C_1^{bl}, \qquad \text{for} \quad a \le y_2 \le 0. \tag{4.102}$$

The variational equation for $-\beta^{a,bl} + \beta^{bl}$ reads

$$\int_{Z_{BL}} \nabla(-\beta^{a,bl} + \beta^{bl}) \nabla\varphi \, dy = \int_0^1 (\varphi_1(y_1, a) - \varphi_1(y_1, 0)) \, dy_1, \qquad \forall \varphi \in V.$$

Testing with $\varphi = -\beta^{a,bl} + \beta^{bl}$ and using (4.102), we obtain

$$\int_{Z_{BL}} |\nabla(\beta^{a,bl} - \beta^{bl})|^2 \, dy = -\int_0^1 (\beta_1^{bl}(y_1, 0) - \beta_1^{bl}(y_1, a)) \, dy_1 = -a.$$

At the other hand

$$\int_{Z_{BL}} |\nabla(\beta^{a,bl} - \beta^{bl})|^2 \, dy = \int_{Z_{BL}} |\nabla\beta^{a,bl}|^2 \, dy + \int_{Z_{BL}} |\nabla\beta^{bl}|^2 \, dy -$$
$$2\int_{Z_{BL}} \nabla\beta^{a,bl} \nabla\beta^{bl} \, dy = -C_1^{bl} + C_1^{a,bl}$$

and (4.101) is proved. $\blacksquare$

This simple result will imply the invariance of the obtained law on the position of the interface.

Remark 4.21. Let $\Omega_{a\varepsilon} = (0, b) \times (a\varepsilon, h)$ for $a < 0$ and let $\{u^{a,eff}, p^{a,eff}\}$ be a solution for (4.91)-(4.95) in $\Omega_{a\varepsilon}$, with (4.95) replaced by

$$u_2^{a,eff} = 0 \qquad \text{and} \qquad u_1^{a,eff} + \varepsilon C_1^{a,bl} \frac{\partial u_1^{a,eff}}{\partial x_2} = 0 \quad \text{on} \quad \Sigma_a = (0, b) \times a\varepsilon \tag{4.103}$$

The unique solution $\{u^{a,eff}, p^{a,eff}\}$ for (4.91)-(4.94), (4.103) is given by

$$u^{a,eff} = \left(\frac{p_b - p_0}{2b\mu} \left((x_2 - a\varepsilon)^2 - (x_2 - a\varepsilon - \varepsilon C_1^{a,bl}) \frac{(h - a\varepsilon)^2}{h - a\varepsilon - \varepsilon C_1^{a,bl}} \right), 0 \right)$$

for $a\varepsilon \le x_2 \le h$ and

$$p^{a,eff} = p^0 = \frac{p_b - p_0}{b} x_1 + p_0 \qquad \text{for} \quad 0 \le x_1 \le b.$$

By Lemma 4.20., $C_1^{a,bl} = C_1^{bl} - a$ and

$$u^{a,eff}(x) = u^{eff}(x) + O(\varepsilon^2).$$

Therefore, a perturbation of the interface position for an $O(\varepsilon)$ implies a perturbation in the solution of $O(\varepsilon^2)$. Consequently, there is a freedom in fixing position of Σ. It influences the result only at the next order of the asymptotic expansion.

Concluding remarks . *Therefore we have found that the velocity field satisfies approximatively the interface condition (4.1) of Beavers and Joseph in the sense of the estimate (4.87). It is important to point out that the parameter α from the expression (4.1) is determined through the auxiliary problems (4.2)-(4.6) and (4.76) by $\alpha = -\dfrac{1}{\varepsilon C_1^{bl}} > 0$.*

In what concerns the pressure approximation, after extending it as in § 1, we get the uniform bound on it. Hence the effective pressure in Ω_2 is $\mu C_\omega^{bl} \frac{\partial v_1^0}{\partial x_2}(0)$. This indicates the discontinuity of the pressure field at Σ and doesn't confirm the law proposed in [ESP]. However, getting the relationship between the two pressures requires additional boundary layers and we don't undertake this task here.

References

References for §1

[Ace92] **E.Acerbi, V.Chiadò Piat, G.Dal Maso, D.Percivale** : An extension theorem from connected sets, and homogenization in general periodic domains, *Nonlinear Anal., TMA*, 18 (1992), pp. 481 - 496.

[ALL89] **G. Allaire** : Homogenization of the Stokes Flow in a Connected Porous Medium, *Asymptotic Analysis* 2 (1989), 203-222.

[All91] **G.Allaire** : Continuity of the Darcy's law in the low-volume fraction limit , *Ann. Scuola Norm. Sup. Pisa* , Vol. 18 (1991), 475-499.

[All92]**G.Allaire** : Homogenization and two-scale convergence, *SIAM J. Math. Anal.*, 23 (1992), pp. 1482 - 1518.

[All97]**G.Allaire** : One-Phase Newtonian Flow, in *Homogenization and Porous Media* , ed. U.Hornung, Springer, New-York, 1997, p. 45-68.

[BeKo95] **A.Yu. Beliaev, S.M.Kozlov** : Darcy equation for random porous media, *Comm. Pure Appl. Math.* , Vol. 49 (1995), 1-34.

[BorSo] **W.Borchers, H.Sohr** : On the equations rot $v = g$ and div $u = f$ with zero boundary conditions, *Hokkaido Mathematical Journal,* Vol. 19 (1990), p. 67-87.

[Bing] **A. Bourgeat, A. Mikelić** : Note on the Homogenization of Bingham Flow through Porous Medium, *J. Math. Pures Appl.* 72 (1993), 405-414.

[BMW] **A.Bourgeat, A.Mikelić, S.Wright** : On the Stochastic Two-Scale Convergence in the Mean and Applications, *Journal für die reine und angewandte Mathematik (Crelles Journal)*, Vol. 456 (1994), pp. 19 – 51.

[CFGM] **Th.Clopeau, J.L.Ferrín, R.P.Gilbert, A.Mikelić** : Homogenizing the Acoustic Properties of the Seabed ,II, to appear in *Math. Comput. Modelling* , 2000.

[ESP] **H.I.Ene, E.Sanchez-Palencia** : Equations et phénomènes de surface pour l'écoulement dans un modèle de milieu poreux, *J. Mécan.* , 14 (1975), pp. 73-108.

[GiMi99] **R.P.Gilbert, A.Mikelić** : Homogenizing the Acoustic Properties of the Seabed, Part I, *Nonlinear Anal.*, , 40 (2000), 185-212.

[Horn 1997] **U.Hornung, ed.** , Homogenization and Porous Media, *Interdisciplinary Applied Mathematics Series* , Vol. 6, Springer, New York, 1997.

[JaMi1] **W.Jäger, A.Mikelić** : On the Flow Conditions at the Boundary Between a Porous Medium and an Impervious Solid, in *" Progress in Partial Differential Equations: the Metz Surveys 3 "*, eds. M.Chipot, J.Saint Jean Paulin et I. Shafrir, πPitman Research Notes in Mathematics no. 314, p. 145-161, Longman Scientific and Technical, London, 1994.

[JaMi2] **W.Jäger, A.Mikelić** : On the Boundary Conditions at the Contact Interface between a Porous Medium and a Free Fluid, *Ann. Sc. Norm. Super. Pisa, Cl. Sci.* - Ser. IV, Vol. XXIII (1996), Fasc. 3, p. 403 - 465.

[JaMi3] **W.Jäger, A.Mikelić** : On the effective equations for a viscous incompressible fluid flow through a filter of finite thickness, *Communications on Pure and Applied Mathematics,* Vol. LI (1998), p. 1073–1121.

[JM5] **W.Jäger, A.Mikelić** : On the boundary conditions at the contact interface between two porous media, in *"Partial differential equations, Theory and numerical solution "* , eds. W. Jäger, J. Nečas, O. John, K. Najzar and J. Stará, π Chapman and Hall/CRC Research Notes in Mathematics no. 406, p. 175-186, CRC Press, London, 1999.

[JKO] **V.V.Jikov, S.Kozlov, O.Oleinik** : Homogenization of Differential Operators and Integral Functionals, Springer Verlag, New York, 1994.

[LiPe] **J. L. Lions** : Some Methods in the Mathematical Analysis of Systems and their Control, *Gordon and Breach, Science Publishers, Inc,* New York, 1981.

[LA] **R. Lipton, M. Avellaneda** : A Darcy Law for Slow Viscous Flow Past a Stationary Array of Bubbles, *Proc. Royal Soc. Edinburgh* 114A, 1990, 71-79.

[MPM] **E.Marušić- Paloka, A.Mikelić** : An Error Estimate for Correctors in the Homogenization of the Stokes and Navier-Stokes Equations in a Porous Medium, *Boll. Unione Mat. Ital.* , A(7) 10 (1996), no. 3, p. 661–671.

[MiPa] **A.Mikelić, L.Paoli** : Homogenization of the inviscid incompressible fluid flow through a 2D porous medium, *Proc. Amer. Math. Soc.* , Vol. 127(1999), pp. 2019-2028.

[Ng89] **G. Nguetseng** : A General Convergence Result for a Functional Related to the Theory of Homogenization, *SIAM J. Math. Analysis,* vol. 20(3), (1989), 608-623.

[SP80] **E. Sanchez-Palencia** : Non-Homogeneous Media and Vibration Theory, *Lecture Notes in Physics* 127, Springer Verlag, 1980.

[Ta1980] **L. Tartar** : Convergence of the Homogenization Process, *Appendix of* [SP80].

[Te1] **R. Temam** : Navier-Stokes Equations, 3rd (revised) edition, *Elsevier Science Publishers B. V.,* Amsterdam, 1984.

[Zh94] **V.V. Zhikov** :On the homogenization of the system of Stokes equations in a porous medium, *Russian Acad. Sci. Dokl. Math.* , Vol. 49 (1994), 52-57.

References for § 2

[BBN] **Bellout, H.; Bloom, F.; Nečas, J.** : Young measure-valued solutions for non-Newtonian incompressible fluids. *Comm. Partial Differential Equations,* 19 (1994), no. 11–12, 1763–1803.

[BSL] **R. B. Bird, W. E. Stewart, E. N. Lightfoot** : Transport Phenomena, *Wiley and Sons,* New York, 1960.

[BAH] **R. B. Bird, R. C. Armstrong, O. Hassager** : Dynamics of Polymeric Liquids, Vol. 1, Fluid Mechanics, *Wiley and Sons,* New York, 1987.

[Bing] **A.Bourgeat, A.Mikelić** : Note on the Homogenization of Bingham Flow through Porous Medium. *Journal de Mathématiques Pures et Appliquées*, Vol. 72 (1993), pp. 405–414.

[BM1] **A.Bourgeat, A.Mikelić** : Homogenization of the Non-Newtonian Flow through Porous Medium, *Nonlinear Anal., TMA*, Vol. 26 (1996), p. 1221 – 1253.

[BMT] **A.Bourgeat, A.Mikelić, R.Tapiéro** : Dérivation des équations moyennées décrivant un écoulement non-newtonien dans un domaine de faible épaisseur, *Comptes Rendus de l'Académie des Sciences, Série I*, t. 316 (1993), p. 965-970.

[CMi] **R. H. Christopher, S. Middleman** : Power-law through a packed tube, *I & EC Fundamentals*, Vol. 4 (1965), p. 422-426.

[Ci] **D. Cioranescu** : Quelques Exemples de Fluides Newtoniens Generalisés, in " *Mathematical topics in fluid mechanics* ", eds. J.F.Rodrigues and A.Sequeira, π Pitman Research Notes in Mathematics Series no. 274, Longman, Harlow, 1992, p. 1-31.

[ClM] **Th. Clopeau, A.Mikelić** : On the non-stationary quasi-Newtonian flow through a thin slab, " *Navier-Stokes Equations, Theory and Numerical Methods* " , ed. R. Salvi, πPitman Research Notes in Mathematics no. 388, Addison Wesley Longman, Harlow, 1998, p. 1-15.

[FMS] **Frehse, J.; Málek, J.; Steinhauer, M.** : An existence result for fluids with shear dependent viscosity-steady flows, to appear in *Proceedings of the second world congress of nonlinear analysis*, preprint n° 482 of SFB 256 Bonn, 1996

[Lno] **J. L. Lions** : Quelques méthodes de résolution des problèmes aux limites non linéaires, *Dunod Gauthier-Villars*, Paris, 1969.

[LiSP] **J. L. Lions, E. Sanchez-Palencia** : Écoulement d'un fluide viscoplastique de Bingham dans un milieu poreux, *J. Math. Pures Appl.* 60 (1981), 341-360.

[LA] **R. Lipton, M. Avellaneda** : A Darcy Law for Slow Viscous Flow Past a Stationary Array of Bubbles, *Proc. Royal Soc. Edinburgh* 114A, 1990, 71-79.

[MaMi] **E.Marušić–Paloka, A.Mikelić** : The Derivation of a Non-Linear Filtration Law Including the Inertia Effects Via Homogenization, *Nonlinear Anal.*, 42 (2000), 97 - 137.

[Mik1] **A. Mikelić** : Homogenization of Nonstationary Navier-Stokes Equations in a Domain with a Grained Boundary, *Annali di Matematica pura ed applicata* (IV), Vol. CLVIII (1991), 167-179.

[MT] **Mikelić, A.; Tapiéro R.** : Mathematical derivation of the power law describing polymer flow through a thin slab. *Modél. Math. Anal. Numér.* 29 (1995), no. 1, 3–21.

[MH] **A.Mikelić** : Non-Newtonian Flow, in " *Homogenization and Porous Media* " ed. U.Hornung, Interdisciplinary Applied Mathematics Series, Vol. 6, Springer, New York, 1997, p. 69–95.

[RS] **Raugel, G.; Sell, G. R.** : Navier-Stokes equations on thin 3D domains. I. Global attractors and global regularity of solutions. *J. Amer. Math. Soc.* 6 (1993), no. 3, 503–568.

[SY] **Ch. Shah, Y. C. Yortsos** : Aspects of Non-Newtonian Flow and Displacement in Porous Media, Topical Report for U.S. DOE, University of Southern California, Los Angeles, 1993.

[Te1] **R. Temam** : Navier-Stokes Equations, 3rd (revised) edition, *Elsevier Science Publishers B. V.*, Amsterdam, 1984.

[WPW] **Y. S. Wu, K. Pruess, A. Witherspoon** : Displacement of a Newtonian Fluid by a Non-Newtonian Fluid in a Porous Medium, *Transport in Porous Media*, Vol. 6, (1991), pp. 115-142.

References for § 3

[BER] **G.I. Barenblatt, V.M. Entov, V.M. Ryzhik** : Theory of Fluid Flows Through Natural Rocks, Kluwer, Dordrecht, 1990.

[BE] **J. Bear**: Hydraulics of Groundwater, McGraw-Hill, Jerusalem, 1979.

[BMM] **A. Bourgeat, E. Marušić-Paloka, A. Mikelić** : The Weak Non -Linear Corrections for Darcy's Law, *Math. Models Methods Appl.Sci. ($M^3 AS$)*, 8 (6) (1996), 1143-1155.

[FG] **M. Firdaouss, J.L. Guermond** : Sur l'homogénéisation des équations de Navier-Stokes à faible nombre de Reynolds, *C.R.Acad. Sci. Paris*, t.320, Série I (1995), 245-251.

[FGL] **M. Firdaouss, J.L. Guermond, P. Le Quéré** : Nonlinear Corrections to Darcy's Law at Low Reynolds Numbers, *J.Fluid Mech.*, Vol. 343 (1997), 331-350.

[GT] **D. Gilbarg, N.S. Trudinger** : Elliptic Partial Differential Equations of Second Order, 2nd edition, Springer-Verlag, Berlin, 1983.

[HG] **S.M. Hassanizadeh, W.G. Gray** : High Velocity Flow in Porous Media, *Transport in Porous Media*, Vol. 2 (1987), 521-531.

[LiPe] **J.L. Lions** : Some Methods in the Mathematical Analysis of Systems and Their Control, Gordon and Breach, New York, 1981.

[MaMi] **E. Marušić-Paloka, A. Mikelić** : The derivation of a non-linear filtration law including the inertia effects via homogenization, *Nonlinear Anal.*, 42 (2000), 97 - 137.

[MA] **C.C. Mei, J.-L. Auriault** : The Effect of Weak Inertia on Flow Through a Porous Medium, *J.Fluid Mech.*, Vol. 222 (1991), 647-663.

[Mik1] **A. Mikelić** : Homogenization of Nonstationary Navier-Stokes Equations in a Domain with a Grained Boundary, *Annali di Mat. pura ed appl. (VI)*, Vol.158 (1991), 167-179.

[Mik2] **A. Mikelić** : Effets inertiels pour un écoulement stationnaire visqueux incompressible dans un milieu poreux, *C.R. Acad. Sci. Paris*, t.320, Série I (1995), 1289-1294.

[RA] **M. Rasoloarijaona, J.-L. Auriault** : Non-linear Seepage Flow Through a Rigid Porous Medium, *Eur.J.Mech., B/Fluids* , 13, No 2, (1994),

177-195.

[RuMa] **D.W. Ruth, H. Ma** : On the Derivation of the Forchheimer Equation by Means of the Averaging Theorem, *Transport in Porous Media* , Vol. 7(1992), 255-264.

[SP80] **E. Sanchez-Palencia** : Non-Homogeneous Media and Vibration Theory, Springer Lecture Notes in Physics 127, Springer-Verlag, Berlin, 1980.

[Sch] **A.E. Scheidegger** : Hydrodynamics in Porous Media, in *Encyclopedia of Physics,* Vol. VIII/2 (Fluid Dynamics II), ed. S. Flügge, Springer-Verlag, Berlin, 1963, 625-663.

[WL] **J.-C. Wodié, T. Levy** : Correction non linéaire de la loi de Darcy, *C.R. Acad. Sci. Paris,* t.312, Série II (1991), 157-161.

References for § 4

[BJ] **G.S. Beavers, D.D. Joseph** : Boundary conditions at a naturally permeable wall, *J. Fluid Mech.* , 30 (1967), pp. 197-207.

[CO] **C. Conca** : Étude d'un fluide traversant une paroi perforée I. Comportement limite près de la paroi, *J. Math. pures et appl.* , 66 (1987), pp. 1-44. II. Comportement limite loin de la paroi, *J. Math. pures et appl.* , 66 (1987), pp. 45-69.

[ESP] **H.I. Ene, E. Sanchez-Palencia** : Equations et phénomènes de surface pour l'écoulement dans un modèle de milieu poreux, *J. Mécan.* , 14 (1975), pp. 73-108.

[DA] **G. Dagan** : The Generalization of Darcy's Law for Nonuniform Flows, *Water Resources Research* , Vol. 15 (1981), p. 1-7.

[JM1] **W.Jäger, A.Mikelić** : Homogenization of the Laplace equation in a partially perforated domain, prépublication no. 157, Equipe d'Analyse Numérique Lyon-St-Etienne, September 1993, published in *"Homogenization, In Memory of Serguei Kozlov* " , eds. V. Berdichevsky, V. Jikov and G. Papanicolaou, p. 259-284, World Scientific, Singapore, 1999.

[JM2] **W.Jäger, A.Mikelić** : On the Flow Conditions at the Boundary Between a Porous Medium and an Impervious Solid, in *" Progress in Partial Differential Equations: the Metz Surveys 3 "*, eds. M.Chipot, J.Saint Jean Paulin et I. Shafrir, πPitman Research Notes in Mathematics no. 314, p. 145-161, Longman Scientific and Technical, London, 1994.

[JM3] **W.Jäger, A.Mikelić** : On the Boundary Conditions at the Contact Interface between a Porous Medium and a Free Fluid, *Ann. Sc. Norm. Super. Pisa, Cl. Sci.* - Ser. IV, Vol. XXIII (1996), Fasc. 3, p. 403 - 465.

[JM4] **W.Jäger, A.Mikelić** : On the effective equations for a viscous incompressible fluid flow through a filter of finite thickness, *Communications on Pure and Applied Mathematics,* Vol. LI (1998), p. 1073–1121.

[JM5] **W.Jäger, A.Mikelić** : On the boundary conditions at the contact interface between two porous media, in *"Partial differential equations, Theory and numerical solution "* , eds. W. Jäger, J. Nečas, O. John, K. Najzar and J.

Stará, π Chapman and Hall/CRC Research Notes in Mathematics no. 406, p. 175-186, CRC Press, London, 1999.

[JM6] **W.Jäger, A.Mikelić** : On the interface boundary conditions by Beavers, Joseph and Saffman, *SIAM J. Appl. Math.* ,60 (2000), 1111- 1127.

[JM7] **W. Jäger, A. Mikelić** : On the roughness-induced effective boundary conditions for a viscous flow , accepted for publication in *Journal of Differential Equations* , 1999.

[JMN] **W.Jäger, A.Mikelić, N.Neuß** : Asymptotic analysis of the laminar viscous flow over a porous bed, preprint, Universität Heidelberg, Germany, January 1999.

[LSP] **Th. Levy, E. Sanchez-Palencia** : On Boundary Conditions for Fluid Flow in Porous Media, *Int. J. Engng. Sci.* , Vol. 13 (1975), p. 923–940.

[LiPe] **J.L. Lions** : Some Methods in the Mathematical Analysis of Systems and Their Control, Gordon and Breach, New York, 1981.

[MPM] **E.Marušić- Paloka, A.Mikelić** : An Error Estimate for Correctors in the Homogenization of the Stokes and Navier-Stokes Equations in a Porous Medium, *Boll. Unione Mat. Ital.* , (7) 10-A (1996), no. 3, p. 661–671.

[SAF] **P.G. Saffman** : On the boundary condition at the interface of a porous medium, *Studies in Applied Mathematics,* 1 (1971), pp. 93-101.

LIST OF C.I.M.E. SEMINARS

83. Matroid theory and its applications

1981	84. Kinetic Theories and the Boltzmann Equation	(LNM 1048)	Springer-Verlag
	85. Algebraic Threefolds	(LNM 947)	"
	86. Nonlinear Filtering and Stochastic Control	(LNM 972)	
1982	87. Invariant Theory	(LNM 996)	"
	88. Thermodynamics and Constitutive Equations (LN Physics 228)		"
	89. Fluid Dynamics	(LNM 1047)	
1983	90. Complete Intersections	(LNM 1092)	"
	91. Bifurcation Theory and Applications	(LNM 1057)	"
	92. Numerical Methods in Fluid Dynamics	(LNM 1127)	"
1984	93. Harmonic Mappings and Minimal Immersions	(LNM 1161)	"
	94. Schrödinger Operators	(LNM 1159)	"
	95. Buildings and the Geometry of Diagrams	(LNM 1181)	
1985	96. Probability and Analysis	(LNM 1206)	"
	97. Some Problems in Nonlinear Diffusion	(LNM 1224)	"
	98. Theory of Moduli	(LNM 1337)	"
1986	99. Inverse Problems	(LNM 1225)	"
	100. Mathematical Economics	(LNM 1330)	"
	101. Combinatorial Optimization	(LNM 1403)	"
1987	102. Relativistic Fluid Dynamics	(LNM 1385)	"
	103. Topics in Calculus of Variations	(LNM 1365)	"
1988	104. Logic and Computer Science	(LNM 1429)	"
	105. Global Geometry and Mathematical Physics	(LNM 1451)	"
1989	106. Methods of nonconvex analysis	(LNM 1446)	"
	107. Microlocal Analysis and Applications	(LNM 1495)	"
1990	108. Geometric Topology: Recent Developments	(LNM 1504)	"
	109. H∞ Control Theory	(LNM 1496)	"
	110. Mathematical Modelling of Industrial Processes	(LNM 1521)	"
1991	111. Topological Methods for Ordinary Differential Equations	(LNM 1537)	"
	112. Arithmetic Algebraic Geometry	(LNM 1553)	"
	113. Transition to Chaos in Classical and Quantum Mechanics	(LNM 1589)	"
1992	114. Dirichlet Forms	(LNM 1563)	"
	115. D-Modules, Representation Theory, and Quantum Groups	(LNM 1565)	"
	116. Nonequilibrium Problems in Many-Particle Systems	(LNM 1551)	"
1993	117. Integrable Systems and Quantum Groups	(LNM 1620)	"
	118. Algebraic Cycles and Hodge Theory	(LNM 1594)	"
	119. Phase Transitions and Hysteresis	(LNM 1584)	"

1994	120. Recent Mathematical Methods in Nonlinear Wave Propagation	(LNM 1640)	Springer-Verlag
	121. Dynamical Systems	(LNM 1609)	"
	122. Transcendental Methods in Algebraic Geometry	(LNM 1646)	
1995	123. Probabilistic Models for Nonlinear PDE's	(LNM 1627)	"
	124. Viscosity Solutions and Applications	(LNM 1660)	"
	125. Vector Bundles on Curves. New Directions	(LNM 1649)	"
1996	126. Integral Geometry, Radon Transforms and Complex Analysis	(LNM 1684)	"
	127. Calculus of Variations and Geometric Evolution Problems	(LNM 1713)	"
	128. Financial Mathematics	(LNM 1656)	"
1997	129. Mathematics Inspired by Biology	(LNM 1714)	"
	130. Advanced Numerical Approximation of Nonlinear Hyperbolic Equations	(LNM 1697)	"
	131. Arithmetic Theory of Elliptic Curves	(LNM 1716)	"
	132. Quantum Cohomology	(LNM)	"
1998	133. Optimal Shape Design	(LNM 1740)	"
	134. Dynamical Systems and Small Divisors	to appear	"
	135. Mathematical Problems in Semiconductor Physics	to appear	"
	136. Stochastic PDE's and Kolmogorov Equations in Infinite Dimension	(LNM 1715)	"
	137. Filtration in Porous Media and Industrial Applications	(LNM 1734)	"
1999	138. Computational Mathematics driven by Industrial Applications	(LNM 1739)	"
	139. Iwahori-Hecke Algebras and Representation Theory	to appear	"
	140. Theory and Applications of Hamiltonian Dynamics	to appear	"
	141. Global Theory of Minimal Surfaces in Flat Spaces	to appear	"
	142. Direct and Inverse Methods in Solving Nonlinear Evolution Equations	to appear	"

Lecture Notes in Mathematics

For information about Vols. 1–1560
please contact your bookseller or Springer-Verlag

Vol. 1603: V. Ene, Real Functions – Current Topics. XIII, 310 pages. 1995.

Vol. 1604: A. Huber, Mixed Motives and their Realization in Derived Categories. XV, 207 pages. 1995.

Vol. 1605: L. B. Wahlbin, Superconvergence in Galerkin Finite Element Methods. XI, 166 pages. 1995.

Vol. 1606: P.-D. Liu, M. Qian, Smooth Ergodic Theory of Random Dynamical Systems. XI, 221 pages. 1995.

Vol. 1607: G. Schwarz, Hodge Decomposition – A Method for Solving Boundary Value Problems. VII, 155 pages. 1995.

Vol. 1608: P. Biane, R. Durrett, Lectures on Probability Theory. Editor: P. Bernard. VII, 210 pages. 1995.

Vol. 1609: L. Arnold, C. Jones, K. Mischaikow, G. Raugel, Dynamical Systems. Montecatini Terme, 1994. Editor: R. Johnson. VIII, 329 pages. 1995.

Vol. 1610: A. S. Üstünel, An Introduction to Analysis on Wiener Space. X, 95 pages. 1995.

Vol. 1611: N. Knarr, Translation Planes. VI, 112 pages. 1995.

Vol. 1612: W. Kühnel, Tight Polyhedral Submanifolds and Tight Triangulations. VII, 122 pages. 1995.

Vol. 1613: J. Azéma, M. Emery, P. A. Meyer, M. Yor (Eds.), Séminaire de Probabilités XXIX. VI, 326 pages. 1995.

Vol. 1614: A. Koshelev, Regularity Problem for Quasilinear Elliptic and Parabolic Systems. XXI, 255 pages. 1995.

Vol. 1615: D. B. Massey, Le Cycles and Hypersurface Singularities. XI, 131 pages. 1995.

Vol. 1616: I. Moerdijk, Classifying Spaces and Classifying Topoi. VII, 94 pages. 1995.

Vol. 1617: V. Yurinsky, Sums and Gaussian Vectors. XI, 305 pages. 1995.

Vol. 1618: G. Pisier, Similarity Problems and Completely Bounded Maps. VII, 156 pages. 1996.

Vol. 1619: E. Landvogt, A Compactification of the Bruhat-Tits Building. VII, 152 pages. 1996.

Vol. 1620: R. Donagi, B. Dubrovin, E. Frenkel, E. Previato, Integrable Systems and Quantum Groups. Montecatini Terme, 1993. Editors:M. Francaviglia, S. Greco. VIII, 488 pages. 1996.

Vol. 1621: H. Bass, M. V. Otero-Espinar, D. N. Rockmore, C. P. L. Tresser, Cyclic Renormalization and Auto-morphism Groups of Rooted Trees. XXI, 136 pages. 1996.

Vol. 1622: E. D. Farjoun, Cellular Spaces, Null Spaces and Homotopy Localization. XIV, 199 pages. 1996.

Vol. 1623: H.P. Yap, Total Colourings of Graphs. VIII, 131 pages. 1996.

Vol. 1624: V. Brinzanescu, Holomorphic Vector Bundles over Compact Complex Surfaces. X, 170 pages. 1996.

Vol.1625: S. Lang, Topics in Cohomology of Groups. VII, 226 pages. 1996.

Vol. 1626: J. Azéma, M. Emery, M. Yor (Eds.), Séminaire de Probabilités XXX. VIII, 382 pages. 1996.

Vol. 1627: C. Graham, Th. G. Kurtz, S. Méléard, Ph. E. Protter, M. Pulvirenti, D. Talay, Probabilistic Models for Nonlinear Partial Differential Equations. Montecatini Terme, 1995. Editors: D. Talay, L. Tubaro. X, 301 pages. 1996.

Vol. 1628: P.-H. Zieschang, An Algebraic Approach to Association Schemes. XII, 189 pages. 1996.

Vol. 1629: J. D. Moore, Lectures on Seiberg-Witten Invariants. VII, 105 pages. 1996.

Vol. 1630: D. Neuenschwander, Probabilities on the Heisenberg Group: Limit Theorems and Brownian Motion. VIII, 139 pages. 1996.

Vol. 1631: K. Nishioka, Mahler Functions and Transcendence. VIII, 185 pages. 1996.

Vol. 1632: A. Kushkuley, Z. Balanov, Geometric Methods in Degree Theory for Equivariant Maps. VII, 136 pages. 1996.

Vol.1633: H. Aikawa, M. Essén, Potential Theory – Selected Topics. IX, 200 pages. 1996.

Vol. 1634: J. Xu, Flat Covers of Modules. IX, 161 pages. 1996.

Vol. 1635: E. Hebey, Sobolev Spaces on Riemannian Manifolds. X, 116 pages. 1996.

Vol. 1636: M. A. Marshall, Spaces of Orderings and Abstract Real Spectra. VI, 190 pages. 1996.

Vol. 1637: B. Hunt, The Geometry of some special Arithmetic Quotients. XIII, 332 pages. 1996.

Vol. 1638: P. Vanhaecke, Integrable Systems in the realm of Algebraic Geometry. VIII, 218 pages. 1996.

Vol. 1639: K. Dekimpe, Almost-Bieberbach Groups: Affine and Polynomial Structures. X, 259 pages. 1996.

Vol. 1640: G. Boillat, C. M. Dafermos, P. D. Lax, T. P. Liu, Recent Mathematical Methods in Nonlinear Wave Propagation. Montecatini Terme, 1994. Editor: T. Ruggeri. VII, 142 pages. 1996.

Vol. 1641: P. Abramenko, Twin Buildings and Applications to S-Arithmetic Groups. IX, 123 pages. 1996.

Vol. 1642: M. Puschnigg, Asymptotic Cyclic Cohomology. XXII, 138 pages. 1996.

Vol. 1643: J. Richter-Gebert, Realization Spaces of Polytopes. XI, 187 pages. 1996.

Vol. 1644: A. Adler, S. Ramanan, Moduli of Abelian Varieties. VI, 196 pages. 1996.

Vol. 1645: H. W. Broer, G. B. Huitema, M. B. Sevryuk, Quasi-Periodic Motions in Families of Dynamical Systems. XI, 195 pages. 1996.

Vol. 1646: J.-P. Demailly, T. Peternell, G. Tian, A. N. Tyurin, Transcendental Methods in Algebraic Geometry. Cetraro, 1994. Editors: F. Catanese, C. Ciliberto. VII, 257 pages. 1996.

Vol. 1647: D. Dias, P. Le Barz, Configuration Spaces over Hilbert Schemes and Applications. VII. 143 pages. 1996.

Vol. 1648: R. Dobrushin, P. Groeneboom, M. Ledoux, Lectures on Probability Theory and Statistics. Editor: P. Bernard. VIII, 300 pages. 1996.

Vol. 1649: S. Kumar, G. Laumon, U. Stuhler, Vector Bundles on Curves – New Directions. Cetraro, 1995. Editor: M. S. Narasimhan. VII, 193 pages. 1997.

Vol. 1650: J. Wildeshaus, Realizations of Polylogarithms. XI, 343 pages. 1997.

Vol. 1651: M. Drmota, R. F. Tichy, Sequences, Discrepancies and Applications. XIII, 503 pages. 1997.

Vol. 1652: S. Todorcevic, Topics in Topology. VIII, 153 pages. 1997.

Vol. 1653: R. Benedetti, C. Petronio, Branched Standard Spines of 3-manifolds. VIII, 132 pages. 1997.

Vol. 1654: R. W. Ghrist, P. J. Holmes, M. C. Sullivan, Knots and Links in Three-Dimensional Flows. X, 208 pages. 1997.

Vol. 1655: J. Azéma, M. Emery, M. Yor (Eds.), Séminaire de Probabilités XXXI. VIII, 329 pages. 1997.

Vol. 1656: B. Biais, T. Björk, J. Cvitanic, N. El Karoui, E. Jouini, J. C. Rochet, Financial Mathematics. Bressanone, 1996. Editor: W. J. Runggaldier. VII, 316 pages. 1997.

Vol. 1657: H. Reimann, The semi-simple zeta function of quaternionic Shimura varieties. IX, 143 pages. 1997.

Vol. 1658: A. Pumarino, J. A. Rodrıguez, Coexistence and Persistence of Strange Attractors. VIII, 195 pages. 1997.

Vol. 1659: V, Kozlov, V. Maz'ya, Theory of a Higher-Order Sturm-Liouville Equation. XI, 140 pages. 1997.

Vol. 1660: M. Bardi, M. G. Crandall, L. C. Evans, H. M. Soner, P. E. Souganidis, Viscosity Solutions and Applications. Montecatini Terme, 1995. Editors: I. Capuzzo Dolcetta, P. L. Lions. IX, 259 pages. 1997.

Vol. 1661: A. Tralle, J. Oprea, Symplectic Manifolds with no Kähler Structure. VIII, 207 pages. 1997.

Vol. 1662: J. W. Rutter, Spaces of Homotopy Self-Equivalences – A Survey. IX, 170 pages. 1997.

Vol. 1663: Y. E. Karpeshina; Perturbation Theory for the Schrödinger Operator with a Periodic Potential. VII, 352 pages. 1997.

Vol. 1664: M. Väth, Ideal Spaces. V, 146 pages. 1997.

Vol. 1665: E. Giné, G. R. Grimmett, L. Saloff-Coste, Lectures on Probability Theory and Statistics 1996. Editor: P. Bernard. X, 424 pages, 1997.

Vol. 1666: M. van der Put, M. F. Singer, Galois Theory of Difference Equations. VII, 179 pages. 1997.

Vol. 1667: J. M. F. Castillo, M. González, Three-space Problems in Banach Space Theory. XII, 267 pages. 1997.

Vol. 1668: D. B. Dix, Large-Time Behavior of Solutions of Linear Dispersive Equations. XIV, 203 pages. 1997.

Vol. 1669: U. Kaiser, Link Theory in Manifolds. XIV, 167 pages. 1997.

Vol. 1670: J. W. Neuberger, Sobolev Gradients and Differential Equations. VIII, 150 pages. 1997.

Vol. 1671: S. Bouc, Green Functors and G-sets. VII, 342 pages. 1997.

Vol. 1672: S. Mandal, Projective Modules and Complete Intersections. VIII, 114 pages. 1997.

Vol. 1673: F. D. Grosshans, Algebraic Homogeneous Spaces and Invariant Theory. VI, 148 pages. 1997.

Vol. 1674: G. Klaas, C. R. Leedham-Green, W. Plesken, Linear Pro-p-Groups of Finite Width. VIII, 115 pages. 1997.

Vol. 1675: J. E. Yukich, Probability Theory of Classical Euclidean Optimization Problems. X, 152 pages. 1998.

Vol. 1676: P. Cembranos, J. Mendoza, Banach Spaces of Vector-Valued Functions. VIII, 118 pages. 1997.

Vol. 1677: N. Proskurin, Cubic Metaplectic Forms and Theta Functions. VIII, 196 pages. 1998.

Vol. 1678: O. Krupková, The Geometry of Ordinary Variational Equations. X, 251 pages. 1997.

Vol. 1679: K.-G. Grosse-Erdmann, The Blocking Technique. Weighted Mean Operators and Hardy's Inequality. IX, 114 pages. 1998.

Vol. 1680: K.-Z. Li, F. Oort, Moduli of Supersingular Abelian Varieties. V, 116 pages. 1998.

Vol. 1681: G. J. Wirsching, The Dynamical System Generated by the 3n+1 Function. VII, 158 pages. 1998.

Vol. 1682: H.-D. Alber, Materials with Memory. X, 166 pages. 1998.

Vol. 1683: A. Pomp, The Boundary-Domain Integral Method for Elliptic Systems. XVI, 163 pages. 1998.

Vol. 1684: C. A. Berenstein, P. F. Ebenfelt, S. G. Gindikin, S. Helgason, A. E. Tumanov, Integral Geometry, Radon Transforms and Complex Analysis. Firenze, 1996. Editors: E. Casadio Tarabusi, M. A. Picardello, G. Zampieri. VII, 160 pages. 1998

Vol. 1685: S. König, A. Zimmermann, Derived Equivalences for Group Rings. X, 146 pages. 1998.

Vol. 1686: J. Azéma, M. Émery, M. Ledoux, M. Yor (Eds.), Séminaire de Probabilités XXXII. VI, 440 pages. 1998.

Vol. 1687: F. Bornemann, Homogenization in Time of Singularly Perturbed Mechanical Systems. XII, 156 pages. 1998.

Vol. 1688: S. Assing, W. Schmidt, Continuous Strong Markov Processes in Dimension One. XII, 137 page. 1998.

Vol. 1689: W. Fulton, P. Pragacz, Schubert Varieties and Degeneracy Loci. XI. 148 pages. 1998.

Vol. 1690: M. T. Barlow, D. Nualart, Lectures on Probability Theory and Statistics. Editor: P. Bernard. VIII, 237 pages. 1998.

Vol. 1691: R. Bezrukavnikov, M. Finkelberg, V. Schechtman, Factorizable Sheaves and Quantum Groups. X, 282 pages. 1998.

Vol. 1692: T. M. W. Eyre, Quantum Stochastic Calculus and Representations of Lie Superalgebras. IX, 138 pages. 1998.

Vol. 1694: A. Braides, Approximation of Free-Discontinuity Problems. XI, 149 pages. 1998.

Vol. 1695: D. J. Hartfiel, Markov Set-Chains. VIII, 131 pages. 1998.

Vol. 1696: E. Bouscaren (Ed.): Model Theory and Algebraic Geometry. XV, 211 pages. 1998.

Vol. 1697: B. Cockburn, C. Johnson, C.-W. Shu, E. Tadmor, Advanced Numerical Approximation of Nonlinear Hyperbolic Equations. Cetraro, Italy, 1997. Editor: A. Quarteroni. VII, 390 pages. 1998.

Vol. 1698: M. Bhattacharjee, D. Macpherson, R. G. Möller, P. Neumann, Notes on Infinite Permutation Groups. XI, 202 pages. 1998.

Vol. 1699: A. Inoue,Tomita-Takesaki Theory in Algebras of Unbounded Operators. VIII, 241 pages. 1998.

Vol. 1700: W. A. Woyczyński, Burgers-KPZ Turbulence,XI, 318 pages. 1998.

Vol. 1701: Ti-Jun Xiao, J. Liang, The Cauchy Problem of Higher Order Abstract Differential Equations, XII, 302 pages. 1998.

Vol. 1702: J. Ma, J. Yong, Forward-Backward Stochastic Differential Equations and Their Applications. XIII, 270 pages. 1999.

Vol. 1703: R. M. Dudley, R. Norvaiša, Differentiability of Six Operators on Nonsmooth Functions and p-Variation. VIII, 272 pages. 1999.

Vol. 1704: H. Tamanoi, Elliptic Genera and Vertex Operator Super-Algebras. VI, 390 pages. 1999.

Vol. 1705: I. Nikolaev, E. Zhuzhoma, Flows in 2-dimensional Manifolds. XIX, 294 pages. 1999.

4. Lecture Notes are printed by photo-offset from the master-copy delivered in camera-ready form by the authors. Springer-Verlag provides technical instructions for the preparation of manuscripts. Macro packages in T_EX, L^AT_EX2e, L^AT_EX2.09 are available from Springer's web-pages at

http://www.springer.de/math/authors/b-tex.html.

Careful preparation of the manuscripts will help keep production time short and ensure satisfactory appearance of the finished book.

The actual production of a Lecture Notes volume takes approximately 12 weeks.

5. Authors receive a total of 50 free copies of their volume, but no royalties. They are entitled to a discount of 33.3 % on the price of Springer books purchase for their personal use, if ordering directly from Springer-Verlag.

Commitment to publish is made by letter of intent rather than by signing a formal contract. Springer-Verlag secures the copyright for each volume. Authors are free to reuse material contained in their LNM volumes in later publications: A brief written (or e-mail) request for formal permission is sufficient.

Addresses:

Professor F. Takens, Mathematisch Instituut,
Rijksuniversiteit Groningen, Postbus 800,
9700 AV Groningen, The Netherlands
E-mail: F.Takens@math.rug.nl

Professor B. Teissier
Université Paris 7
UFR de Mathématiques
Equipe Géométrie et Dynamique
Case 7012
2 place Jussieu
75251 Paris Cedex 05
E-mail: Teissier@math.jussieu.fr

Springer-Verlag, Mathematics Editorial, Tiergartenstr. 17,
D-69121 Heidelberg, Germany,
Tel.: *49 (6221) 487-701
Fax: *49 (6221) 487-355
E-mail: lnm@Springer.de